Biomembrane Si

Series in Computational Biophysics

Series Editor: Nikolay Dokholyan

Molecular Modeling at the Atomic Scale: Methods and Applications in Quantitative Biology
Ruhong Zhou

Coarse-Grained Modeling of Biomolecules
Garegin A. Papoian

Computational Approaches to Protein Dynamics: From Quantum to Coarse-Grained Methods
Monika Fuxreiter

Modeling the 3D Conformation of Genomes
Guido Tiana, Luca Giorgetti

Biomembrane Simulations: Computational Studies of Biological Membranes
Max L. Berkowitz

For more information about this series, please visit:
[www.crcpress.com/Series-in-Computational-Biophysics/book-series/CRCSERCOMBIO]

Biomembrane Simulations

Computational Studies of
Biological Membranes

Edited by
Max L. Berkowitz

CRC Press
Taylor & Francis Group
Boca Raton London New York

CRC Press is an imprint of the
Taylor & Francis Group, an **informa** business

CRC Press
Taylor & Francis Group
6000 Broken Sound Parkway NW, Suite 300
Boca Raton, FL 33487-2742
© 2019 by Taylor & Francis Group, LLC

First issued in paperback 2020

CRC Press is an imprint of Taylor & Francis Group, an Informa business

No claim to original U.S. Government works

ISBN-13: 978-1-4987-9979-9 (hbk)
ISBN-13: 978-0-367-77964-1 (pbk)

Library of Congress Cataloging-in-Publication Data

Names: Berkowitz, Max L., editor.
Title: Biomembrane simulations : computational studies of biological membranes / edited by Max L. Berkowitz.
Other titles: Series in computational biophysics.
Description: Boca Raton, FL : CRC Press, Taylor & Francis Group, [2019] |
Series: Series in computational biophysics
Identifiers: LCCN 2018048428| ISBN 9781498799799 (hardback ; alk. paper) | ISBN 1498799795 (hardback ; alk. paper)
Subjects: LCSH: Membranes (Biology)–Computer simulation. | Membranes (Biology)–Mathematical models. | Computational biology.
Classification: LCC QH601 .B5258 2019 | DDC 571.6/40285–dc23
LC record available at https://lccn.loc.gov/2018048428

Visit the Taylor & Francis Web site at
http://www.taylorandfrancis.com

and the CRC Press Web site at
http://www.crcpress.com

To all the girls in my family: Paula, Ruth,

Jessica, Noa, Ariella, and Annika.

Contents

Series Preface

Computer modeling techniques, such as Monte–Carlo and molecular dynamics, were initially used to study systems consisting of hard-sphere particles, providing valuable insights into behavior of these systems. Nevertheless, even at the early stages of techniques development, researchers quickly realized that they offer a powerful tool for study of biological macromolecules such as proteins, DNA, and also molecular assemblies like membranes. Biological membranes contain mixtures of lipid molecules arranged in bilayers and proteins embedded in them. Due to the complexity of such mixtures, the initial simulations of biological membranes considered assemblies of lipid monolayers only. First simulations that used Monte–Carlo method to study configurational properties of lipid molecules in monolayers already appeared in 1960s, early 1970s. To study dynamical properties of monolayers, molecular dynamics simulations appeared in mid-1970s. These Monte–Carlo and molecular dynamics simulations of monolayers used a simplified description of lipid molecules. For example, the initial molecular dynamics simulation of a monolayer containing lipid molecules with two carbon chains described the molecules as dumbbells; in today's language, we can classify this model as a coarse-grained one. The more detailed and sophisticated description of lipid molecules arranged in a bilayer was reported in early 1980s, while persistent activity in molecular detailed simulations of lipid bilayers started only in early 1990s. The then available force fields (CHARMM, AMBER, GROMOS) for simulations of molecules such as proteins were extended to include parameters for the description of lipids. Since detailed simulations required large computational resources to study membranes, the simulations were performed on patches of membranes containing a rather small number of lipid molecules, around a hundred. Although the sampling in the simulations was performed over small in size bilayer patches and over limited number of configurations, they could provide information about the structure and dynamics of lipid bilayers that complemented experimental findings obtained from spectroscopic measurements.

To simulate biological membranes composed of a mixture of different lipids and proteins, one needs to simulate processes occurring on multitude of temporal and spatial scales. Even today, in spite of the impressive development of hardware and software, one needs to use coarse-grained description of membrane if one wants to describe its morphological transitions or the interaction between proteins in it. Coarse graining of membranes can be performed on different levels: using particle description or using continuum description. The first Chapter contributed by Lyubartsev and Rabinovich presents a brief review of force fields used today in membrane simulations: it considers detailed all-atom force fields and particle-based coarse-grained force fields, such as very popular for membrane simulations MARTINI force field. More detailed description of different particle-based coarse-grained fields and examples of their applications are presented in the Chapter contributed by Laradji and Sperotto. Continuum description, that can be very useful when considering elastic properties of membranes, is presented in the Chapter contributed by Sodt.

Biological membranes need to separate an aqueous environment outside the cell from the aqueous solution in the cell interior and this explains why lipids have bilayer architecture in the membrane. Since the self-assembly of lipids into bilayers is controlled by the hydrophobic effect, it is important to study the interaction of lipids with water. How far into the water propagates the influence of membrane surface? Chapter by Samatas et al. describes the information provided by simulations that helps answering this question. Fusion of cells represents a very important process in biology and it consists of a number of steps. Prior to fusion cells need to approach each other closely. Experiments show that when lipid bilayers approach each other, they are exposed to the action of repulsive forces, called hydration forces, because of the importance of water in determining their nature. The molecular origin of the hydration force acting between phospholipid membranes was probed by simulations performed in the R. Netz's group; the review of this work is presented in the chapter by Kanduc et al. At a certain stage of fusion, lipids rearrange and create a pore in the bilayer and the process of pore creation can be described as a reaction. In general, to explain the energetics and kinetics of reactions, one needs to know how the free energy changes along the reaction coordinate(s). What is/are proper reaction coordinate(s) and their dimensionality is a complicated question, often answered using intuition. It is also important to make sure that the free-energy change calculated along the reaction coordinate is converged in the simulations. The issues of reaction coordinate(s) that can be used to describe the process of pore creation in lipid bilayer and convergence of free energy along the coordinate(s) are considered in the contribution by Awasthi and Hub. Although membranes separate the exterior from cell's interior, a constant traffic across membranes takes place to guarantee the functioning of living cells. Often protein channels or transporters assist in molecular transport, but some molecules can directly diffuse across lipid bilayer. Again, one needs to calculate the free-energy profile along the reaction coordinate(s) that describe(s) the permeation of molecules across the bilayer and different methods exist for these calculations. In their contribution, Pokhrel and Maibaum analyze the issue of free-energy convergence, the detection and ways to overcome failures when using different methods. Permeation of molecules across membranes occurs on a time scale ranging from microseconds to hours, while typical molecular dynamics simulations cover molecular motion on a substantially shorter time scales. How can one extract information from these shorter simulations to describe the kinetics of molecular motion across membranes? Cardenas and Elber present a review that provides answers to this question. Among many molecules that can permeate membrane, of special interest to us is the ability of drug molecules to do this. Today, when nanotechnology undergoes an active development, nanoparticles are used to assist in drug delivery. How nanoparticles interact with biological membranes is the topic considered in the review by Rossi et al. Another issue that is very important in medical and pharmaceutical sciences is the action of general anesthetics, which is connected to interaction of certain molecules with biological membranes. This issue is reviewed in the contribution presented by Jedlovszky. Biological membranes contain mixtures of different lipid molecules and these mixtures are not homogeneous. Domains of liquid-ordered lipids, called lipid rafts, appear in the otherwise liquid-disordered lipid mixture. Can ions mediate the creation of nanodomains? Ganesan et al. address this question in their contribution.

For many years, membranes were considered to be passive observers of the processes occurring in cells. Latest findings clearly show that due to their complexity, membranes are playing an important role in the cell dynamics and the interaction of lipids in membranes with peptides and proteins. Of interest to us is the interaction of the so-called antimicrobial peptides that can damage bacterial membranes; such peptides can serve as effective antibiotics. To study computationally these interactions, we need to be able to simulate more realistically bacterial membranes, and progress in this area is reviewed in the contribution by Khalid et al.

A brief description of the contributions to this volume demonstrates how broad, interesting, and important is the field of biological membrane simulations. I enjoyed reading the chapters and learned many things from this reading and hope that all readers of this book will share my feelings. I am grateful to all the contributors for the wonderful work they did.

Max Berkowitz

See the book's dedicated webpage at the publisher's site for online supplements, including full-color images available to download: www.crcpress.com/9781498799799.

About the Editor

Max L. Berkowitz is a Professor in the Department of Chemistry at the University of North Carolina, Chapel Hill, NC, USA. He earned his PhD from the Weizmann Institute of Science, Rehovot, Israel. His research interests include studies of the structural and dynamical properties of aqueous ionic solutions, structure and dynamics of biomembranes, and influence of cavitation effect on biomembranes. He has given numerous invited talks and presentations and is an author or a co-author of more than 150 peer-reviewed journal publications. He is a Fellow of the American Physical Society.

Contributors

Mohamed Laradji
The University of Memphis, Department
of Physics and Materials Science
Memphis, Tennessee

Alfredo E. Cardenas
The University of Texas at Austin
Oden Institute for Computational
Engineering and Sciences
Austin, Texas

Ron Elber
The University of Texas at Austin
Oden Institute for Computational
Engineering and Sciences
Department of Chemistry
Austin, Texas

Pál Jedlovszky
Eszterhazy Karoly University
Department of Chemistry
Eger, Hungary

Nihit Pokhrel
Department of Chemistry
University of Washington
Seattle, Washington

Lutz Maibaum
Department of Chemistry
University of Washington
Seattle, Washington

Matej Kanduč
Jozef Stefan Institute
Department of Theoretical Physics
Ljubljana, Slovenia

Alexander Schlaich
Université Grenoble Alpes
Laboratoire Interdisciplinaire
de Physique
Grenoble, France

Bartosz Kowalik
Freie Universität Berlin
Fachbereich Physik
Berlin, Germany

Amanuel Wolde-Kidan
Freie Universität Berlin
Fachbereich Physik
Berlin, Germany

Roland R. Netz
Freie Universität Berlin
Fachbereich Physik
Berlin, Germany

Emanuel Schneck
Max Planck Institute of
Colloids and Interfaces
Department of Biomaterials
Potsdam, Germany

Alexander P. Lyubartsev
Stockholm University
Department of Materials and
Environmental Chemistry
Stockholm, Sweden

Alexander L. Rabinovich
Institute of Biology of the Karelian
Research Centre of the Russian
Academy of Sciences
Laboratory of ecological biochemistry
Petrozavodsk, Russian Federation

Maria Maddalena Sperotto
Technical University of Denmark
Kgs. Lyngby, Denmark

Alexander J. Sodt
Eunice Kennedy Shriver National
Institute of Child Health and
Human Development
National Institutes of Health
Bethesda, Maryland

Hongcheng Xu
University of Maryland, College Park
Fischell Department of Bioengineering,
Biophysics Program
College Park, Maryland

Sai J. Ganesan
University of Maryland, College Park

Fischell Department of Bioengineering
College Park, Maryland

Silvina Matysiak
University of Maryland, College Park
Fischell Department of Bioengineering
College Park, Maryland

Neha Awasthi
University of Goettingen
Institute for Microbiology and Genetics
Göttingen, Germany

Jochen S Hub
Saarland University
Theoretical Physics
Saarbrücken, Germany

Syma Khalid
University of Southampton
School of Chemistry
Southampton, United Kingdom

Graham Saunders
University of Southampton
School of Chemistry
Southampton, United Kingdom

Sotiris Samatas
Universitat de Barcelona
Department, Company:
Secciò de Fisica Estadistica i
    Interdisciplinaria—Departament de
Física de la Matèria Condensada & Institute
    of Nanoscience and
Nanotechnology (IN2UB)
Barcelona, Spain

Carles Calero
Universitat de Barcelona
Department, Company:
Secciò de Fisica Estadistica i
    Interdisciplinaria—Departament de
Física de la Matèria Condensada & Institute
    of Nanoscience and
    Nanotechnology (IN2UB)
Barcelona, Spain

Fausto Martelli
IBM Research
Darsebury, United Kingdom
The Hartree Centre, Daresbury
United Kingdom

Giancarlo Franzese
Universitat de Barcelona
Department, Company:
Secciò de Fisica Estadistica i
    Interdisciplinaria—Departament de
Física de la Matèria Condensada & Institute
    of Nanoscience and
Nanotechnology (IN2UB)
Barcelona, Spain

Giulia Rossi
Department of Physics
University of Genoa
Genoa, Italy

Sebastian Salassi
Department of Physics
University of Genoa
Genoa, Italy

Federica Simonelli
Department of Physics
University of Genoa
Genoa, Italy

Alessio Bartocci
Department of Physics
University of Genoa
Genoa, Italy

Luca Monticelli
Molecular Microbiology and
    Structural Biochemistry
UMR 5086, CNRS & University of Lyon
Lyon, France

Taylor Haynes
University of Southampton
School of Chemistry
Southampton, United Kingdom

1

Force Fields for Biomembranes Simulations

Alexander P. Lyubartsev
Department of Materials and Environmental Chemistry, Stockholm University, SE 106 91, Stockholm, Sweden

Alexander L. Rabinovich
Institute of Biology of the Karelian Research Centre of the Russian Academy of Sciences, Pushkinskaya 11, Petrozavodsk, 185910, Russian Federation

1 Introduction

Biomembranes form outer shells in living cell and various intracellular structures. They play an active part in the life of the cell, regulating transport of various molecules and cell signaling. Biomembranes are very complex heterogeneous systems consisting of many different types of lipids and membrane-associated molecules. Lipid molecules forming bilayers provide a structural framework in which other biomembrane components, such as proteins, polypeptides, and carbohydrates are inserted or attached to. The lipids constituting membranes differ with respect to the type of hydrophilic head-group and occur with a wide variety of hydrophobic hydrocarbon chains of fatty acids (FAs). The most abundant phospholipid in animals and plants is phosphatidylcholine (PC); other common lipid types are phosphatidylethanolamine (PE), phosphatidylserine (PS), phosphatidylglycerol (PG), and sphingomyelin (SM). The most commonly occurring FA chains consist of 14–22 carbon atoms and may contain 1–6 carbon–carbon double bonds of the cis configuration in different positions. In most cases, at least half of the FA chains are unsaturated. The double bonds of polyunsaturated (PU) chains are, as a rule, methylene-interrupted. The PU FA tails of lipids are of great importance for the structure and functioning of biomembranes especially in such crucially important tissues of higher organisms as brain white and gray matter or eye retina (Dratz & Deese, 1986; Rabinovich & Ripatti, 1994; Gawrisch et al., 2003; Stillwell & Wassall, 2003; Rabinovich et al., 2003; Valentine & Valentine, 2004; Feller & Gawrisch, 2005; Gawrisch et al., 2008; Stillwell, 2008; Wassall & Stillwell, 2009). Evidently, the basis (and primary cause) of these phenomena is the specific chemical structure of PU FA chains having methylene-interrupted cis double bonds, which results in their specific physical properties, which in turn remains the cause for their specific functioning in living organisms. Nevertheless, the molecular mechanisms of many biological functions of PU FAs remain a subject of much debate. Investigation of the interplay of various lipid types in bilayers is a key element of our general understanding of biomembrane functioning, which is one of the greatest challenging problems in biophysical and biomedical sciences.

Much insight into the biomembrane properties can be gained by molecular simulations (molecular dynamics or Monte–Carlo), which provide detailed three-dimensional imaging of the system with atomistic resolution, and hence give essential information which otherwise is hardly accessible by any other experimental method. Compared to experimental studies, computer simulations have the advantage that, within various levels of approximation, allow one to separately probe different physical mechanisms and chemical pathways, and assess, both qualitatively and quantitatively, their importance for the studied phenomena. The first works on computer simulations of biomembranes appeared in already 1980s when attempts of modeling of bilayers composed of amphiphilic molecules were made (Kox et al., 1980; van der Ploeg & Berendsen, 1982). A decade later, molecular dynamics simulations of fully hydrated bilayers built by lipids typical for biological membranes have been performed (Heller et al., 1993; Marrink & Berendsen, 1994; Tieleman & Berendsen, 1996). Since then the amount of works on simulations of lipid membranes has been increasing tremendously, which is reflected in a number of reviews accounting for this published in the past decades (Damodaran & Merz, 1994; Mouritsen & Jorgensen, 1994; Pastor, 1994; Mouritsen et al., 1996; Jacobsson, 1997; Merz, 1997; Tieleman et al., 1997; Tobias et al., 1997; Merz, 1997; Berendsen & Tieleman, 1998; Feller & MacKerell, Jr., 2000; Feller, 2000; Forrest & Sansom, 2000), and more recently by Feller (2001), Tobias (2001), Scott (2002), Hansson et al. (2002), Saiz & Klein (2002), Saiz et al. (2002), Vigh et al. (2005), Chan and Boxer (2007), Vermeer et al. (2007), Feller (2008), Marrink et al. (2009), Pandit & Scott (2009), Gurtovenko et al. (2010), Lyubartsev & Rabinovich (2011), Yanga & Ma (2012), Bennett & Tieleman (2013), Rabinovich & Lyubartsev (2013), Pluhackova and Böckmann (2015), Baoukina & Tieleman (2016), Bunker et al. (2016), Kirsch & Böckmann (2016), Lyubartsev & Rabinovich (2016), Pasenkiewicz-Gierula et al. (2016), and Pöyry & Vattulainen (2016).

The value of information obtained by computer simulations depends crucially on the force field (FF) defining molecular interactions. Proper parametrization of the FF is an ongoing problem in molecular simulations. A good FF should provide agreement with all available experimental data, within the simulation and experimental uncertainty. As simulations are becoming longer, uncertainties caused by the equilibration stage and statistical error are decreasing. Experimental techniques are also improving. At some point, the FF which earlier provided satisfactory agreement with experimental data may begin to show discrepancies. Furthermore, FFs optimized for pure components, such as lipids, proteins, and small molecules, may not work optimally or even fail when used in description of complex inhomogeneous systems including lipid membrane in contact with proteins, drugs, nanoparticles, etc. This development initiates further improvements of the FF leading to better description of the molecular interactions and better agreement between computer simulations and experimental results.

In this chapter, we give an overview of the current state of development of FFs for biomembrane simulations. We describe principles of the FF parametrization, different classes of FFs, validation issues, as well as show some examples of applications. In the following, we will use the numerical notations for description of FA chains of lipids having the form: $N : k(n - j)$ cis, where N refers to the total number of carbon atoms in the chain; k is the number of the methylene-interrupted double bonds (i.e., one methylene group is localized between each pair of double bonds); the letter n means that the so-called 'n minus' nomenclature is used, i.e., the position j of the first double bond is counted from the methyl terminus of the chain (with the methyl carbon as number 1). The first double bond extends from the j-th carbon to the $(j + 1)$-th carbon from the end, 'cis' refers to the conformation around the double bonds. For brevity, the fragment $(n - j)$ cis in the notation is frequently omitted. Notation for a lipid thus consists of description of the FA chains (separated by a slash symbol) and followed by notation of the lipid head, e.g. 16:0/18:1 (n-9)cis PC describes 1-palmitoyl-2-oleoyl-sn-glycero-3-PC (POPC) lipid.

2 Force Fields

2.1 General Principles of Force Field Organization

In classical molecular dynamics simulations, the atoms move according to the Newtonian equation of motion. The forces acting on atoms are determined as gradients of the potential energy, which is a function of atom coordinates, and often called FF. In Monte–Carlo simulations, the same energy function determines the probabilities of the Monte–Carlo steps. The results of simulations depend crucially on the quality of the FF, that is why accurate parameterization and validation of the FFs is an ongoing subject of research in molecular modeling.

In principle, a rather accurate way to compute forces acting on atoms located in certain positions (known as Born–Oppenheimer approximation) is to do high-quality quantum chemical calculation of the electronic ground state and ground state energy $E_0(r_1, \ldots r_N)$ where r_i are atom coordinates (this function is also called *energy surface*). The force acting on atom i can then be computed as gradient of the energy surface:

$$F_i = -\frac{\partial}{\partial r_i} E_0(r_1, \ldots r_N) \tag{1}$$

Good quality quantum-chemical computations are extremely time-consuming and in practice can be done for systems consisting of no more than a few hundred atoms. Moreover, quantum-chemical methods of electron structure calculations scales with the number of atoms as N^3 or even faster, which makes considerable increase of the tractable system size unlikely. That is why practically all biomolecular simulations use molecular-mechanics FF approach when the quantum-mechanical energy surface is approximated by a sum of simple potential energy functions representing some specific interatomic interactions. The most often used expression (the so-called standard molecular-mechanical FF) which provides a good compromise between ability to describe molecular interactions accurately enough and computational efficiency is:

$$\begin{aligned}
E_0(r_1, \ldots r_N) \approx U(r_1, \ldots r_N) = \sum_{ij\,\text{bond}} k_{b,ij}(r_{ij} - r_{0,ij})^2 \;\; + \sum_{ijk\,\text{ang}} k_{\theta,ijk}(\theta_{ijk} - \theta_{0,ijk})^2 \\[1em]
+ \sum_{ijkl\,\text{tors}} \frac{1}{2} V_{\varphi,ijkl}\left[1 + \cos\left(n\varphi_{ijkl} - \varphi_{0,ijkl}\right)\right] \\[1em]
+ \sum_{\substack{ij\,\text{nonbonded} \\ (\text{excl})}} 4\varepsilon_{ij}\left[\left(\frac{\sigma_{ij}}{r_{ij}}\right)^{12} - \left(\frac{\sigma_{ij}}{r_{ij}}\right)^6\right] \;\; + \sum_{\substack{ij\,\text{nonbonded} \\ (\text{excl})}} \frac{1}{4\pi\varepsilon_0} \frac{q_i q_j}{r_{ij}}
\end{aligned} \tag{2}$$

The first two terms represent energy change due to small deviation of covalent bond lengths and angles between the bonds from the equilibrium values ($r_{0,ij}, \theta_{0,ijk}$); the torsion term (which may contain several cosine functions with different multiplicities n) describes rotation around torsion angles $\varphi_{0,ijkl}$. The non-bonded term is taken over all atom pairs excluding those involved in the bond and angular interactions. The repulsive (r^{-12}) part of the Lennard–Jones potential approximates the short-distance repulsion of atoms when their electron orbitals become strongly overlapping. The attractive (r^{-6}) term describes van der Waals interaction which has a physical origin in the London dispersion force due to correlation of dipole moment fluctuations. The electrostatic energy is presented as interaction of fixed point charges located on atoms. The whole FF is determined by the sets of force constants (k_b, k_θ and V_φ) for bond, angular, and torsion interactions, the corresponding equilibrium bond distances and angles (r_0, θ_0 and φ_0), Lennard–Jones parameters (ε and σ) and partial charges (q) in (2) for all chemically different atom types involved.

The principle limitation of expression for the FF (2) is that all its terms are independent on each other. Fixed partial charges can be a matter of special concern. In some FFs, additional terms can be assigned for various cross-interactions (e.g. Urey–Bradley term coupling bond and angular energies), hydrogen bonds, dipole interactions, etc. Still, despite a number of limitations, expression for the FF (2) has proved through the years its robustness in simulations of a very wide range of biomolecular and soft matter systems. Particularly, for biomembrane systems, the vast majority of simulations is done using the standard molecular-mechanical FFs described by expression (2).

The work on the FF development consists in finding the values of the FF parameters which describe the targeted class of systems in the 'best' possible way. Other important requirement for an FF is computational efficiency (one should avoid additional terms in the FF if possible), transferability (parameters depend only on the chemical type of atom and its bonding, but not based on which molecule the atom is present), and representativity; the same set of FF parameters should describe accurately different properties, such as structure (molecular conformations, various distribution functions), thermodynamics (enthalpies, free energies, binding, partitioning), and dynamics (diffusion, time correlation functions). In the next section, we consider several families of the FFs used in recent years in atomistic biomembrane simulations: GROMOS, CHARMM, Slipids, AMBER, OPLS-AA critically accessing their performance relative to these criteria.

2.2 GROMOS

GROMOS ('GROningen MOlecular Simulation package') (Hermans et al., 1984; Schuler et al., 2001; Chandrasekhar et al., 2003) has been widely used in simulations of lipid bilayers from 1990s (Marrink & Berendsen, 1994; Tieleman & Berendsen, 1996). It employs a united atoms approach representing each of nonpolar CH, CH_2, and CH_3 groups of hydrocarbons as a single particle. The GROMOS FF was parametrized primarily from experimental thermodynamic properties of model compounds, such as enthalpies and free energies of solvation of small molecules representing biomolecular fragments in polar and apolar environments. There exist several versions of the GROMOS FF used in simulations of lipid bilayers, the most popular being the so-called Berger modification Berger et al. (1997) (or Berger lipid FF). Other variations are 45A3 parameter set (Chandrasekhar et al., 2003; with original GROMOS parameters for non-bonded interactions) as well as further developments such as G53A (Oostenbrink et al., 2004), 43A1-S3 (Chiu et al., 2009), and G54A (Poger et al., 2010) parameter sets.

A number of variations of GROMOS FF were tested by Chandrasekhar et al. (2003) in simulations of 16:0/16:0 PC bilayer at $T = 325$ K, by comparison with experimentally known average membrane area per lipid, NMR bond order parameters, and lipid lateral diffusion. In Benz et al. (2005), comparison of simulated and measured (by X-ray or neutron diffraction) structure factor was made for 18:1(n-9)cis/18:1(n-9)cis PC lipids described within Berger lipid FF. In Högberg & Lyubartsev (2006), 14:0/14:0 PC bilayer was simulated using Berger parameter set (Berger et al., 1997) at 303 K and 323 K and comparison with similar set of experimental data has been made. A common conclusion from these as well as several other studies (Anezo et al., 2003; Siu et al., 2008; Prakash & Sankararamakrishnan, 2010) can be made that while giving a fair representation of the bilayer structure and dynamics, the GROMOS FF still has some small, but going beyond possible computational or experimental error differences for a number of experimentally measured properties.

Recent updates of GROMOS parameters (43A1-S3 parameter set (Chiu et al., 2009) and G53A6 parameter set (Poger et al., 2010) resulted in improved agreement with experiment for the area per lipid for a number of lipids in comparison with previous versions of the GROMOS FF. The ability of G53A6 parameter set to reproduce the structural and hydration properties of common phospholipids of varying length and degree of unsaturation of the acyl chains, i.e., pure bilayers of 12:0/12:0 PC, 14:0/14:0 PC, 18:1(n-9)cis/18:1(n-9)cis PC, and 16:0/18:1(n-9)cis PC in a liquid crystal phase was examined in Poger and Mark (2010). In a more recent update, the united atom GROMOS model of diacyl (DMPC and DPPC) lipids was further optimized using ab initio computations to obtain partial charges of the headgroup atoms and torsional parameters of the acyl chains (Tjörnhammar & Edholm, 2015). This has led to substantial improvement of the description of the bilayer gel phase and gel-to-liquid crystalline transition, while keeping a good description of the fluid phase.

The united atom approximation used in GROMOS FF can be considered reasonable for saturated lipid tails. For methyls at the choline group, the united atom representation may be questioned because of the polarity of this group and ability of choline group hydrogens to form weak hydrogen bonds (Pohle et al., 2004). Also, there is weak polarity of hydrogens at double bonds of unsaturated lipids, which is becoming stronger in the case of PU lipids (Ermilova & Lyubartsev, 2016). Explicit hydrogen representation is also important for accurate modeling of various solutes in lipid membranes (Paloncýová et al., 2014). By these reasons, fully atomistic FFs have received more attention in the recent years.

2.3 CHARMM

The CHARMM ('Chemistry at HARvard Macromolecular Mechanics') FF (Mackerell et al., 1995, 1998) describes all hydrogens explicitly. Additionally, it has a more detailed description of intramolecular interactions, including Urey–Bradley term for covalent angles and a richer variety of parameters for dihedral angles, many of which were developed on the basis of quantum-chemical calculations for small model compounds which are then transferred to larger molecules with eventual further optimization. CHARMM parameters for lipids were introduced first in Feller et al. (1997) (within the CHARMM22 parameter set, often denoted also as C22) and then were updated in Feller and MacKerell, Jr. (2000) (CHARMM27, or C27 parameter set), again in Klauda et al. (2005) (C27r parameter set), and in Klauda et al. (2010) (C36 parameter set).

Detailed investigations of the earlier versions of CHARMM FF have shown non-negligible disagreements with experiment (Benz et al., 2005). The most essential drawback was underestimation of average area per lipid in constant-pressure simulations, and transition of fully saturated bilayers to gel phase at conditions corresponding to the liquid crystalline phase (Jensen et al., 2004; Hyvönen & Kovanen, 2005; Sonne et al., 2007; Högberg et al., 2008). This was the reason that many bilayer simulations employing CHARMM27 FF were done either in the NVT ensemble with a fixed area per lipid, or under non-zero surface tension (Gullingsrud & Schulten, 2004; Benz et al., 2005; Klauda et al., 2006b; Prakash & Sankararamakrishnan, 2009; Roark & Feller, 2009). One of the reasons of such behavior can be traced to too strong preference for trans-conformations in the saturated alkane chains of lipid tails described by the CHARMM27 FF (Klauda et al., 2005; Högberg et al., 2008; Prakash & Sankararamakrishnan, 2010).

Several variations of the CHARMM FF were suggested. In works by Sonne et al. (2005, 2007), the partial atom charges were computed within the ab initio Hartree–Fock approach for an ensemble of typical lipid conformations taken from a molecular dynamics trajectory, and then averaged. Though recalculation of charges has brought results closer to the experiment, the resulting area per lipid was still about 4–5 Å^2 too low, both for DPPC Sonne et al. (2007) and DMPC Pedersen et al. (2007) lipids. In Taylor et al. (2009), simulations of bilayers composed of DPPC, POPC, and PDPC lipids, with atomic charges derived in Sonne et al. (2007) and with alkane torsion parameters described by C27r parameter set provided good agreement with experimental data for these types of lipids. In Högberg et al. (2008), an empirical way of gradual change of the energy difference between trans and gauche conformations of alkane chains, by scaling of the 1–4 electrostatic interactions (between atoms separated by exact three covalent bonds) was suggested. Together with recomputation of atomic charges in the same manner as in work by Sonne et al. (2007), simulation of DMPC bilayer at 303 K has shown an area per lipid, as well as bond order parameters, electron density and the structure factor in a perfect agreement with experiment by Högberg et al. (2008). The FF of Högberg et al. (2008) has been applied also to DPPC and POPC bilayers (Andoh et al., 2012a), and to a number of PU lipids (Rabinovich & Lyubartsev, 2014), and the calculated membrane area per lipid molecule and order parameters showed a good agreement with the experimental values.

The most recent update of the CHARMM FF, the C36 parameter set, was presented in Klauda et al. (2010) and validated on six lipid types: 12:0/12:0 PC, 14:0/14:0 PC, 16:0/16:0 PC, 18:1/18:1 PC, 16:0/18:1 PC, and 16:0/18:1 PE. The changes included reparameterization of partial atom charges and torsion potentials on the basis of ab initio computations, as well as revision of some Lennard–Jones parameters. Properties such as average area per lipid at zero tension, structure factors, NMR order parameters, dipole electrostatic potential, X-ray and neutron form factors showed certain improvements relative to the previous C27r parameter set (Klauda et al., 2010; Kang & Klauda, 2015). Thus, the C36 FF is able to accurately represent bilayer properties of fully saturated and monounsaturated lipid molecules in the NPT ensemble. However, MD simulations using C27-based parameters or C36-based parameters with the NPT ensemble on 18:0/22:6(*n*-3)cis PC bilayers of the PU FA chains resulted in inaccurate descriptions of the surface area per lipid, deuterium order parameters, and X-ray form factors, demonstrating the need to improve this portion of the FF. Therefore, the C36 FF was further modified, and the corresponding FF was referred to as C36p (Klauda et al., 2012).

An extension of the C36 FF to cholesterol (CHOL), called C36c, has also been reported (Lim et al., 2012). The new parameters in the C36c modification should enable more accurate simulations of lipid bilayers with CHOL, especially with respect to the free energy of lipid flip/flop or transfer of phospholipids and/or CHOL (Lim et al., 2012). Furthermore, the C36 lipid FF has been extended (Venable et al., 2014) to include sphingolipids, via a combination of high-level quantum mechanical calculations on small molecule fragments, and validation by extensive MD simulations on *N*-palmitoyl and *N*-stearoylsphingomyelin. A set of CHARMM-based parameters for articaine, a potent and widely used local anesthetic, was presented in Prates et al. (2011).

As another line of modification of the CHARMM FF (in particular, the C27 and C27r versions) for lipids, it was suggested to use a united atom (UA) description of hydrocarbons in lipid tails similar to the Ryckaert–Bellemans torsion potential but with some other parameters (Henin et al., 2008). Recently, a UA-chain model has been developed (Lee et al., 2014) based on the C36 all-atom lipid parameters, termed C36-UA, and agreed well with bulk, lipid membrane, and micelle formation of a surfactant. MD simulations of heptane and pentadecane were used to test the validity of C36-UA on density, heat of vaporization, and liquid self-diffusion constants. Simulations using C36-UA resulted in accurate surface area per lipid, X-ray and neutron form factors, and chain order parameters of DMPC, DPPC, POPC, DOPC, and DMPC/CHOL bilayers (Lee et al., 2014). The C36-UA FF offers a useful alternative to the all-atom C36 lipid FF by requiring less computational cost while still maintaining similar level of accuracy, which may prove useful for large systems with proteins.

Up-to-date overviews of the CHARMM FF family were given in Pastor & MacKerell, Jr. (2011), Zhu et al. (2012), Vanommeslaeghe and MacKerell, Jr. (2015), where a brief presentation on the historical aspects of FFs was given, including underlying methodologies and principles, along with a description of the strategies used for parameter development.

2.4 Slipids

Following ideas discussed in papers by Sonne et al. (2007) and Högberg et al. (2008), a new all-atomistic FF called Slipids (Stockholm lipids) has recently been developed for fully saturated phospholipids (Jämbeck & Lyubartsev, 2012a), for lipids containing a single double bond in one or two tails (Jämbeck & Lyubartsev, 2012b), and then has been extended to PU lipids (Ermilova & Lyubartsev, 2016). Parameters for a number of lipid heads (PE, PG, PS, and SM) and cholesterol have also been reported (Jämbeck & Lyubartsev, 2013a). The CHARMM36 FF was used as a starting point of parameters optimization, and parameters for covalent bonds and angles, as well some of torsion angles and Lennard–Jones parameters of Slipids coincide with those of the CHARMM36 FF, while all partial charges, Lennard–Jones parameters, and torsion parameters of lipid tails were updated.

The parametrization has been largely based on high-level ab initio calculations in order to keep the empirical input to a minimum. The FF's ability to simulate lipid bilayers in the liquid crystalline phase in a tensionless ensemble was tested in simulations of DLPC, DMPC, and DPPC (Jämbeck & Lyubartsev, 2012a), as well as POPC, SOPC, POPE, and DOPE bilayers (Jämbeck & Lyubartsev, 2012b), and for lipid-CHOL mixtures (Jämbeck & Lyubartsev, 2013a). The new FF reproduces many experimentally measurable properties of lipid bilayers such as area per lipid, NMR order parameters, and structure factors, including their temperature dependence.

These parameters are able to reproduce important structural properties of single- and double-component membranes without applying a surface tension.

All parameters of the Slipids FF (Jämbeck & Lyubartsev, 2012a, 2012b, 2013a; Ermilova & Lyubartsev, 2016) are consistent and fully compatible, and therefore it is possible to study complex systems containing many different lipids and proteins in a fully atomistic resolution in the isothermic–isobaric (NPT) ensemble, which is the proper ensemble for membrane simulations. Since parametrization of atomic charges in Slipid FF, based on ab initio computations using restricted electrostatic potential (RESP) scheme and use of 0.83 scaling factor for 1–4 interactions, is similar to those adopted in the Amber family of FF, Slipids FF is supposed to be compatible with Amber FF. This was demonstrated for amino acids (often used to model membrane proteins) (Jämbeck & Lyubartsev, 2012b) and for a number of drug-related compounds (Jämbeck & Lyubartsev, 2013b; Paloncýová et al., 2014; Saeedi et al., 2017), by computations of partitioning of these molecules in membrane bilayers and comparing with experimental data.

2.5 AMBER

AMBER (Assisted Model Building and Energy Refinement) FF was initially devised for proteins and nucleic acids (Weiner et al., 1984), and though it was used in some earlier studies of lipid bilayers (Damodaran et al., 1992; Essmann et al., 1995), there has been limited focus on lipid simulations for a long time. Simulations carried out within the standard Amber94 set of FF parameters (Moore et al., 2001; Lopez et al., 2004), as well as using its version known as GAFF (Generalized Amber Force Field; Wang et al. (2004)) that was extended to include lipid parameters (Jójárt & Martinek, 2007), provided average lipid area below the experimental value for DMPC and DOPC lipids (Rosso & Gould, 2008). Some additional modifications of the GAFF FF, including recomputations of atomic charges, were made in a paper by Siu et al. (2008). However, the average area per lipid in constant pressure simulations of DOPC bilayer still remained below experimental, and additional surface tension should be applied to maintain the correct area.

Substantial progress in development of AMBER-compatible FF parameters for lipids was made in development 'LIPID11' FF (Skjevik et al., 2012). Several MD simulations of phospholipid bilayers including DOPC, POPE, and POPC were carried out. The authors (Skjevik et al., 2012) concluded that LIPID11 can be a flexible starting point for the development of a comprehensive, Amber-compatible lipid FF. At the same time, the GAFF Lennard–Jones parameters for the simulation of acyl chains were corrected (Dickson et al., 2012) to allow the accurate and stable simulation of pure lipid bilayers composed of 12:0/12:0 PC, 14:0/14:0 PC, 16:0/16:0 PC, 18:1 (*n*-9)cis/18:1(*n*-9)cis PC, 16:0/18:1(*n*-9)cis PC, and 16:0/18:1(*n*-9)cis PE phospholipid types; the resulting FF is GAFFlipid. Continuation of these lines resulted two years later in LIPID14 FF (Dickson et al., 2014). The Lennard–Jones and torsion parameters of both the head and tail groups have been revised and updated partial charges calculated. The modular nature of LIPID14 FF allows numerous combinations of head and tail groups to create different lipid types, enabling an easy insertion of new lipid species. The LIPID14 FF has been validated by simulating bilayers of six lipid types, 12:0/12:0 PC, 14:0/14:0 PC, 16:0/16:0 PC, 18:1(*n*-9)cis/18:1(*n*-9)cis PC, 16:0/18:1(*n*-9)cis PC, and 16:0/18:1(*n*-9)cis PE for a total of 500 ns each without applying a surface tension, with favorable comparison to experiment for properties such as area per lipid, volume per lipid, bilayer thickness, NMR order parameters, scattering data, and lipid lateral diffusion. LIPID14 is compatible with the AMBER protein, nucleic acid, carbohydrate, and small molecule FFs.

It was next shown by Slingsby et al. (2015) that partial charge optimization at a high level of theory is a simple and effective way to improve phospholipid structure properties and solvent and ion interactions for the GAFF (note that the same parameterization strategy is used in Slipids FF). The ab initio calculated charge distributions results in a more realistic electrostatic interactions between the lipid molecules, allowing for an accurate simulation in an NPT ensemble without the need for an external surface tension parameter.

In another development, GAFF parameters for phospholipids (Wang et al., 2004) and GAF-Flipid (Dickson et al., 2012) were improved in work by Ogata and Nakamura (2015) in order to describe the thermal phase transition using MD simulations for six PC bilayers, 16:0/16:0 PC, 16:0/16:0 PE, 18:0/18:0 PC, 18:1(*n*-9)cis/18:1(*n*-9)cis PC, 16:0/18:1(*n*-9)cis PC, and 16:0/18:1(*n*-9)cis PE. These modified parameters (mod-Gaff2) were also found to be useful for performing MD simulation of transmembrane proteins with membrane models. The simulation using mod-Gaff2 parameters showed similar results as the set LIPID14.

2.6 OPLS-AA

The OPLS-AA (Optimized Parameters for Liquid Simulations All-Atom) FF was originally developed for organic liquids (Jorgensen et al., 1996) with emphasis on reproduction of experimental thermodynamical and partitioning properties. This FF was extended many times to include a large set of parameters for proteins, nucleic acids, carbohydrates, and various drug molecules, but before 2014, lipids have not been fully parametrized in OPLS-AA FF (only parts of the molecules including long hydrocarbons (Siu et al., 2012), *n*-pentadecane, methyl acetate, and dimethyl phosphate anion were reparameterized; Murzyn et al., 2013). In work by Maciejewski et al. (2014), parametrization of DPPC bilayer was reported: the parameters for torsion angles in the PC and glycerol moieties and in the acyl chains, as well the partial atomic charges, were determined. The surface area per DPPC lipid was within the range of the experimental values; agreement between form factors from the X-ray experiment and the MD simulations was good. The order parameter S_{CD} profiles of C–D bonds along the DPPC *sn*-2 chain showed overall good agreement with a small deviation at the beginning of the chain; S_{CD} parameters calculated for the carbon atoms in the glycerol backbone and the choline moiety agreed well with those obtained experimentally, with an exception for one carbon atom. Small deviations between simulated and experimental data indicate that there is still space for further improvement in the OPLS-AA FF parameters for DPPC lipid. Besides, it is needed to develop OPLS-AA parameters for other lipids.

2.7 Coarse-Grained Force Fields

Investigation of certain properties of lipid bilayers requires length and time scales not reachable (or not affordable) in atomistic simulations. Examples are: undulations of membrane surface, membrane bending, formation of different aggregates as micelles, vesicles, lammelar, or hexagonal phase transformations, problems related to domain formations at the membrane surfaces, and insertion/penetrations of larger macromolecules (proteins) or nanoparticles (NPs). In principle, there are examples of studies of these phenomena using all-atom or united-atom approaches (for example, aggregation of lipids into smaller micelles or bilayers; Lee et al. (2014), Skjevik et al. (2015)) but these require very large computer resources. In order to make large-scale membrane simulations practically feasible, it becomes necessary to simplify description of individual lipids, grouping atoms into pseudo-particles in a coarse-grained (CG) description of a bilayer. CG models emerged as a practical alternative to all-atom simulations in cases when detailed atomistic representation is not necessary while it is important to observe membrane phenomena over long distances or time scales. Several reviews and discussions appeared describing different aspects of CG, or multiscale modeling of bilayers (Nielsen et al., 2004; Brannigan et al., 2006; Müller et al., 2006; Venturoli et al., 2006; Ayton et al., 2007; Shillcock & Lipowsky, 2007; Klein & Shinoda, 2008; Southern et al., 2008; Ayton & Voth, 2009b; Bennun et al., 2009;, Cascella & Peraro, 2009; Marrink et al., 2009; Pandit & Scott, 2009; Ayton et al., 2010; Kamerlin et al., 2011; Riniker et al., 2012; Baaden & Marrink, 2013; Bradley & Radhakrishnan, 2013; Noid, 2013; Marrink & Tieleman, 2013; Ingólfsson et al., 2014; Vicatos et al., 2014; Pluhackova & Böckmann, 2015).

There exist a large variety of CG models of lipids differing by the level of details, account for the solvent, and the way how interaction potentials are defined. It is rather common to unite groups consisting of 3–5 heavy atoms into a single CG site, though even more simple models (e.g. a generic 3-site lipid model; Cooke & Deserno, 2005) were used. Further, while some of CG models use explicit 'coarse-grained' water, other models are formulated to use an implicit solvent, where the effect of solvent (water) is described by effective potentials.

One of the most widely used explicit solvent CG lipid models is based on the MARTINI FF Marrink et al. (2004, 2007). The MARTINI FF uses predominantly a four-to-one mapping, i.e., on average, four heavy atoms are represented by a single interaction center. Solvent (water) is also modeled by particles each representing four water molecules. The interaction potential consists of Lennard–Jones and eventually electrostatic terms, which are tuned to reproduce experimental partitioning data. There are several particle types in the MARTINI FF representing particles on the hydrophobic–hydrophilic scale, and CG sites corresponding to fragments of various lipid molecules are assigned to one of such predetermined types, which provides possibility to easily build CG models of various lipids. MARTINI FF was generalized to include protein and peptide models (Monticelli et al., 2008), carbohydrates (Lopez et al., 2009), and glycolipids (López et al., 2013). The current state of the MARTINI model, recent highlights as well as shortcomings, and ideas on the further development of the model were analyzed by the developers (Marrink & Tieleman, 2013). An implicit-solvent version of the CG MARTINI model, nicknamed 'Dry Martini,' was introduced by Arnarez et al. (2015).

A CG model for lipids and proteins in water solution, built initially on similar principles as MARTINI FF, and refined by atomistic simulations, is described in a series of works by Schulten group (Shih et al., 2006; Arkhipov et al., 2008). This model was used to study self-assembly of lipoprotein systems (Shih, Arkhipov, et al., 2007; Shih, Freddolino, et al., 2007) and the effect of proteins on membrane curvature (Arkhipov et al., 2008; Yin et al., 2009; Arkhipov et al., 2009). In.Orsi et al. (2008, 2010), another CG lipid model was considered, in which CG sites of lipids were presented as ellipsoidal particles interacting by Gay–Berne potential with embedded charges on the head group and dipoles at the ester groups. Water was presented as a one-site spherical particle with a dipole. This model provides a CG description of hydrated bilayer in somewhat more details than MARTINI FF. This model was used to study membrane properties such as pressure distribution, spontaneous curvature, water permeation in 14:0/14:0 PC and 18:1/18:1 PC bilayers (Orsi et al., 2010). One more CG lipid FF, built on principles similar to MARTINI FF, was developed by Shinoda et al. (2010) and called SDK-model. The interactions between non-bonded CG sites were described by Lennard–Jones-like potentials with powers 12–4 and 9–6, with parameters fitted to a number of experimentally observable properties of lipid bilayers and atomistic simulations.

Within the so-called 'bottom-up' approach, parameters of a CG FF are deduced from the atomistic simulations. There exist several practical schemes to implement the bottom-up coarse-graining. The force-matching procedure (called also Multi Scale Coarse Graining, MS-CG) fits a pairwise expression for the force to reproduce the n-body potential of mean force for the CG sites in the atomistic system (Izvekov & Voth, 2005; Noid et al., 2008). Within the structure-based coarse-graining, the effective potentials between CG sites are derived from the structural properties determined in atomistic simulations. Particularly, radial distribution functions between CG sites, as well as distributions of intramolecular bond lengths and angles, are used for parametrization of inter- and intramolecular potentials, which can be done by the inverse Monte–Carlo (Lyubartsev & Laaksonen, 1995) or inverse Boltzmann (Reith et al., 2003) methods.

Several studies have employed 'bottom-up' approaches to parametrize CG models of lipids that accurately describe the ensembles sampled by all-atom models. The force matching approach was used for multiscale simulations of DMPC bilayers (Izvekov & Voth, 2006), DMPC–CHOL lipid mixtures in plain bilayers as well as liposomes (Izvekov & Voth, 2006, 2009), and DOPC–DOPE lipid mixtures (Lu & Voth, 2009). A hybrid algorithm with a part of CG interaction potential presented by the Gay–Berne potential was also considered (Ayton & Voth, 2009a). Effective potentials, computed in a work by Lyubartsev (2005), from atomistic simulations of DMPC lipids using the inverse Monte–Carlo technique, were used to describe processes of spontaneous formation of bicells, micelles, and multilammelar structures (Lyubartsev et al., 2009, 2010).

Transferability of the inverse Monte–Carlo-derived CG potentials for DMPC lipids with respect to lipid/water molar ratio has been investigated by Mirzoev and Lyubartsev (2014). In Hadley and McCabe (2010), the iterative inverse Boltzmann approach was used to derive effective potentials for CG 16:0/16:0 PC lipid model which are suitable for description of both liquid crystalline (liquid disordered) and amorphous states of this lipid. In Murtola et al. (2007), the inverse Monte–Carlo method was used to derive effective potentials for two-dimensional model of lipid–CHOL domain-forming mixtures.

---

3 Validation and Comparisons of Force Fields

3.1 Methodological Issues

While discussing the question on how well an FF describes real systems, a question of other than FF factors affecting simulation results should not been missed. Equilibration and sampling issues is an ongoing problem of all types of simulation studies. For pure lipid bilayers, it is now accepted that several hundred ns simulations provide typically sufficient time for equilibration and average collection. However, for more complicated systems such as mixed bilayers and larger penetrants, longer simulation times may be needed and this question needs to be investigated in each specific case.

In certain cases, use of advanced sampling methods helps in solving problems requiring longer simulation scale. For instance, the preferred conformations of the glycerol region of DPPC have been explored using replica exchange MD simulations and compared with the results of standard MD approaches and with experiment (Vogel & Feller, 2012). It was found that due to slow isomerization rates in key torsions, standard MD is not able to produce accurate equilibrium conformer distributions from reasonable trajectory lengths while replica exchange MD provides quite efficient sampling due to the rapid increase in isomerization rate with temperature (Vogel & Feller, 2012). Another example is accelerated MD; it is an enhanced sampling technique that expedites conformational space sampling by reducing the barriers separating various low-energy states of a system. An application of the accelerated MD method on POPC and DMPC lipid membranes was presented by Wang et al. (2011). In Bochicchio et al. (2015), transfer of polyethylene and polypropylene oligomers through POPC membrane using umbrella sampling and metadynamics has been studied.

Finite system-size effects is another methodological issue which is subject to discussions. In Castro-Román et al. (2006), system-size effects on the structure of a DOPC bilayer are investigated by performing MD simulations of small and large single bilayer patches (72 and 288 lipids, respectively), as well as an explicitly multilamellar system consisting of a stack of five 72-lipid bilayers, demonstrating that finite-size effects are negligible in the considered cases. A similar MD simulations study was performed for DPPC saturated bilayers composed of 72 and 288 lipids to examine system size dependence on dynamical properties (Klauda et al., 2006a). These studies showed that a typical size of MD simulations of lipid bilayers with about 100 lipids is enough for representation of structural properties of large bilayer fragments. However, for lateral diffusion, the effect of system size can be substantial, which was demonstrated, in a paper by Camley et al. (2015), by simulations within MARTINI FF and by a hydrodynamic model.

Treatment of long-range interactions outside the cut-off distance affects also results of the simulations. While importance of correct treatment of the electrostatic forces is well recognized (Patra et al., 2004; Venable et al., 2009), the role of long-range corrections to the Lennard–Jones forces is less appreciated. Many of MD simulations employ a short force cutoff distance of 10–12 Å (and even 8 Å in some cases), out of which van der Waals interactions are neglected. Though the attractive part of the Lennard–Jones potential may seem to be small at such distances, its total contributions to the energy and especially pressure are not negligible. This may noticeably affect average lipid area, and through this other properties. It is sometimes claimed that the cutoff distance is a part of the FF. Such an approach is however inconsistent since the attractive part of the Lennard–Jones potential has a clear physical origin in the London dispersion force. The simplest way to take into account the long-range part of the Lennard–Jones potential is to use the isotropic long-range correction (Allen & Tildesley, 1987). A more accurate isotropic periodic sum (IPS) approach (Wu & Brooks, 2005; Klauda et al., 2007) takes into account long-range corrections to the Lennard–Jones potential in the similar way as Ewald summation of the electrostatic interactions, which may be more appropriate for inhomogeneous systems. Different treatments of out-cutoff corrections may also explain some difference in results for the simulated bilayers computed in different works implementing the same FF.

Finally, it should be mentioned (Zhu et al., 2012) that even the most accurate FF is only as good as the computational design with which it is applied. Issues related to software-introduced artifacts in biomembrane simulations are discussed by Wong-Ekkabut & Karttunen (2016).

3.2 Validation of the Force Fields

As it was mentioned above, a good FF should describe available experimental data within experimental and simulation errors. The average area per lipid defined in constant pressure – zero tension simulations, is a parameter which is often used as a first check to define the quality of the FF, and one of the most common ways to determine whether the bilayer system has reached equilibrium. When the area per lipid reaches a stable value, other structural properties (density distributions, NMR order parameters) do not show noticeable trends either.

Experimental values of the area per lipid are most often obtained from X-ray or neutron diffraction and volumetric data; some other techniques including NMR have been also used. As an example, in Table 7 of paper by Rabinovich et al. (2017) (and in Table 1 of paper by Lyubartsev & Rabinovich (2016)), available experimental average areas per lipid A_{pl} of liquid-crystalline phase mixed-chain PC bilayers are collected. A detailed analysis and comparison of data from MD simulations with different FFs and experiment for different lipids is given in our review (Lyubartsev & Rabinovich, 2016). Another collection of average lipid areas for several fluid-phase PC bilayers, three fully saturated (12:0/12:0 PC, 14:0/14:0 PC, 16:0/16:0 PC) and two monounsaturated (16:0/18:1(n-9)cis PC and 18:1(n-9)cis/18:1(n-9)cis PC), computed from different FFs, as well as corresponding collection of experimental areas, is available in Tables of papers by Poger and Mark (2010) and Poger et al. (2016).

Experimental data presented in Table 7 of paper by Rabinovich et al. (2017) (and Table 1 of paper by Lyubartsev and Rabinovich (2016)) show a wide spread of the average A_{pl} values associated with the same PC bilayer (for which the data are available). For example, for 18:0/18:1 PC lipid at 303 K, the experimental data are in the range 0.61–0.71 nm². The FF used in simulations has also an influence on the computed MD area (Lyubartsev & Rabinovich, 2016). Clearly, validation of an FF against experimental lipid area should necessarily include discussion on the kind of experiment and accuracy of the experimental value. It is important to have in mind that while the statistical errors are usually easily evaluated both in simulations and experiment, systematic errors caused by the artifacts of the simulation setup or by the interpretation of experiment are not easy to discover and they are often not known. Many experimental structural methods were reviewed by Nagle and Tristram-Nagle (2000a, 2000b) and Nagle (2013), and considerable quantitative uncertainty was demonstrated in structural results for lipid bilayers, including the value of area A_{pl} in the fluid phase. The differences across experiments in Table 7 of Rabinovich et al. (2017) (and in Table 1 of Lyubartsev & Rabinovich (2016)) obtained by methods of different nature may be connected to the different experimental approaches.

A structural correction based on fluctuations that have not been included in previous studies was introduced in works by Nagle and Tristram-Nagle (2000a, 2000b). A hybrid zero-baseline structural model was developed by Klauda et al. (2006b) for the electron density profile for the purpose of interpreting X-ray diffraction data. An NMR database for simulations of membrane dynamics was reviewed in Leftin & Brown (2011). There are some difficulties in creating a firm base for comparison of computer simulation results and experimental data because it is not improbable that the latter will also be reanalyzed and corrected. Still it seems at the moment that the combination of X-ray and small angle neutron scattering (Kučerka et al., 2011; Nagle, 2013) provides the most consistent results for the average area per lipid.

A new computational approach to quantify the area per lipid of membranes was presented by Chacón et al. (2015), which relies on the analysis of the membrane fluctuations using coupled undulatory mode (Tarazona et al., 2013). The inclusion of undulation effects is only important for very large bilayer fragments. Unlike the projected area, widely used in computer simulations of membranes, the coupled undulatory area is thermodynamically consistent. This new area definition makes it possible to accurately estimate the area of the undulating bilayer, and the area per lipid, by excluding any contributions related to the phospholipid protrusions (Chacón et al., 2015). It should also be mentioned that as it can be seen from Table 7 of Rabinovich et al. (2017) (and Table 1 of Lyubartsev and Rabinovich (2016)), availability of experimental data on average areas varies between the bilayers. For PC bilayers with PU chains (the same can be said about other than PC lipids), experimental data are scarce and scattering data are often not available. Therefore, a care should be always taken while using experimental data from a single experiment for validation of the FF.

More reliable validation of an FF can be done by comparison of simulated and experimental structure factors of the system and, via Fourier reconstruction, the overall transbilayer scattering-density profiles (Benz et al., 2005). The advantage of validation against scattering factors is that this property is directly accessible from the experiment not subject to model interpretations and approximations. Such comparison can be done both for X-ray scattering data (determined by the electron density) and for neutron scattering (determined by the mass density) which allows to get additional insight into bilayer structure. Comparison of the simulated structure factors with experimental scattering data has become a necessary component of the validation of new FFs and their variations.

Additional important source of data for validation of an FF used in lipid bilayer simulations is NMR bond order parameters. Comparison of experimental and simulated order parameters for lipid tails is usually included into validation section of most of papers presenting new FFs and their variations, and characteristic features of order parameter profile for saturated and unsaturated tails are typically well reproduced. Validation against order parameters of lipid headgroups is less common. This can be partially explained by the relative lack of such experimental data. In a recent paper by Botan et al. (2015), 13 different FFs have been compared against experimentally available NMR C-H order parameters of the glycerol and choline headgroups moiety of several PC lipids, with variation of hydration levels and the presence of CHOL. CHARMM- and GAFF-based lipid FF, as well as FF from work by Maciejewski et al. (2014), were found to perform better for glycerol backbone than the GROMOS-based models.

3.3 Partitioning and Penetration of Small Molecules

One of the key membrane functions is the regulation of the transport of small molecules across the membrane. Behavior of small solutes in lipid membranes is widely discussed by Bemporad et al. (2004a, 2004b, 2005). The accurate description of not only pure lipid bilayers but also molecular interactions between guest molecules and membranes is another challenge for FF development. An FF well tuned for pure lipid bilayer simulations may not be compatible with FF describing penetrants. For more complicated goals such as membrane–protein and membrane–peptide studies, FF should also achieve a properly balanced description of structural and dynamical features of proteins, peptides, etc. interacting with lipid bilayers, as well as their correct partitioning between hydrophobic and headgroup regions of the bilayer and surrounding water.

While the membrane transport, as a rule, involves special channel-forming peptides and proteins, various small, uncharged molecules, such as O_2 and CO_2, water, as well as some drugs, can permeate in small amounts the cell membrane without the aid of any transmembrane protein. Partitioning of small molecules across the lipid bilayer is directly related to the free energy profile and adsorption (binding) free energies, which also determines the rate of their passive penetration through the membrane. These properties, often accessible experimentally, is a good benchmark of compatibility of the FFs describing lipids and small molecules.

MD simulations of DPPC, POPC, and DAPC membranes were performed to explore the energetics and mechanism of passive CHOL flip-flop and its dependence on chain saturation (Jo et al., 2010). The resulting paths indicate that CHOL prefers to tilt first and then move to the bilayer center where the free energy barrier exists. The barrier is lower in DAPC than in DPPC or POPC, and the calculated flip-flop rates show that CHOL flip-flop in a PU bilayer is faster than in more saturated bilayers. Free energy profile of a pair of CHOL molecules in a leaflet of POPC bilayers in the liquid-crystalline phase has been calculated in Andoh et al. (2012b) as a function of their lateral distance using a combination of NPT-constant atomistic MD calculations and the thermodynamic integration method.

To elucidate the molecular mechanism of the reduction in water leakage across the membranes by the addition of CHOL, water permeability of DPPC and palmitoyl-SM bilayers in the absence and presence of CHOL (0, 10, 20, 30, 40, and 50 mol%) has been studied by MD simulations (Saito & Shinoda, 2011). An enhanced free energy barrier was observed in these membranes with increased CHOL concentration, and this was explained by the reduced cavity density around the CHOL in the hydrophobic membrane core and this was found to be the main reason to reduce the water permeability.

Systematic MD simulations were done by Wennberg et al. (2012) to study the partitioning of solutes between water and membranes. Potentials of mean force were derived for six different solutes (ethanol, ammonia, nitric oxide, propane, benzene, and neopentane) permeating across 20 different lipid membranes containing one out of four types of phospholipids (DMPC, DPPC, POPC, POPE) plus a CHOL content of 0, 20, 30, 40, and 50 mol%. The simulations showed that the partitioning is more sensitive to CHOL (i) for larger solutes, (ii) in membranes with saturated as compared to membranes with unsaturated lipid tails, and (iii) in membranes with smaller lipid head groups.

In Sugii et al. (2005), the effects of the hydrocarbon chain length of lipid molecules on the permeation process of small molecules (O_2, CO, NO, and water) through lipid bilayers were investigated. MD simulations of three saturated lipid bilayer systems were performed: DLPC, DMPC, DPPC.

Six molecules of $(CF_3)_2$–benzoic acid with deprotonated carboxyl groups were inserted into the MD simulation box containing a lipid bilayer with DMPC molecules (Dürr et al., 2012). MD simulations confirmed the intuitive expectation that the $(CF_3)_2$–benzoic acid molecules are oriented in the lipid bilayer according to their amphiphilic properties, which also allows for favorable hydrogen bonding within the lipid headgroup region. MD simulations have also been carried out to scan the interdigitation effect at the (5S)-1-benzylo-5-(1H-benzimidazol-1-ylo-methylo)-2-pyrrolidinone/DMPC system (Mavromoustakos et al., 2011); partial interdigitation was observed. Behavior of fluorescent probes (Loura & Ramalho, 2011) in a lipid bilayer from computer simulations is also widely discussed.

A rather large amount of works studying passive membrane transport were done with drug molecules. To reach their biological target, drugs have to cross cell membranes, and understanding passive membrane permeation of drug-like molecules (Ulander & Haymet, 2003; Mukhopadhyay et al., 2004; Bemporad et al., 2005) is therefore very important. Behavior of anesthetics (e.g., lidocaine, benzocaine, articaine, halothane, hexafluoroethane, short chain alcohols like methanol, ethanol, 1-alkanols) (Koubi et al., 2003; Patra et al., 2006; Högberg et al., 2007; Castro et al., 2008; Högberg & Lyubartsev, 2008; Terama et al., 2008; Bernardi et al., 2009; Porasso et al., 2009; Vemparala et al., 2010; Prates et al., 2011) in lipid bilayer is interesting not only from the permeation point of view, but also because possible modulation of membrane properties by anesthetics is considered as possible mechanism of their action (Castro et al., 2008). It was demonstrated by Porasso et al. (2009) that addition of local anesthetic benzocaine increases disorder in the membrane. A thermodynamic study of benzocaine insertion into DPPC and DPPS bilayers by means of MD was carried out by Cascales et al. (2011). It was shown that an increase in the DPPS fraction of the lipid bilayer facilitates the insertion of the benzocaine into the bilayer, an observation that could be related with the activity of certain drugs that depend on the lipid composition of the cell membrane. It was shown (Bernardi & Pascutti, 2012) for benzocaine, lidocaine, and tetracaine, that the charged form of these drugs are oriented at the interface as one of the lipids, while the neutral form can easily cross the interface, entering the membrane (DPPC) in agreement with most experimental results. Overviews of the MD simulation studies used for different drug–membrane interactions (Lopes et al., 2017) and molecular simulations of nonfacilitated membrane permeation (Awoonor-Williams & Rowley, 2016) were recently presented.

MD simulations and X-ray diffraction analysis study of several 18:1(n-9)trans/18:1(n-9)trans PE membranes containing free FAs was performed in Cordomí et al. (2010). The study was aimed at understanding the interactions of several structurally related FAs with biomembranes, which is necessary for further rational lipid drug design in membrane–lipid therapy. FAs able to affect biophysical properties of cell membranes in turn will also alter localization and/or function of membrane protein involved in the regulation of cellular processes. For the above reasons, FAs or other lipids could be a tool to modulate pathophysiological conditions via cell membrane properties (Cordomí et al., 2010).

An important contributor to the thermodynamic driving force is the available free volume across a membrane. Thus, the diffusion properties of the penetrants are obviously related to the properties of the free volume clusters (e.g., their size, shape, orientation, etc.) present in the membrane, and therefore a detailed analysis of the voids can also provide some information on the permeability properties of the membrane. Such a study was performed for bilayers composed of 18:0/18:1(n-9)cis PC, 18:0/18:2(n-6)cis PC, 18:0/18:3(n-3)cis PC, 18:0/20:4(n-6)cis PC, and 18:0/22:6(n-3)cis PC molecules in Rabinovich et al. (2005). It was found that the preformed cross-membrane channels are not broad enough to let small molecules, such as water, go readily through them; however, they are likely to facilitate the permeation of such molecules across the membrane.

Effects of various nanostructured materials (nanoparticles) on lipid membranes is also widely discussed (Makarucha et al., 2011), including carbon nanotubes (Li et al., 2012; Lee & Kim, 2012; Skandani et al., 2012), gold (Lin et al., 2011), and other nanocrystals (Song et al., 2011), fullerenes Jusufi et al. (2011), etc. Many hydrophobic nanoparticles are found to be able to transverse a membrane, with some nanoparticles even causing damage to the membrane, thus potentially leading to cytotoxic effects. Though lipid membranes have been very intensively studied by computer simulations during last decade, in general, modeling translocation of nanoparticles through a lipid membrane is a significant challenge. The physical mechanisms of the effect of some nanomaterials on lipid membranes as well as other functional biomolecules on the basis of recent computational studies were discussed in a review by Ding and Ma (2018).

3.4 Comparisons of Force Fields

While most of work introducing new FF or FF updates concentrate on validation of the introduced FF, comparison of results obtained with different FFs provides another source of very important, valuable information.

The free energy profiles of 11 different molecules through a model DMPC bilayer using five FFs (Berger (Berger et al., 1997), Slipids (Jämbeck & Lyubartsev, 2012a, 2012b, 2013a), CHARMM36 (Klauda et al., 2010; Pastor & MacKerell, Jr., 2011), GAFFlipid (Dickson et al., 2012), and GROMOS 43A1-S3 (Chiu et al., 2009)) were calculated (Paloncýová et al., 2014). Molecules used in the study were glycerol, methanol, acetone, 1-butanol, benzylalcohol, aniline, 2-nitrotoluene, p-xylene, 4-chloro-3-methylphenol, 2,4,5-trichloroaniline, and hexachlorobenzene. A relevant result of the study is that Slipids and CHARMM36 performed well in the prediction of free energy barriers; GAFFlipid predicted mean differences of water/lipids barriers very well. To study hydrophilic molecules, CHARMM36 is the only FF able to predict a correct ranking of lipophilicity. Finally, Slipids is recommended as the versatile FF for simulations of complex molecular systems containing lipid bilayers (Paloncýová et al., 2014).

Using two different all-atom lipid FFs, AMBER LIPID14 (Dickson et al., 2014) and CHARMM36 (Klauda et al., 2010), MD simulations started from random mixtures of lipids and water were performed (Skjevik et al., 2015) in which four different types of phospholipids (DPPC, POPC, DOPC, and POPE) self-assembled into organized bilayers under 1 microsecond. In all of the simulations, the lipids self-assembled into bilayers via the same general pathway. The CHARMM36 PC lipids were shown to self-assemble faster than their LIPID14 equivalents, whereas the POPE bilayer formation times are quite similar when comparing the two FFs. In a work by Skjevik et al. (2016), the above MD simulations were expanded: three all-atom FFs, AMBER LIPID14 (Dickson et al., 2014), CHARMM36 (Klauda et al., 2010), and Slipids (Jämbeck & Lyubartsev, 2012a, 2012b, 2013a) were used and eight different types of phospholipids (DPPC, POPC, DOPC, POPE, POPS, POPG, DOPS, and DOPG) self-assembled into organized bilayers. Irrespective of the underlying FF, the lipids were shown to spontaneously form stable lamellar bilayer structures within 1 microsecond, the majority of which display properties in satisfactory agreement with the experimental data.

Four lipid FFs, including the united-atom GROMOS54a7 (Schmid et al., 2011) and the all-atom FFs CHARMM36 (Klauda et al., 2010), Slipids (Jämbeck & Lyubartsev, 2012a), and Lipid14 (Dickson et al., 2014), for a broad range of structural and dynamical properties of saturated and monounsaturated PC bilayers (DMPC and POPC) as well as for monounsaturated PE bilayers (POPE) were compared (Pluhackova et al., 2016). Additionally, in this work, the ability of the different FFs to describe the gel–liquid crystalline phase transition was compared and their computational efficiency estimated. Moreover, membrane properties like the water flux across the lipid bilayer and lipid acyl chain protrusion probabilities were compared. All studied FFs were found to satisfactorily describe the overall structural characteristics of phospholipid bilayers; the lipid diffusion and water permeability were well described by the three all-atom FFs. However, also deficiencies in reproducing different observables were still seen for all lipid FFs (Pluhackova et al., 2016). For example, the GROMOS54a7 FF was found to underestimate the lipid self-diffusion and showed some weakness in describing the melting characteristics of membranes and is therefore less suitable for studies where dynamic lipid behavior is expected to play an important role. It was also found that Slipids is very well suited for membrane studies closely above the melting temperature; Lipid14 and CHARMM36 showed well-ordered membrane structures for the gel phase and a highly cooperative transition during melting. The acyl chain order and X-ray and neutron scattering form factors were best reproduced by Slipids and Lipid14; CHARMM36 yielded improved values for the lipid volume and bilayer thickness as compared to the other all-atom FFs and described most realistically the lipid diffusion. According to Pluhackova et al. (2016), the computationally most efficient all-atom FF was Lipid14.

In a work by Papadimitriou et al. (2015), a series of MD simulations of a fully hydrated lipid bilayer consisting of pure ceramide (CER NS 24:0) have been presented; ceramides comprise the major component of the lipid phase of human skin. The simulations were performed by using five different FFs: OPLS (Jorgensen & Tirado-Rives, 1985), GROMOS 54A7 (Schmid et al., 2011), Berger (Berger et al., 1997), CHARMM C36 (Klauda et al., 2010), and GAFFlipid (Dickson et al., 2012), in order to evaluate and compare their performance in modeling lipid systems that contain ceramides. The examined properties include bilayer thickness, chain tilt, density profiles, order parameters, chain conformation, area per lipid, and (intermolecular or intramolecular) hydrogen bonding between the head groups. In general, it was observed that an FF may be more accurate in predicting certain structural properties while it may not be so successful in the prediction of others. For a few properties (e.g. intramolecular hydrogen bonding), there was some discrepancy between the FFs but the lack of respective experimental data did not allow an unambiguous conclusion regarding which FF was the most reliable. In any case, all considered FFs gave similar qualitative results.

4 Conclusions

Rapid development of the modern computer hardware makes simulations of larger molecular systems during longer time possible, which leads to higher requirements for the FF quality. This connection can be drawn for all molecular systems, but it is especially relevant for lipid bilayers which are soft systems with properties defined by a delicate balance of different interactions between hydrophilic and hydrophobic parts of the molecules and water as a solvent. In this review, we have discussed the recent progress in the development of FF for lipid membranes simulations, considering main classes of the atomistic FFs such as GROMOS, CHARMM, AMBER, Slipids, and their variations, and gave comparative analysis of the FFs and their ability to describe experimental behavior of the bilayers.

ACKNOWLEDGEMENTS

This work has been supported by the Swedish Science Research Council (Vetenskapsrådet, grant 621-2013-4260, to AL), the state order (project 0221-2017-0050, regist. no. AAAA-A17-117031710039-3) of Russian Federation (to AR), and EU FP7 MembraneNanoPart project (to AL and AR).

REFERENCES

Allen, M. P. & Tildesley, D. J. (1987), *Computer simulations of liquids*, 2nd edn, Oxford, Clarendon.

Andoh, Y., Ito, T. & Okazaki, S. (2012a), 'An application of improved force field to fully hydrated DPPC and POPC bilayers in a tensionless NPT ensemble: A test of CHARMM 27-based new force field by Högberg et al.', *Mol. Simul.* **38**(5), 414–418.

Andoh, Y., Oono, K., Okazaki, S. & Hatta, I. (2012b), 'A molecular dynamics study of the lateral free energy profile of a pair of cholesterol molecules as a function of their distance in phospholipid bilayers', *J. Chem. Phys.* **136**, 155104.

Anezo, C., de Vries, A. H., Höltje, H.-D., Tieleman, D. P. & Marrink, S.-J. (2003), 'Methodological issues in lipid bilayer simulations', *J. Phys. Chem. B* **107**, 9424–9433.

Arkhipov, A., Yin, Y. & Schulten, K. (2008), 'Four-scale description of membrane sculpting by BAR domains', *Biophys. J.* **95**, 2806–2821.

Arkhipov, A., Yin, Y. & Schulten, K. (2009), 'Membrane-bending mechanism of amphiphysin N-BAR domains', *Biophys. J.* **97**, 2727–2735.

Arnarez, C., Uusitalo, J. J., Masman, M. F., Ingólfsson, H. I., de Jong, D. H., Melo, M. N., Periole, X., de Vries, A. H. & Marrink, S. J. (2015), 'Dry Martini, a coarse-grained force field for lipid membrane simulations with implicit solvent', *J. Chem. Theory Comput.* **11**(1), 260–275.

Awoonor-Williams, E. & Rowley, C. N. (2016), 'Molecular simulation of non-facilitated membrane permeation', *Biochim. Biophys. Acta, Biomembranes* **1858**(7), 1672–1687.

Ayton, G. S., Lyman, E. & Voth, G. A. (2010), 'Hierarchical coarse-grained strategy for protein-membrane systems to access mesoscopic scales', *Faraday Discuss.* **144**, 347–358.

Ayton, G. S., Noid, W. G. & Voth, G. A. (2007), 'Multiscale modeling of biomolecular systems: In serial and in parallel', *Curr. Opin. Struct. Biol.* **17**, 192–198.

Ayton, G. S. & Voth, G. A. (2009a), 'Hybrid coarse-graining approach for lipid bilayers at large length and time scales', *J. Phys. Chem. B* **113**, 4413–4424.

Ayton, G. S. & Voth, G. A. (2009b), 'Systematic multiscale simulation of membrane protein systems', *Curr. Opin. Struct. Biol.* **19**, 138–144.

Baaden, M. & Marrink, S. J. (2013), 'Coarse-grain modelling of protein-protein interactions', *Curr. Opin. Struct. Biol.* **23**(6), 878–886.

Baoukina, S. & Tieleman, D. P. (2016), 'Computer simulations of lung surfactant', *Biochim. Biophys. Acta* **1858**(10), 2431–2440.

Bemporad, D., Essex, J. W. & Luttmann, C. (2004a), 'Computer simulation of small molecule permeation across a lipid bilayer: Dependence on bilayer properties and solute volume, size and cross – sectional area', *Biophys. J.* **87**, 1–13.

Bemporad, D., Essex, J. W. & Luttmann, C. (2004b), 'Permeation of small molecules through a lipid bilayer: A computer simulation study', *J. Phys. Chem. B* **108**, 4875–4884.

Bemporad, D., Essex, J. W. & Luttmann, C. (2005), 'Behaviour of small solutes and large drugs in a lipid bilayer from computer simulations', *Biochim. Biophys. Acta* **1718**, 1–21.

Bennett, W. F. D. & Tieleman, D. P. (2013), 'Computer simulations of lipid membrane domains', *Biochim. Biophys. Acta* **1828**(8), 1765–1776.

Bennun, S. V., Hoopes, M. I., Xing, C. & Faller, R. (2009), 'Coarse-grained modeling of lipids', *Chem. Phys. Lipids* **159**, 59–66.

Benz, R. W., Castro-Román, F., Tobias, D. J. & White, S. H. (2005), 'Experimental validation of molecular dynamics simulations of lipid bilayers: A new approach', *Biophys. J.* **88**(2), 805–817.

Berendsen, H. J. C. & Tieleman, D. P. (1998), 'Molecular dynamics: Studies of lipid bilayers', in *Encyclopedia of computational chemistry*, R. Schleyer, ed. Chichester, NY et al., J. Wiley & Sons, pp. 1639–1650.

Berger, O., Edholm, O. & Jähnig, F. (1997), 'Molecular dynamics simulations of a fluid bilayer of dipalmitoylphosphatidylcholine at full hydration, constant pressure and constant temperature', *Biophys. J.* **72**(5), 2002–2013.

Bernardi, R. C., Gomes, D. E., Gobato, R., Taft, C. A., Ota, A. T. & Pascutti, P. G. (2009), 'Molecular dynamics study of biomembrane/local anesthetics interactions', *Mol. Phys.* **107**(14), 1437–1443.

Bernardi, R. C. & Pascutti, P. G. (2012), 'Hybrid QM/MM molecular dynamics study of benzocaine in a membrane environment: How does a quantum mechanical treatment of both anesthetic and lipids affect their interaction', *J. Chem. Theory Comput.* **8**, 2197–2203.

Bochicchio, D., Panizon, E., Ferrando, R., Monticelli, L. & Rossi, G. (2015), 'Calculating the free energy of transfer of small solutes into a model lipid membrane: Comparison between metadynamics and umbrella sampling', *J. Chem. Phys.* **143**, 144108.

Botan, A., Fernando, F., Francois, P., Fuchs, J., Javanainen, M., Kanduc, M., Kulig, W., Lamberg, A., Loison, C., Lyubartsev, A. P., Miettinen, M. S., Monticelli, L., Määttä, J., Ollila, S. O. H., Retegan, M., Rog, T., Santuz, H. & Tynkkynen, J. P. (2015), 'Towards atomistic resolution structure of phosphatidylcholine headgroup and glycerol backbone at different ambient conditions', *J. Phys. Chem. B* **119**, 15075–15088.

Bradley, R. & Radhakrishnan, R. (2013), 'Coarse-grained models for protein-cell membrane interactions', *Polymers* **5**(3), 890–936.

Brannigan, G., Lin, L. C. L. & Brown, F. L. H. (2006), 'Implicit solvent simulation models for biomembranes', *Biophys. J.* **35**, 104–124.

Bunker, A., Magarkar, A. & Viitala, T. (2016), 'Rational design of liposomal drug delivery systems, a review: Combined experimental and computational studies of lipid membranes, liposomes and their pegylation', *Biochim. Biophys. Acta* **1858**(10), 2334–2352.

Camley, B. A., Lerner, M. G., Pastor, R. W. & Brown, F. L. H. (2015), 'Strong influence of periodic boundary conditions on lateral diffusion in lipid bilayer membranes', *J. Chem. Phys.* **143**, 243113.

Cascales, J. J. L., Costa, S. D. O. & Porasso, R. D. (2011), 'Thermodynamic study of benzocaine insertion into different lipid bilayers', *J. Chem. Phys.* **135**, 135103.

Cascella, M. & Peraro, M. D. (2009), 'Challenges and perspectives in biomolecular simulations: From the atomistic picture to multiscale modeling', *Chimia* **63**(1–2), 14–18.

Castro, V., Stevensson, B., Dvinskikh, S. V., Högberg, C.-J., Lyubartsev, A. P., Zimmermann, H., Sandström, D. & Maliniak, A. (2008), 'NMR investigations of interactions between anesthetics and lipid bilayers', *Biochim. Biophys. Acta* **1778**, 2604–2611.

Castro-Román, F., Benz, R. W., White, S. H. & Tobias, D. J. (2006), 'Investigation of finite system-size effects in molecular dynamics simulations of lipid bilayers', *J. Phys. Chem. B* **110**, 24157–24164.

Chacón, E., Tarazona, P. & Bresme, F. (2015), 'A computer simulation approach to quantify the true area and true area compressibility modulus of biological membranes', *J. Chem. Phys.* **143**(3), 034706.

Chan, Y.-H. M. & Boxer, S. G. (2007), 'Model membrane systems and their applications', *Curr. Opin. Chem. Biol.* **11**, 581–587.

Chandrasekhar, I., Kastenholz, M., Lins, R. D., Oostenbrink, C., Schuler, L. D., Tieleman, D. P. & van Gunsteren, W. F. (2003), 'A consistent potential energy parameter set for lipids: Dipalmitoyl-phosphatidylcholine as a benchmark of the GROMOS96 45A3 force field', *Eur. Biophys. J.* **32**(1), 67–77.

Chiu, S. W., Pandit, S. A., Scott, H. L. & Jakobsson, E. (2009), 'An improved united atom force field for simulations of mixed lipid bilayers', *J. Phys. Chem. B* **113**, 2748–2763.

Cooke, I. R. & Deserno, M. (2005), 'Solvent-free model for self-assembling fluid bilayer membranes: Stabilization of the lipid phase based on broad attractive tail potentials', *J. Chem. Phys.* **123**, 224710.

Cordomí, A., Prades, J., Frau, J., Vögler, O., Funari, S. S., Perez, J. J., Escribá, P. V. & Barceló, F. (2010), 'Interactions of fatty acids with phosphatidylethanolamine membranes: X-ray diffraction and molecular dynamics studies', *J. Lipid Res.* **51**, 1113–1124.

Damodaran, K. V. & Merz, K. M. (1994), 'Computer simulation of lipid systems', in *Reviews in computational chemistry*, Vol. 5, K. B. Lipkowitz & D. B. Boyd, eds. New York, VCH Publishers, Inc, pp. 269–298.

Damodaran, K. V., Merz, K. M. & Gaber, B. P. (1992), 'Structure and dynamics of the dilauroylphosphatidylethanolamine lipid bilayer', *Biochem.* **31**, 7656–7664.

Dickson, C. J., Madej, B. D., Skjevik, A. A., Betz, R. M., Teigen, K., Gould, I. R. & Walker, R. C. (2014), 'Lipid14: The Amber lipid force field', *J. Chem. Theory Comput.* **10**(2), 865–879.

Dickson, C. J., Rosso, L., Betz, R. M., Walker, R. C. & Gould, I. R. (2012), 'GAFFlipid: A General Amber Force Field for the accurate molecular dynamics simulation of phospholipid', *Soft Matter* **8**(37), 9617–9627.

Ding, H. M. & Ma, Y. Q. (2018), 'Computational approaches to cell – nanomaterial interactions: Keeping balance between therapeutic efficiency and cytotoxicity', *Nanoscale Horiz* **3**(1), 6–27.

Dratz, E. A. & Deese, A. J. (1986), 'The role of docosahexaenoic acid (22:6ω3) in biological membranes: Examples from photoreceptors and model membrane bilayers', in *The health effects of polyunsaturated fatty acids in seafoods*, A. P. Simopoulous, R. R. Kifer & R. E. Martin, eds. Washington, Academic Press, Inc., pp. 319–351.

Dürr, U. H. N., Afonin, S., Hoff, B., de Luca, G., Emsley, J. W. & Ulrich, A. S. (2012), 'Alignment of druglike compounds in lipid bilayers analyzed by solid-state ^{19}F − NMR and molecular dynamics, based on dipolar couplings of adjacent CF 3 groups', *J. Phys. Chem. B* **116**, 4769–4782.

Ermilova, I. & Lyubartsev, A. P. (2016), 'Extension of the slipids force field to polyunsaturated lipids', *J. Phys. Chem. B* **120**(50), 12826–12842.

Essmann, U., Perera, L. & Berkowitz, M. L. (1995), 'The origin of the hydration interaction of lipid bilayers from MD simulation of dipalmitoylphosphatidylcholine membranes in gel and liquid crystalline phases', *Langmuir* **11**, 4519–4531.

Feller, S. E. (2000), 'Molecular dynamics simulations of lipid bilayers', *Curr. Opin. Colloid Interf. Sci.* **5**, 217–223.

Feller, S. E. (2001), 'Molecular dynamics simulation of phospholipid bilayers', in *Lipid Bilayers. Structure and Interactions*, John Katsaras & Thomas Gutberlet, eds. Berlin, Heidelberg, N.Y, Springer-Verlag, pp. 89–107.

Feller, S. E., ed. (2008), *Computational modeling of membrane bilayers. Current topics in membranes*, Vol. 60, San Diego, Elsevier Inc., 448 p.

Feller, S. E. & Gawrisch, K. (2005), 'Properties of docosahexaenoic-acidcontaining lipids and their influence on the function of rhodopsin', *Curr. Opin. Struct. Biol.* **15**, 416–422.

Feller, S. E. & MacKerell, Jr., A. D. (2000), 'An improved empirical potential energy function for molecular simulations of phospholipids', *J. Phys. Chem. B* **104**, 7510–7515.

Feller, S. E., Yin, D., Pastor, R. W. & MacKerell, Jr., A. D. (1997), 'Molecular dynamics simulation of unsaturated lipid bilayers at low hydration: Parametrization and comparison with diffraction studies', *Biophys. J.* **73**(5), 2269–2279.

Forrest, L. R. & Sansom, M. S. P. (2000), 'Membrane simulations: Bigger and better?', *Curr. Opin. Struct. Biol.* **10**, 174–181.

Gawrisch, K., Eldho, N. V. & Holte, L. L. (2003), 'The structure of DHA in phospholipid membranes', *Lipids* **38**(4), 445–452.

Gawrisch, K., Soubias, O. & Mihailescu, M. (2008), 'Insights from biophysical studies on the role of polyunsaturated fatty acids for function of g-protein coupled membrane receptors', *Prostagl., Leukotr. & Essent. Fatty Acids* **79**, 131–134.

Gullingsrud, J. & Schulten, K. (2004), 'Lipid bilayer pressure profiles and mechanosensitive channel gating', *Biophys. J.* **86**, 3496–3509.

Gurtovenko, A. A., Anwar, J. & Vattulainen, I. (2010), 'Defect-mediated trafficking across cell membranes: Insights from in silico modeling', *Chem. Rev.* **110**, 6077–6103.

Hadley, K. R. & McCabe, C. (2010), 'A coarse-grained model for amorphous and crystalline fatty acids', *J. Chem. Phys.* **132**(13), 134505.

Hansson, T., Oostenbrink, C. & van Gunsteren, W. F. (2002), 'Molecular dynamics simulations', *Curr. Opin. Struct. Biol.* **12**, 190–196.

Heller, H., Schaefer, M. & Schulten, K. (1993), 'Molecular dynamics simulations of a bilayer of 20 lipids in the gel and in the liquid-crystall phase', *J. Phys. Chem.* **97**, 8343–8360.

Henin, J., Shinoda, W. & Klein, M. L. (2008), 'United atom acyl chain for CHARMM phospholipids', *J. Phys. Chem. B* **112**, 7008–7015.

Hermans, J., Berendsen, H. J. C., van Gunsteren, W. F. & Postma, J. P. M. (1984), 'A consistent empirical potential for water-protein interactions', *Biopolymers* **23**, 1513–1518.

Högberg, C.-J. & Lyubartsev, A. P. (2006), 'A molecular dynamics investigation of the influence of hydration and temperature on structural and dynamical properties of a dimyristoylphosphatidylcholine bilayer', *J. Phys. Chem. B* **110**, 14326–14336.

Högberg, C.-J. & Lyubartsev, A. P. (2008), 'Effect of local anesthetic lidocaine on electrostatic properties of a lipid bilayer', *Biophys. J.* **94**, 525–531.

Högberg, C.-J., Maliniak, A. & Lyubartsev, A. P. (2007), 'Dynamical and structural properties of charged and uncharged lidocaine in a lipid bilayer', *Biophys. Chem.* **125**, 416–424.

Högberg, C.-J., Nikitin, A. M. & Lyubartsev, A. P. (2008), 'Modification of the CHARMM force field for DMPC lipid bilayer', *J. Comp. Chem.* **29**, 2359–2369.

Hyvönen, M. T. & Kovanen, P. T. (2005), 'Molecular dynamics simulations of unsaturated lipid bilayers: Effects of varying the number of double bonds', *Eur. Biophys. J.* **34**, 294–305.

Ingólfsson, H. I., Lopez, C. A., Uusitalo, J. J., de Jong, D. H., Gopal, S. M., Periole, X. & Marrink, S. J. (2014), 'The power of coarse graining in biomolecular simulations', *WIREs Comput. Mol. Sci.* **4**(3), 225–248.

Izvekov, S. & Voth, G. A. (2005), 'A multiscale coarse-graining method for biomolecular systems', *J. Phys. Chem. B* **109**, 2469–2473.

Izvekov, S. & Voth, G. A. (2006), 'Multiscale coarse-graining of mixed phospholipid/cholesterol bilayers', *J. Chem. Theory Comput.* **2**, 637–648.

Izvekov, S. & Voth, G. A. (2009), 'Solvent-free lipid bilayer model using multi-scale coarse-graining', *J. Phys. Chem. B* **113**, 4443–4455.

Jacobsson, E. (1997), 'Computer simulation studies of biological membranes: Progress promise and pitfalls', *Trends Biochem Sci* **22**(9), 339–344.

Jämbeck, J. P. M. & Lyubartsev, A. P. (2012a), 'Derivation and systematic validation of a refined all-atom force field for phosphatidylcholine lipids', *J. Phys. Chem. B* **116**, 3164–3179.

Jämbeck, J. P. M. & Lyubartsev, A. P. (2012b), 'An extension and further validation of an all-atomistic force field for biological membranes', *J. Chem. Theory Comput.* **8**, 2938–2948.

Jämbeck, J. P. M. & Lyubartsev, A. P. (2013a), 'Another piece of the membrane puzzle: Extending Slipids further', *J. Chem. Theory Comput.* **9**, 774–784.

Jämbeck, J. P. M. & Lyubartsev, A. P. (2013b), 'Exploring the free energy landscape of solutes embedded in lipid bilayers', *J. Phys. Chem. Lett.* **4**, 1781–1787.

Jensen, M. Ø., Mouritsen, O. G. & Peters, G. H. (2004), 'Simulations of a membrane-anchored peptide: Structure, dynamics, and influence on bilayer properties', *Biophys. J.* **86**, 3556–3575.

Jo, S., Rui, H., Lim, J. B., Klauda, J. B. & Im, W. (2010), 'Cholesterol flip-flop: Insights from free energy simulation studies', *J. Phys. Chem. B* **114**(42), 13342–13348.

Jójárt, B. & Martinek, T. A. (2007), 'Performance of the general Amber force field in modeling aqueous POPC membrane bilayers', *J. Comput. Chem.* **28**(12), 2051–2058.

Jorgensen, W. L., Maxwell, D. S. & Tirado-Rives, J. (1996), 'Developing and testing of the OPLS all-atom force field on conformational energetics and properties of organic liquids', *J. Am. Chem. Soc.* **118**(45), 11225–11236.

Jorgensen, W. L. & Tirado-Rives, J. (1985), 'The OPLS potential functions for proteins: Energy minimizations for crystals of cyclic peptides and crambin', *J. Am. Chem. Soc.* **110**, 1657–1666.

Jusufi, A., DeVane, R. H., Shinoda, W. & Klein, M. L. (2011), 'Nanoscale carbon particles and the stability of lipid bilayers', *Soft Matter* **7**, 1139–1146.

Kamerlin, S. C. L., Vicatos, S., Dryga, A. & Warshel, A. (2011), 'Coarsegrained (multiscale) simulations in studies of biophysical and chemical systems', *Annu. Rev. Phys. Chem.* **62**, 41–64.

Kang, H. & Klauda, J. B. (2015), 'Molecular dynamics simulations of palmitoyloleoylphosphatidylgly-cerol bilayers', *Mol. Simul.* **41**(10–12), 948–954.

Kirsch, S. A. & Böckmann, R. A. (2016), 'Membrane pore formation in atomistic and coarse-grained simulations', *Biochim. Biophys. Acta* **1858**(10), 2266–2277.

Klauda, J. B., Brooks, B. R., MacKerell, A. D., Venable, R. M. & Pastor, R. W. (2005), 'An ab initio study on the torsional surface of alkanes and its effect on molecular simulations of alkanes and a DPPC bilayer', *J. Phys. Chem. B* **109**, 5300–5311.

Klauda, J. B., Brooks, B. R. & Pastor, R. W. (2006a), 'Simulation-based methods for interpreting x-ray data from lipid bilayers', *J. Chem. Phys.* **125**, 144710.

Klauda, J. B., Kučerka, N., Brooks, B. R., Pastor, R. W. & Nagle, J. F. (2006b), 'Simulation-based methods for interpreting x-ray data from lipid bilayers', *Biophys. J.* **90**, 2796–2807.

Klauda, J. B., Monje, V., Kim, T. & Im, W. (2012), 'Improving the CHARMM force field for poly-unsaturated fatty acid chains', *J. Phys. Chem. B* **116**, 9424–9431.

Klauda, J. B., Venable, R. M., Freites, J. A., O'Connor, J. W., Tobias, D. J., Mondragon-Ramires, C., Vorobyov, I., MacKerell, A. D. & Pastor, R. W. (2010), 'Update of the CHARMM all-atom additive force field for lipids: Validation on six lipid types', *J. Phys. Chem. B* **114**, 7830–7843.

Klauda, J. B., Wu, X., Pastor, R. W. & Brooks, B. R. (2007), 'Long-range Lennard-Jones and electrostatic interactions in interfaces: Application of the isotropic periodic sum method', *J. Phys. Chem. B* **111**, 4393–4400.

Klein, M. L. & Shinoda, W. (2008), 'Large-scale molecular dynamics simulations of self-assembling systems', *Science* **321**, 798–800.

Koubi, L., Saiz, L., Tarek, M., Scharf, D. & Klein, M. L. (2003), 'Influence of anesthetic and nonimmobilizer molecules on the physical properties of a polyunsaturated lipid bilayer', *J. Phys. Chem. B* **107**(51), 14500–14508.

Kox, A. J., Michels, J. P. J. & Wiegel, F. W. (1980), 'Simulation of a lipid monolayer using molecular dynamics', *Nature* **287**, 317–319.

Kučerka, N., Nieh, M.-P. & Katsaras, J. (2011), 'Fluid phase lipid areas and bilayer thicknesses of commonly used phosphatidylcholines as a function of temperature', *Biochim. Biophys. Acta* **1808**, 2761–2771.

Lee, H. & Kim, H. (2012), 'Self-assembly of lipids and single-walled carbon nanotubes: Effects of lipid structure and PEGylation', *J. Phys. Chem. C* **116**, 9327–9333.

Lee, S., Tran, A., Allsopp, M., Lim, J. B., Hénin, J. & Klauda, J. B. (2014), 'CHARMM36 United Atom chain model for lipids and surfactants', *J. Phys. Chem. B* **118**, 547–556.

Leftin, A. & Brown, M. F. (2011), 'An NMR database for simulations of membrane dynamics', *Biochim. Biophys. Acta, Biomembranes* **1808**, 818–839.

Li, X., Shi, Y., Miao, B. & Zhao, Y. (2012), 'Effects of embedded carbon nanotube on properties of biomembrane', *J. Phys. Chem. B* **116**, 5391–5397.

Lim, J. B., Rogaski, B. & Klauda, J. B. (2012), 'Update of the cholesterol force field parameters in CHARMM', *J. Phys. Chem. B* **116**, 203–210.

Lin, J.-Q., Zheng, Y.-G., Zhang, H.-W. & Chen, Z. (2011), 'A simulation study on nanoscale holes generated by gold nanoparticles on negative lipid bilayers', *Langmuir* **27**, 8323–8332.

Lopes, D., Jakobtorweihen, S., Nunes, C., Sarmentoc, B. & Reis, S. (2017), 'Shedding light on the puzzle of drug-membrane interactions: Experimental techniques and molecular dynamics simulations', *Prog. Lipid Res.* **65**, 24–44.

Lopez, C. A., Rzepiela, A. J., de Vries, A. H., Dijkhuizen, L., Hunenberger, P. H. & Marrink, S. J. (2009), 'MARTINI coarse-grained force field: Extension to carbohydrates', *J. Chem. Theory Comput.* **5**, 3195–3210.

López, C. A., Sovova, Z., van Eerden, F. J., de Vries, A. H. & Marrink, S. J. (2013), 'MARTINI force field parameters for glycolipids', *J. Chem. Theory Comput.* **9**, 1694–1708.

Lopez, C. F., Nielsen, S. O., Klein, M. L. & Moore, P. B. (2004), 'Hydrogen bonding structure and dynamics of water at the dimyristoylphospatidylcholine lipid bilayer surface from a molecular dynamics simulation', *J. Phys. Chem. B* **108**, 6603–6610.

Loura, L. M. S. & Ramalho, J. P. P. (2011), 'Recent developments in molecular dynamics simulations of fluorescent membrane probes', *Molecules* **16**, 5437–5452.

Lu, L. & Voth, G. A. (2009), 'Systematic coarse-graining of a multicomponent lipid bilayer', *J. Phys. Chem. B* **113**, 1501–1510.

Lyubartsev, A., Mirzoev, A., Chen, L. J. & Laaksonen, A. (2010), 'Systematic coarse-graining molecular models by the Newton inversion method', *Faraday Discuss.* **144**, 43–56.

Lyubartsev, A., Tu, Y. & Laaksonen, A. (2009), 'Hierarchical multiscale modelling scheme from first principles to mesoscale', *J. Comp. Theory Nanosci.* **6**, 951–959.

Lyubartsev, A. P. (2005), 'Multiscale modeling of lipids and lipid bilayers', *Eur. Biophys. J.* **35**, 53–61.

Lyubartsev, A. P. & Laaksonen, A. (1995), 'Calculation of effective interaction potentials from radial distribution functions: A reverse Monte Carlo approach', *Phys. Rev. E* **52**(4), 3730–3737.

Lyubartsev, A. P. & Rabinovich, A. L. (2011), 'Recent development in computer simulations of lipid bilayers', *Soft Matter* **7**, 25–39.

Lyubartsev, A. P. & Rabinovich, A. L. (2016), 'Force field development for lipid membrane simulations', *Biochim. Biophys. Acta* **1858**(10), 2483–2497.

Maciejewski, A., Pasenkiewicz-Gierula, M., Cramariuc, O., Vattulainen, I. & Rog, T. (2014), 'Refined OPLS all-atom force field for saturated phosphatidylcholine bilayers at full hydration', *J. Phys. Chem. B* **118**(17), 4571–4581.

MacKerell, A. D., Bashford, D., Bellott, M., Dunbrack, R. L., Evanseck, J. D., Field, M. J., Fisher, S., Gao, J., Ha, S., Joseph-McCarthy, D., Kuchnir, L., Kuczera, K., Lau, F. T. K., Mattos, C., Michnick, S., Ngo, T., Nguyen, D. T., Prodhom, B., Reiher, III, W. E., Roux, B., Schlenkrich, M., Smith, J. C., Stote, R., Straub, J., Watanabe, M., Wiórkiewicz-Kuczera, J., Yin, D. & Karplus, M. (1998), 'All-atom empirical potential for molecular modeling and dynamics studies of proteins', *J. Phys. Chem. B* **102**, 3586–3616.

Mackerell, A. D., Wiórkiewicz-Kuczera, J. & Karplus, M. (1995), 'An all-atom empirical energy function for the simulation of nucleic acids', *J. Am. Chem. Soc.* **117**, 11946–11975.

Makarucha, A. J., Todorova, N. & Yarovsky, I. (2011), 'Nanomaterials in biological environment: A review of computer modelling studies', *Eur. Biophys. J.* **40**, 103–115.

Marrink, S. J. & Berendsen, H. J. C. (1994), 'Simulation of water transport through a lipid membrane', *J. Phys. Chem.* **98**, 4155–4168.

Marrink, S. J., de Vries, A. H. & Mark, A. E. (2004), 'Coarse grained model for semiquantitative lipid simulations', *J. Phys. Chem. B* **108**, 750–760.

Marrink, S. J., de Vries, A. H. & Tieleman, D. P. (2009), 'Lipids on the move: Simulations of membrane pores, domains, stalks and curves', *Biochim. Biophys. Acta* **1788**, 149–168.

Marrink, S. J., Risselada, H. J., Yefimov, S., Tieleman, D. P. & de Vries, A. H. (2007), 'The MARTINI force field: Coarse grained model for biomolecular simulations', *J. Phys. Chem. B* **111**(27), 7812–7824.

Marrink, S. J. & Tieleman, D. P. (2013), 'Perspective on the Martini model', *Chem. Soc. Rev.* **42**(16), 6801–6822.

Mavromoustakos, T., Chatzigeorgiou, P., Koukoulitsa, C. & Durdagi, S. (2011), 'Partial interdigitation of lipid bilayers', *Int. J. Quantum Chem.* **111**, 1172–1183.

Merz, M. (1997), 'Molecular dynamics simulations of lipid bilayers', *Curr. Opin. Struct. Biol.* **7**(11), 511–517.

Mirzoev, A. & Lyubartsev, A. P. (2014), 'Systematic implicit solvent coarse graining of dimyristoylphosphatidylcholine lipids', *J. Comput. Chem.* **35**, 1208–1218.

Monticelli, L., Kandasamy, S. K., Periole, X., Larson, R. G., Tieleman, D. P. & Marrink, S. J. (2008), 'The MARTINI coarse-grained force field: Extension to proteins', *J. Chem. Theory Comput.* **112**, 819–834.

Moore, P. B., Lopez, C. F. & Klein, M. L. (2001), 'Dynamical properties of a hydrated lipid bilayer from a multinanosecond molecular dynamics simulation', *Biophys. J.* **81**, 2484–2494.

Mouritsen, O. G. & Jorgensen, K. (1994), 'Dynamical order and disorder in lipid bilayers', *Chem. Phys. Lipids* **73**, 3–25.

Mouritsen, O. G., Sperotto, M. M., Risbo, J., Zhang, Z. & Zuckermann, M. J. (1996). 'Computational approach to lipid – protein interactions in membranes', in *Advances in computational biology*, Vol. 2, Amsterdam, JAI Press, Elsevier, pp. 15–64.

Mukhopadhyay, P., Vogel, H. J. & Tieleman, D. P. (2004), 'Distribution of pentachlorophenol in phospholipid bilayers: A molecular dynamics study', *Biophys. J.* **86**, 337–345.

Müller, M., Katsov, K. & Schick, M. (2006), 'Biological and synthetic membranes: What can be learned from a coarse-grained description?', *Phys. Rep* **434**, 113–176.

Murtola, T., Falck, E., Karttunen, M. & Vattulainen, I. (2007), 'Coarse-grained model for phospholipid/ cholesterol bilayer employing inverse Monte Carlo with thermodynamic constraints', *J. Chem. Phys.* **126**(7), 075101.

Murzyn, K., Bratek, M. & Pasenkiewicz-Gierula, M. (2013), 'Refined OPLS all-atom force field parameters for n-pentadecane, methyl acetate, and dimethyl phosphate', *J. Phys. Chem. B* **117**, 16388–16396.

Nagle, J. F. (2013), 'Introductory lecture: Basic quantities in model biomembranes', *Faraday Discuss.* **161**, 11–29.

Nagle, J. F. & Tristram-Nagle, S. (2000a), 'Lipid bilayer structure', *Curr. Opin. Struct. Biol.* **10**, 474–480.

Nagle, J. F. & Tristram-Nagle, S. (2000b), 'Structure of lipid bilayers', *Biochim. Biophys. Acta* **1469**, 159–195.

Nielsen, S. O., Lopez, C. F., Srinivas, G. & Klein, M. L. (2004), 'Coarse grain models and the computer simulation of soft materials', *J. Phys.: Condens. Matter* **16**, R481–R512.

Noid, W. G. (2013), 'Perspective: Coarse-grained models for biomolecular systems', *J. Chem. Phys.* **139** (9), 090901.

Noid, W. G., Chu, J.-W., Ayton, G. S., Krishna, V., Izvekov, S., Voth, G. A., Das, A. & Andersen, H. C. (2008), 'The multiscale coarse-graining method I. a rigorous bridge between atomistic and coarse-grained models', *J. Chem. Phys.* **128**, 244114.

Ogata, K. & Nakamura, S. (2015), 'Improvement of parameters of the AMBER potential force field for phospholipids for description of thermal phase transitions', *J. Phys. Chem. B* **119**, 9726–9739.

Oostenbrink, C., Villa, A., Mark, A. E. & Gunsteren, W. F. V. (2004), 'A biomolecular force field based on the free enthalpy of hydration and solvation: The GROMOS force-field parameter sets 53a5 and 53a6', *J. Comput. Chem.* **25**(13), 1656–1676.

Orsi, M., Haubertin, D. Y., Sanderson, W. E. & Essex, J. W. (2008), 'A quantitative coarse-grained model for lipid bilayers', *J. Phys. Chem. B* **112**, 802–815.

Orsi, M., Michel, J. & Essex, J. W. (2010), 'Coarse-grain modelling of DMPC and DOPC lipid bilayers', *J. Phys.: Condens. Matter* **22**, 155106.

Paloncýová, M., Fabre, G., DeVane, R. H., Trouillas, P., Berka, K. & Otyepka, M. (2014), 'Benchmarking of force fields for molecule-membrane interactions', *J. Chem. Theory Comput.* **10**(9), 4143–4151.

Pandit, S. A. & Scott, H. L. (2009), 'Multiscale simulations of heterogeneous model membranes', *Biochim. Biophys. Acta* **1788**, 136–148.

Papadimitriou, N. I., Kainourgiakis, M. E., Karozis, S. N. & Charalambopoulou, G. C. (2015), 'Studying the structure of single-component ceramide bilayers with molecular dynamics simulations using different force fields', *Mol. Simul.* **41**(13), 1122–1136.

Pasenkiewicz-Gierula, M., Baczynski, K., Markiewicz, M. & Murzyn, K. (2016), 'Computer modelling studies of the bilayer/water interface', *Biochim. Biophys. Acta* **1858**(10), 2305–2321.

Pastor, R. W. (1994), 'Molecular dynamics and Monte Carlo simulations of lipid bilayers', *Curr. Opin. Struct. Biol.* **4**, 486–492.

Pastor, R. W. & MacKerell, Jr., A. D. (2011), 'Development of the CHARMM force field for lipids', *J. Phys. Chem. Lett.* **2**, 1526–1532.

Patra, M., Karttunen, M., Hyvönen, M., Falck, E. & Vattulainen, I. (2004), 'Lipid bilayers driven to a wrong lane in molecular dynamics simulations by subtle changes in long-range electrostatic interactions', *J. Phys. Chem. B* **108**, 4485–4494.

Patra, M., Salonen, E., Terama, E., Vattulainen, I., Faller, R., Lee, B. W., Holopainen, J. & Karttunen, M. (2006), 'Under the influence of alcohol: The effect of ethanol and methanol on lipid bilayers', *Biophys. J.* **90**, 1121–1135.

Pedersen, U. R., Peters, G. H. & Westh, P. (2007), 'Molecular packing in 1hexanol-DMPC bilayers studied by molecular dynamics simulations', *Biophys. Chem.* **125**(1), 104–111.

Pluhackova, K. & Böckmann, R. A. (2015), 'Biomembranes in atomistic and coarse-grained simulations', *J. Phys.: Condens. Matter* **27**, 323103.

Pluhackova, K., Kirsch, S. A., Han, J., Sun, L., Jiang, Z., Unruh, T. & Böckmann, R. A. (2016), 'A critical comparison of biomembrane force fields: Structure and dynamics of model dmpc, popc, and pope bilayers', *J. Phys. Chem. B* **120**(16), 3888–3903.

Poger, D., Carona, B. & Mark, A. E. (2016), 'Validating lipid force fields against experimental data: Progress, challenges and perspectives', *Biochim. Biophys. Acta* **1858**(7), 1556–1565.

Poger, D. & Mark, A. E. (2010), 'On the validation of molecular dynamics simulations of saturated and cis-monounsaturated phosphatidylcholine lipid bilayers: A comparison with experiment', *J. Chem. Theory Comput.* **6**, 325–336.

Poger, D., van Gunsteren, W. F. & Mark, A. E. (2010), 'A new force field for simulating phosphatidylcholine bilayers', *J. Comput. Chem.* **31**, 1117–1125.

Pohle, W., Gauger, D. R., Bohl, R., Mrazkova, E. & Hobza, P. (2004), 'Lipid hydration: Headgroup CH moieties are involved in water binding', *Biopolymers* **74**, 27–31.

Porasso, R. D., Bennett, W. F. D., Oliveira-Costa, S. D. & Cascales, J. J. L. (2009), 'Study of the benzocaine transfer from aqueous solution to the interior of a biological membrane', *J. Phys. Chem. B* **113**(29), 9988–9994.

Pöyry, S. & Vattulainen, I. (2016), 'Role of charged lipids in membrane structures -insight given by simulations', *Biochim. Biophys. Acta* **1858**(10), 2322–2333.

Prakash, P. & Sankararamakrishnan, R. (2009), 'Force field dependence of phospholipid headgroup and acyl chain properties: Comparative molecular dynamics simulations of DMPC bilayers', *J. Comp. Chem.* **31**, 266–277.

Prakash, P. & Sankararamakrishnan, R. (2010), 'Force field dependence of phospholipid headgroup and acyl chain properties: Comparative molecular dynamics simulations of DMPC bilayers', *J. Comp. Chem.* **31**, 266–277.

Prates, E. T., Souza, P. C. T., Pickholz, M. & Skaf, M. S. (2011), 'CHARMMbased parameterization of neutral articaine -a widely used local anesthetic', *Int. J. Quantum Chem.* **111**(4), 1339–1345.

Rabinovich, A. L., Balabaev, N. K., Alinchenko, M. G., Voloshin, V. P., Medvedev, N. N. & Jedlovszky, P. (2005), 'Computer simulation study of intermolecular voids in unsaturated phosphatidylcholine lipid bilayers', *J. Chem. Phys.* **122**, 084906.

Rabinovich, A. L. & Lyubartsev, A. P. (2013), 'Computer simulation of lipid membranes: Methodology and achievements', *Polym. Sci. Ser. C* **55**, 162–180.

Rabinovich, A. L. & Lyubartsev, A. P. (2014), 'Bond orientation properties in lipid molecules of membranes: Molecular dynamics simulations', *J. Phys.: Conf. Ser.* **510**, 012022.

Rabinovich, A. L., Lyubartsev, A. P. & Zhurkin, D. V. (2017), 'Unperturbed hydrocarbon chains and liquid phase bilayer lipid chains: A computer simulation study', *Eur. Biophys. J.* DOI 10.1007/s00249–017–1231–9.

Rabinovich, A. L. & Ripatti, P. O. (1994), 'Polyunsaturated hydrocarbon chains of lipids: Structure, properties, functions (in Russian)', *Uspekhi sovremennoi biologii (Progress in modern biology)* **114**, 581–594.

Rabinovich, A. L., Ripatti, P. O., Balabaev, N. K. & Leermakers, F. A. M. (2003), 'Molecular dynamics simulations of hydrated unsaturated lipid bilayers in the liquid-crystal phase and comparison to self-consistent field modeling', *Phys. Rev. E* **67**(1), 011909.

Reith, D., Pütz, M. & Müller-Plathe, F. (2003), 'Deriving effective mesoscale potentials from atomistic simulations', *J. Comp. Chem.* **24**, 1624–1636.

Riniker, S., Allison, J. R. & van Gunsteren, W. F. (2012), 'On developing coarse-grained models for biomolecular simulation: A review', *Phys. Chem. Chem. Phys.* **14**(36), 12423–12430.

Roark, M. & Feller, S. E. (2009), 'Molecular dynamics simulation study of correlated motions in phospholipid bilayer membranes', *J. Phys. Chem. B* **113**, 13229–13234.

Rosso, L. & Gould, I. R. (2008), 'Structure and dynamics of phospholipid bilayers using recently developed general all-atom force fields', *J. Comp. Chem.* **29**(1), 24–37.

Saeedi, M., Lyubartsev, A. P. & Jalili, S. (2017), 'Anesthetics mechanism on a dmpc lipid membrane model: Insights from molecular dynamics simulations', *Biophys. Chem.* **226**, 1–13.

Saito, H. & Shinoda, W. (2011), 'Cholesterol effect on water permeability through DPPC and PSM lipid bilayers: A molecular dynamics study', *J. Phys. Chem. B* **115**, 15241–15250.

Saiz, L., Bandyopadhyay, S. & Klein, M. L. (2002), 'Towards an understanding of complex biological membranes from atomistic molecular dynamics simulations', *Biosci. Rep.* **22**(2), 151–173.

Saiz, L. & Klein, M. L. (2002), 'Computer simulation studies of model biological membranes', *Acc. Chem. Res.* **35**, 482–489.

Schmid, N., Eichenberger, A. P., Choutko, A., Riniker, S., Winger, M., Mark, A. E. & van Gunsteren, W. F. (2011), 'Definition and testing of the GROMOS force-field versions 54A7 and 54B7', *Eur. Biophys. J.* **40**, 843–856.

Schuler, L. D., Daura, X. & van Gunsteren, W. F. (2001), 'An improved GROMOS96 force field for aliphatic hydrocarbons in the condensed phase', *J. Comp. Chem.* **22**, 1205–1218.

Scott, H. L. (2002), 'Modeling the lipid component of membranes', *Curr. Opin. Struct. Biol.* **12**, 495–502.

Shih, A. Y., Arkhipov, A., Freddolino, P. L. & Schulten, K. (2006), 'Coarse grained protein-lipid model with application to lipoprotein particles', *J. Phys. Chem. B* **110**, 3674–3684.

Shih, A. Y., Arkhipov, A., Freddolino, P. L., Sligar, S. G. & Schulten, K. (2007), 'Assembly of lipids and proteins into lipoprotein particles', *J. Chem. Phys. B* **111**, 11095–11104.

Shih, A. Y., Freddolino, P. L., Arkhipov, A. & Schulten, K. (2007), 'Assembly of lipoprotein particles revealed by coarse-grained molecular dynamics simulations', *J. Struct. Biol.* **157**, 579–592.

Shillcock, J. & Lipowsky, R. (2007), 'Visualizing soft matter: Mesoscopic simulations of membranes, vesicles and nanoparticles', *Biophys. Rev. Lett.* **2**(1), 33–55.

Shinoda, W., DeVane, R. & Klein, M. L. (2010), 'Zwitterionic lipid assemblies: Molecular dynamics studies of monolayers, bilayers, and vesicles using a new coarse grain force field', *J. Phys. Chem. B* **114**, 6836–6849.

Siu, S. W. I., Pluhackova, K. & Böckmann, R. A. (2012), 'Optimization of the OPLS-AA force field for long hydrocarbons', *J. Chem. Theory Comput.* **8**(4), 1459–1470.

Siu, S. W. I., Vácha, R., Jungwirth, P. & Böckmann, R. A. (2008), 'Biomolecular simulations of membranes: Physical properties from different force fields', *J. Chem. Phys.* **128**, 125103.

Skandani, A. A., Zeineldin, R. & Al-Haik, M. (2012), 'Effect of chirality and length on the penetrability of single-walled carbon nanotubes into lipid bilayer cell membranes', *Langmuir* **28**, 7872–7879.

Skjevik, A. A., Madej, B. D., Dickson, C. J., Lin, C., Teigen, K., Walker, R. C. & Gould, I. R. (2016), 'Simulation of lipid bilayer self-assembly using all-atom lipid force fields', *Phys. Chem. Chem. Phys.* **18**(15), 10573–10584.

Skjevik, A. A., Madej, B. D., Dickson, C. J., Teigen, K., Walker, R. C. & Gould, I. R. (2015), 'All-atom lipid bilayer self-assembly with the AMBER and CHARMM lipid force fields', *Chem. Commun.* **51**(21), 4402–4405.

Skjevik, A. A., Madej, B. D., Walker, R. C. & Teigen, K. (2012), 'LIPID11: A modular framework for lipid simulations using Amber', *J. Phys. Chem. B* **116**(36), 11124–11136.

Slingsby, J. G., Vyas, S. & Maupin, C. M. (2015), 'A charge-modified general amber force field for phospholipids: Improved structural properties in the tensionless ensemble', *Mol. Simul.* **41**(18), 1449–1458.

Song, B., Yuan, H., Jameson, C. J. & Murad, S. (2011), 'Permeation of nanocrystals across lipid membranes', *Mol. Phys.: Int. J. at the Interface Between Chem. & Phys.* **109**(11), 1511–1526.

Sonne, J., Hansen, F. Y. & Peters, G. H. (2005), 'Methodological problems in pressure profile calculations for lipid bilayers', *J. Chem. Phys.* **122**, 124903.

Sonne, J., Jensen, M. Ø., Hansen, F. Y., Hemmingsen, L. & Peters, G. H. (2007), 'Reparametrization of all-atom dipalmitoylphosphatidylcholine lipid parameters enables simulation of fluid bilayers at zero tension', *Biophys. J.* **92**, 4157–4167.

Southern, J., Pitt-Francis, J., Whiteley, J., Stokeley, D., Kobashi, H., Nobes, R., Kadooka, Y. & Gavaghan, D. (2008), 'Multi-scale computational modelling in biology and physiology', *Prog. in Biophys. and Mol. Biol.* **96**, 60–89.

Stillwell, W. (2008), 'Docosahexaenoic acid: A most unusual fatty acid', *Chem. Phys. Lipids* **153**(1), 1–2.

Stillwell, W. & Wassall, S. R. (2003), 'Docosahexaenoic acid: Membrane properties of a unique fatty acid', *Chem. Phys. Lipids* **126**, 1–27.

Sugii, T., Takagi, S. & Matsumoto, Y. (2005), 'A molecular dynamics study of lipid bilayers: Effects of the hydrocarbon chain length on permeability', *J. Chem. Phys.* **123**, 184714.

Tarazona, P., Chacón, E. & Bresme, F. (2013), 'Thermal fluctuations and bending rigidity of bilayer membranes', *J. Chem. Phys.* **139**(9), 094902.

Taylor, J., Whiteford, N. E., Bradley, G. & Watson, G. W. (2009), 'Validation of all-atom phosphatidylcholine lipid force fields in the tensionless NPT ensemble', *Biochim. Biophys. Acta* **1788**, 638–649.

Terama, E., Ollila, O. H. S., Salonen, E., Rowat, A. C., Trandum, C., Westh, P., Patra, M., Karttunen, M. & Vattulainen, I. (2008), 'Influence of ethanol on lipid membranes: From lateral pressure profiles to dynamics and partitioning', *J. Phys. Chem. B* **112**, 4131–4139.

Tieleman, D. P. & Berendsen, H. J. C. (1996), 'Molecular dynamics simulations of a fully hydrated dipalmitoylphosphatidylcholine bilayer with different macroscopic boundary conditions and parameters', *J. Chem. Phys.* **105**(11), 4871–4880.

Tieleman, D. P., Marrink, S. J. & Berendsen, H. J. C. (1997), 'A computer perspective of membranes: Molecular dynamics studies of lipid bilayer systems', *Biochim. Biophys. Acta* **1331**, 235–270.

Tjörnhammar, R. & Edholm, O. (2015), 'Reparameterized united atom model for molecular dynamics simulations of gel and fluid phosphatidylcholine bilayers', *J. Chem. Theory Comput.* **10**, 5706–5715.

Tobias, D. J. (2001), 'Membrane simulations', in *Computational biochemistry and biophysics*, O. H. Becker A. D., MacKerell, Jr., B. Roux & M. Watanabe, eds. New York, Dekker, pp. 465–496.

Tobias, D. J., Tu, K. & Klein, M. L. (1997), 'Atomic-scale molecular dynamics simulations of lipid membranes', *Curr. Opin. Colloid Interf. Sci.* **2**, 15–26.

Ulander, J. & Haymet, A. D. J. (2003), 'Permeation across hydrated DPPC lipid bilayers: Simulation of the titrable amphiphilic drug valproic acid', *Biophys. J.* **85**, 3475–3484.

Valentine, R. C. & Valentine, D. L. (2004), 'Omega-3 fatty acids in cellular membranes: A unified concept', *Prog. Lipid Res.* **43**, 383–402.

van der Ploeg, P. & Berendsen, H. J. C. (1982), 'Molecular dynamics simulation of a bilayer membrane', *J. Chem. Phys.* **76**, 3271–3276.

Vanommeslaeghe, K. & MacKerell, Jr., A. D. (2015), 'CHARMM additive and polarizable force fields for biophysics and computer-aided drug design', *Biochim. Biophys. Acta* **1850**, 861–871.

Vemparala, S., Domene, C. & Klein, M. L. (2010), 'Computational studies on the interactions of inhalational anesthetics with proteins', *Acc. Chem. Res.* **132**, 103–110.

Venable, R. M., Chen, L. E. & Pastor, R. W. (2009), 'Comparison of the extended isotropic periodoc sum and particle mesh Ewald methods for simulations of lipid bilayers and monolayers', *J. Phys. Chem. B* **113**, 5855–5862.

Venable, R. M., Sodt, A. J., Rogaski, B., Rui, H., Hatcher, E., MacKerell, Jr., A. D., Pastor, R. W. & Klauda, J. B. (2014), 'CHARMM all-atom additive force field for sphingomyelin: Elucidation of hydrogen bonding and of positive curvature', *Biophys. J.* **107**(1), 134–145.

Venturoli, M., Sperotto, M. M., Kranenburg, M. & Smit, B. (2006), 'Mesoscopic models of biological membranes', *Phys. Rep.* **437**, 1–54.

Vermeer, L. S., de Groot, B. L., Réat, V., Milon, A. & Czaplicki, J. (2007), 'Acyl chain order parameter profiles in phospholipid bilayers: Computation from molecular dynamics simulations and comparison with ^{2}H NMR experiments', *Eur. Biophys. J.* **36**, 919–931.

Vicatos, S., Rychkova, A., Mukherjee, S. & Warshel, A. (2014), 'An effective coarse-grained model for biological simulations: Recent refinements and validations', *Proteins – Struct., Funct., Bioinf.* **82**(7), 1168–1185.

Vigh, L., Escribá, P. V., Sonnleitner, A., Sonnleitner, M., Piotto, S., Maresca, B., Horváth, I. & Harwood, J. L. (2005), 'The significance of lipid composition for membrane activity: New concepts and ways of assessing function', *Prog. Lipid Res.* **44**, 303–344.

Vogel, A. & Feller, S. E. (2012), 'Headgroup conformations of phospholipids from molecular dynamics simulation: Sampling challenges and comparison to experiment', *J. Membrane Biol.* **245**, 23–28.

Wang, J., Wolf, R. M., Caldwell, J. W., Kollman, P. A. & Case, D. A. (2004), 'Development and testing of a general Amber force field', *J. Comput. Chem.* **25**, 1157–1174.

Wang, Y., Markwick, P. R. L., de Oliveira, C. A. F. & McCammon, J. A. (2011), 'Enhanced lipid diffusion and mixing in accelerated molecular dynamics', *J. Chem. Theory Comput.* **7**, 3199–3207.

Wassall, S. R. & Stillwell, W. (2009), 'Polyunsaturated fatty acid -cholesterol interactions: Domain formation in membranes', *Biochim. Biophys. Acta* **1788**, 24–32.

Weiner, S. J., Kollman, P. A., Case, D. A., Singh, U. C., Ghio, C., Alagona, G., Profeta, S. & Weiner, P. (1984), 'A new force field for molecular mechanical simulation of nucleic acids and proteins', *J. Am. Chem. Soc.* **106**, 765–784.

Wennberg, C. L., van der Spoel, D. & Hub, J. S. (2012), 'Large influence of cholesterol on solute partitioning into lipid membranes', *J. Am. Chem. Soc.* **134**, 5351–5361.

Wong-Ekkabut, J. & Karttunen, M. (2016), 'The good, the bad and the user in soft matter simulations', *Biochim. Biophys. Acta* **1858**(10), 2529–2538.

Wu, X. W. & Brooks, B. R. (2005), 'Isotropic periodic sum: A method for the calculation of long-range interactions', *J. Chem. Phys.* **122**, 044107.

Yanga, K. & Ma, Y. (2012), 'Computer simulations of fusion, fission and shape deformation in lipid membranes', *Soft Matter* **8**, 606–618.

Yin, Y., Arkhipov, A. & Schulten, K. (2009), 'Simulations of membrane tubulation by lattices of amphiphysin N-BAR domains', *Structure* **17**, 882–892.

Zhu, X., Lopes, P. E. M. & MacKerell, Jr., A. D. (2012), 'Recent developments and applications of the CHARMM force fields', *WIREs Comput. Mol. Sci.* **2**, 167–185.

2

Mesoscopic Particle-Based Modeling of Self-Assembled Lipid Membranes

Mohamed Laradji
Department of Physics and Materials Science, The University of Memphis, Memphis, TN 38152, USA

Maria Maddalena Sperotto[1]
Technical University of Denmark, 2800 Kgs. Lyngby, Denmark

I Introduction

Biomembranes are heterogeneous systems (Nicolson 2014) that display dynamic and structural properties on many length and time scales (Konig and Sackmann 1996). The main constituent of biomembranes is a few-nanometers thick and flexible lipid bilayer, which from the lateral point of view is fluid, i.e., does not exhibit shear restoring forces, and has a curvature-energy of the order of $10 k_B T$. A large number of experimental, phenomenological, and computational studies have been performed on both biomembrane and biomimetic membranes (Fahy et al. 2009, van Meer et al. 2008, van Meer et al. 2011), where the latter are membranes reconstituted with one or more molecular components extracted from natural biomembranes.

Since the 1970s, a wide range of models have been proposed to investigate biomembrane properties at different length and/or time scales. These include phenomenological lattice models, continuum elasticity models, atomistic models, and mesoscopic particle-based models.

In phenomenological lattice models, the interactions between the lipids of the two leaflets of the bilayer are usually neglected. Also, each lipid acyl-chain occupies a lattice point and, depending on the temperature, it may sample a large number of states (Doniach 1978, Mouritsen et al. 1983, Nagle 1973, Pink et al. 1980). During the 1970s and up to the 1990s, lattice models were extensively used to investigate the thermodynamics of lipid membranes because they can be studied by mean field theories and Monte–Carlo simulations. Such models have the drawback to be phenomenological and view lipid membranes as quasi-two-dimensional systems. Thus, lattice models are not suitable for studying phenomena where curvature deformations of the bilayer play a major role, or for inferring transport and non-equilibrium properties of lipid membranes.

Both long-wavelength and microscopic continuum elasticity models were proposed for the study of lipid membranes. Among these, the models based on the long-wavelength Helfrich–Canham Hamiltonian (Canham 1970, Helfrich 1973) view the membrane as an infinitesimally thin sheet, without internal structure, and with a free energy dominated by curvature deformations. The long-wavelength Helfrich–Canham Hamiltonian has been commonly used to investigate properties of lipid membranes through analytical techniques (Seifert 1997), Langevin simulations (Lin and Brown 2004,

---

[1] Corresponding author.

Reister-Gottfried et al. 2007, Taniguchi 1996), and Monte–Carlo simulations (Gompper and Kroll 1994, Kumar et al. 2001, Laradji and Kumar 2004, Noguchi and Gompper 2005, Ramakrishnan et al. 2014, Sreeja et al. 2015). Microscopic elasticity models that account for the bending of both leaflets of the membrane, as well as the local orientation of the lipid chains and local thickness of the bilayer, were also developed to investigate local, small length-scale phenomena such as membrane pores and stalks (Hamm and Kozlov 2000, Kozlovsky and Kozlov 2002, May 2002, Terzi and Deserno 2017, Watson et al. 2012). Such models are mostly investigated analytically and are not very tractable through numerical simulations (Watson et al. 2012).

Since the early 1990s, molecular-scale structure and molecular transport of lipid membranes have been extensively studied by atomistic molecular dynamics (AMD) simulations (Berkowitz et al. 2006, Marrink et al. 1993, Smondyrev and Berkowitz 1999). AMD simulations of lipid membranes typically involve few thousand lipid molecules and are executed over timescales up to few 100 ns. Thus AMD simulations can only probe short time-scale rearrangements of the lipid molecules and are not suitable for studies of large-scale phenomena in lipid membranes. Mesoscopic modeling can therefore bridge the length and time scales that can be probed by atomistic models and continuum elasticity models (Venturoli et al. 2006). Since both the atomistic modeling approach and the continuum elasticity approach are described in detail elsewhere in this book, the focus of this chapter is on mesoscopic particle-based modeling of self-assembled lipid membranes.

In Section II, the basic ideas behind the particle-based mesoscopic modeling are presented. The two following subsections, Section III and Section IV, introduce the explicit and implicit solvent modeling approach, respectively. In Section III, explicit solvent models are reviewed, and subsections IIIA and IIIB are devoted to the description of two of these models, namely the Martini model (Marrink et al. 2004) and the dissipative particle dynamics (DPD) model (Hoogerbrugge and Koelman 1992). In Section IV, the advantages of the approach based on implicit solvent modeling are discussed. In subsections IVA, IVB, IVC, and IVD, the results from some illustrative studies based on a particularly efficient implicit solvent model are presented (Revalee et al. 2008). The selected studies regard the thermal behavior of supported lipid bilayers (SLBs), the effect of system-size on the thermal behavior of liposomes, the cytoskeleton-induced blebbing, and the wrapping and endocytosis of nanoparticles by lipid membranes. Advantages and shortcomings of both the explicit and implicit solvent particle-based mesoscopic modeling are discussed in Section V.

II The Particle-Based Mesoscopic Approach

In 2006, Brannigan et al. (2006) estimated that an AMD simulation of a small patch of a cell membrane for just 1 ms would require 46 years. Although nowadays computers are considerably faster than in 2006, the mesoscopic approach enables to investigate phenomena that occur much beyond the typical AMD's time and length scales that can be sampled nowadays. Mesoscopic modeling implies that some molecular details of the system under investigation can be dismissed as irrelevant to the process or phenomenon to be investigated. This means an a priori evaluation of molecular information and an understanding of which aspects are relevant. Therefore, for ensuring that the predictions from mesoscopic simulations are reliable, it is a priori necessary to evaluate molecular information and understand which have to be accounted for in the mesoscopic description of a system.

Within the mesoscopic particle-based description, a system is coarse-grained into beads, each representing a cluster of atoms whose details can be neglected for the phenomenon or process under investigation. The solvent can be modeled either explicitly, by coarse-graining a number of solvent molecules into beads, or implicitly, by replacing the hydrophobic interactions, which are responsible for the self-assembly of lipids, with effective attractive forces between the lipids tail groups. The mesoscopic particle-based description is based on the assumption that many atomistic aspects of the system are a priori known and effectively accounted for. As this depends on the system and specific phenomenon under investigation, there is no unique way to develop particle-based coarse-grained models. Many particle-based coarse-grained models have been developed to predict generic properties

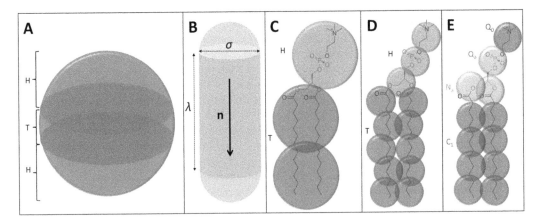

FIGURE 1   Degrees of coarse-graining of the lipid bilayer: (A) Coarse-graining of a patch of a bilayer into a sphere with diameter about the thickness of the bilayer (Drouffe et al. 1991). The middle section of the sphere represents the hydrophobic core of the bilayer, while the top and bottom spherical caps represent the hydrophobic groups of the lipids in the top and bottom leaflets, respectively. (B) Coarse-graining of a lipid into a spherocylinder (Brannigan et al. 2004). $\mathbf{n}$ is a normal vector pointing toward the hydrophobic tail of the lipid. (C) Generic coarse-graining of a lipid into semi-flexible chains of few beads (Cooke and Deserno 2005; Laradji and Kumar 2004; Revalee et al. 2008). (D) Coarse-graining of a dimyristoyl phosphatidylcholine (DMPC) lipid into 13 beads in the model of Venturoli et al. (2005). In (C) and (D), H and T correspond to hydrophilic and hydrophobic beads, respectively. (E) Coarse-graining of DMPC into 12 beads in the Martini model (Marrink et al. 2004). Q_0 and Q_a denote positively and negatively charged beads, respectively. N_a beads are nonpolar and C_1 beads are apolar. Note that in (B) to (E), lipid models have about the same scale. Model (A) is about twice as large.

of lipid membranes, with different levels of coarse-graining (Brannigan et al. 2004, Cooke and Deserno 2005, Drouffe et al. 1991, Farago 2003, Goetz and Lipowsky 1998, Kranenburg et al. 2003, Laradji and Kumar 2004, Morikawa and Saito 1994, Noguchi and Takasu 2001, Revalee et al. 2008, Sintes and Baumgärtner 1997, Whitehead et al. 2001).

Among the coarse-grained models aimed at describing specific lipid membranes, some were developed using the iterative Boltzmann inversion method (Izvekov and Voth 2009, Lyubartsev et al. 2009, McGreevy and Pusztai 1988, Moore et al. 2014), in which the parameters of the models are optimized such that they reproduce average structural properties obtained from AMD simulations. The parameters of other coarse-grained models, also developed to study specific lipid membranes, are optimized by reproducing thermodynamic quantities (Marrink et al. 2004, Shelley et al. 2001). In Fig. 1, examples of different levels of lipid coarse-graining are shown. The solvent is treated explicitly in most of the coarse-grained models that are aimed at specific lipid membranes (Izvekov and Voth 2009, Marrink et al. 2004, Shelley et al. 2001). However, both explicit solvent models (Goetz et al. 1999, Goetz and Lipowsky 1998, Kranenburg et al. 2003, Laradji and Kumar 2004, Shillcock and Lipowsky 2002) and implicit solvent models (Brannigan et al. 2004, Cooke and Deserno 2005, Drouffe et al. 1991, Farago 2003, Morikawa and Saito 1994, Revalee et al. 2008, Sintes and Baumgärtner 1997) were developed to describe generic lipid membranes. There were also attempts to develop particle-based implicit solvent models for specific lipid bilayers (Arnarez et al. 2015, Curtis and Hall 2013, Izvekov and Voth 2009, Sodt and Head-Gordon 2010, Wang and Deserno 2010).

III   Explicit Solvent Modeling

A number of particle-based mesoscopic models of self-assembled lipid membranes with explicit solvent were proposed in the past (Goetz et al. 1999, Goetz and Lipowsky 1998, Laradji and

Kumar 2004, Marrink et al. 2004, Shillcock and Lipowsky 2002). According to these models, lipid molecules are coarse-grained into short chains of beads, representing both the hydrophilic lipid head-group and the hydrophobic lipid chains. Likewise, solvent molecules are coarse-grained into solvent beads. Here, self-assembly of the lipids into membranes is achieved through repulsive interactions between the lipids hydrophobic beads and the solvent beads. Goetz et al. performed some of the earliest coarse-grained molecular dynamics of lipid membranes (Goetz et al. 1999, Goetz and Lipowsky 1998). In their approach, a lipid molecule is modeled as an amphiphile with one or few hydrophilic beads attached to one or two chains of hydrophobic beads and interacting via Lennard–Jones forces (see Fig. 1(C) and (D)). Beads within a lipid molecule are connected via harmonic interactions, to ensure connectivity, and via three-body interactions, to provide stiffness to the chains. They investigated the elastic properties of the self-assembled membrane from the stress tensor and the spectrum of out-of-plane height fluctuations, and showed that the elasticity of a tensionless bilayer is dominated by bending modes (Goetz and Lipowsky 1998).

Similar, but more refined coarse-grained models aimed at describing specific lipid membranes were later developed (Izvekov and Voth 2009, Marrink et al. 2004, Shelley et al. 2001, Shinoda et al. 2010), which have been extensively used for a wide range of studies of lipid membranes. Another widely used explicit solvent model for lipid membranes is based on DPD (Kranenburg et al. 2003, Laradji and Kumar 2004, Shillcock and Lipowsky 2002). Both the Martini and the DPD models are reviewed in the following two subsections.

III.A The Martini Model

Marrink et al. (2004, 2007) proposed a coarse-grained model, known as the Martini model, to describe specific lipid membranes in explicit solvent. The Martini approach (Marrink and Tieleman, 2013) is based on a four-to-one coarse-graining of heavy atoms (e.g., carbon, oxygen, nitrogen, and phosphorus) with bonded hydrogen atoms into one bead. For more complex structures, such as that of cholesterol, a two-to-one coarse-graining is used instead. Four main classes of beads are defined in the Martini model, corresponding to polar, nonpolar, apolar, and charged beads. For example, as depicted in Fig. 1(E), a phospholipid molecule such as dimyristoyl phosphatidylcholine (DMPC) is coarse-grained into twelve beads: Each of the two hydrocarbon flexible chains are coarse-grained by four hydrophobic, i.e., apolar beads (C_1), and the lipid hydrophilic head-group by two charged beads (Q_0 and Q_a) and two nonpolar (N_a) beads. Non-bonded beads interact via the 12–6 Lennard–Jones potential, in addition to Coulomb forces between charged beads. Bonded beads interact via harmonic, three-body, and if needed dihedral interactions. In the Martini model, four water molecules are coarse-grained into one isotropic solvent bead. A trial-and-error procedure is used to optimize the Martini parameters in order to reproduce the experimental values of hydration free energies and the partitioning coefficient of specific molecules in water and selected alkane solvents. The Martini model was subsequently improved to account for the polarization of water (Yesylevskyy et al. 2010).

The Martini model has become very popular due to several reasons, not least its relative simplicity: It is based on a limited number of bead types and a time step that is orders of magnitude larger than that of AMD simulations. Furthermore, an extensive library of the Martini force field parameters has been developed for various biomolecules, including a plethora of lipid molecules (Marrink et al. 2007), amino acids (Monticelli et al. 2008), carbohydrates (Lopez et al. 2009), DNA (Uusitalo et al. 2015), polymers (Panizon et al. 2015, Rossi et al. 2012), dendrimers (Lee and Larson 2008), etc. As a result, the Martini model has been used in various investigations of biomembranes, including multi-component lipid membranes (Davis et al. 2013, Risselada and Marrink 2008), membrane–protein interactions (Baoukina and Tieleman 2010, Periole et al. 2007), structure of membrane tethers (Baoukina et al. 2012), membrane fusion (Baoukina and Tieleman 2010), conformational changes of transmembrane proteins (Louhivuori et al. 2010), and interactions between soft nanoparticles, such as dendrimers, or hard nanoparticles, such as Au-nanoparticles, with lipid membranes (Da Rocha et al. 2013, Gupta and Rai 2017, Lee and Larson 2009, Lin et al. 2011, Oroskar et al. 2016). Fig. 2 illustrates examples of lipid membrane systems studied by the Martini force field: Fig. 2(A), which is taken from Risselada et al. (2011), depicts a time series of snapshots during SNARE-mediated fusion of a vesicle with a planar lipid membrane. Fig. 2(B), which is taken from Louhivuori et al. (2010), shows a time series of snapshots of a mechano-sensitive channel in a liposome undergoing hypo-osmotic shock; the authors found that the increased tension of the liposome (achieved by adding more water beads inside the liposome) leads to conformational changes of the mechano-sensitive channel and reaches a conducting state in a time scale of microseconds. Fig. 2(C), which is taken from Parton et al. (2011), depicts a liposome with transmembrane proteins; the authors found that, in the case of no hydrophobic mismatch between the hydrophobic thicknesses of the bilayer and the proteins hydrophobic length, proteins remain as monomers (top snapshot), while they aggregate in the case of hydrophobic mismatch (bottom snapshot). Fig. 2(D), which is taken from Simonelli et al. (2015), shows an Au-nanoparticle coated with hydrophobic and anionic ligands interacting with a lipid bilayer. The mix of these two types of ligands allows the nanoparticle to intercalate in the bilayer.

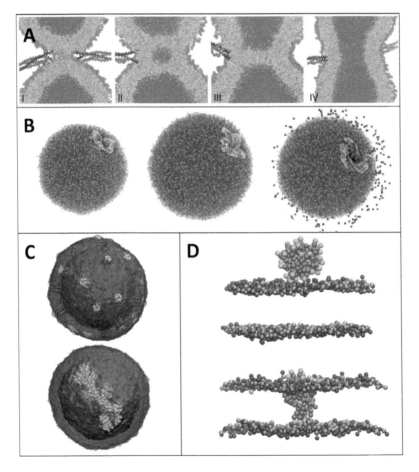

FIGURE 2 (A) Two vesicles undergoing fusion mediated by SNARE proteins (Risselada et al. 2011). Lipids are shown in orange (heads) and gray (tails), and the SNARE proteins are shown in red, green, and blue. The solvent inside the vesicles is shown in blue. (B) Snapshots of a liposome undergoing hypo-osmotic shock due to the interaction with a mechano-sensitive channel (Louhivuori et al. 2010). (C) Aggregation of transmembrane proteins induced by hydrophobic mismatch (Parton et al. 2011). The lipid bilayer is shown in blue and the proteins are shown in yellow. (D) A 2-nm gold nanoparticle coated with hydrophobic octanethiol ligands (red beads) and 11-mercaptoundecanesulfonate ligands (terminating with green anionic beads) that interacts with a lipid bilayer (Simonelli et al. 2015). (See color insert for the color version of this figure)

Despite its wide use, several studies have demonstrated limitations of and, in some cases, erroneous results predicted by Martini force field simulations. For example, simulations with the Martini model predict that ternary mixtures of dipalmitoyl phosphatidylcholine (DPPC), dioleoyl phosphatidylcholine (DOPC), and cholesterol do not exhibit phase separation (Davis et al. 2013); this is in contrast to experimental outcomes which showed that DPPC–DOPC–cholesterol mixtures can exhibit phase coexistence between a liquid-ordered (LO) phase (Ipsen et al. 1987), rich in DPPC and cholesterol, and a liquid-disordered (LD) phase, rich in DOPC (Leventhal and Veatch 2016, Veatch and Keller 2002, 2003, Yanagisawa et al. 2007). In fact, the unsaturated lipid has to be polyunsaturated, such as dilinoleyl phosphatidylcholine (DUPC) or diarachidonoyl phosphatidyl-choline (DAPC), for the Martini model to predict coexistence between the LO and the LD phases.

Systematic simulation studies by Davis et al. (2013) of ternary mixtures composed of DPPC, cholesterol, and unsaturated PC, with different degrees of unsaturation, indicated that phase separation in these mixtures is driven by enthalpy rather than by conformational differences between the saturated DPPC and the unsaturated PC lipids. Fig. 3 shows, for example, that phase separation in DPPC/DOPC/cholesterol can be observed if the C1–C3 interaction strength (where C1 is a hydrophobic fully saturated bead of DPPC and C3 is the partially unsaturated hydrophobic bead of DOPC) is modified to be as strong as that between the C1–C4 interaction (where C4 is the partially unsaturated hydrophobic bead of polyunsaturated lipids such as DAPC and DUPC). In contrast, phase separation in DPPC/DAPC/cholesterol as well as DPPC/DUPC/cholesterol mixtures can be prevented if the C1–C4 interaction, between DPPC and DUPC or DAPC, is modified

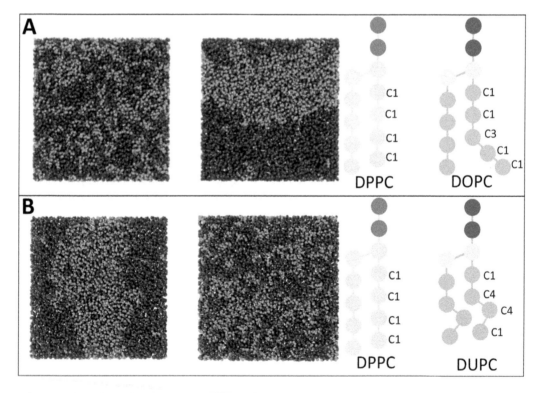

FIGURE 3 Phase changes in lipid–cholesterol bilayers. (A) Left snapshot: Equilibrium configuration of a ternary bilayer composed of DPPC (green), DOPC (blue) and cholesterol (red), with current Martini parameterization. Right snapshot: Equilibrium configuration of the same ternary bilayer, but with a more repulsive C1-C3 interaction. (B) Left snapshot: Equilibrium configuration of a ternary bilayer composed of DPPC (green), DUPC (blue) and cholesterol, with current Martini parameterization. Right snapshot: Equilibrium configuration of the same ternary bilayer, but with a less repulsive C1–C4 interaction. From Davis et al. (2013). (See color insert for the color version of this figure)

to be as weak as the C1–C3 interaction, between DPPC and DOPC or POPC, without modifying the conformational structure of the lipid tail groups. Similar concerns regarding the Martini model parametrization were also raised in recent studies (Javanainen et al. 2017, Stark et al. 2013), which showed that the Martini-interactions between water-soluble proteins are too strong, leading to their aggregation even at concentrations below the experimental solubility limit.

III.B The Dissipative Particle Dynamics Model

Another explicit solvent approach that has been very popular in investigating lipid membranes is based on DPD. DPD was developed in the 1990s by Hoogerbrugge and Koelman (1992) and then cast in its present form by Español and Warren (1995). The DPD approach makes use of soft pairwise conservative, dissipative, and random forces between beads. Due to their pairwise nature, the dissipative and random forces in DPD collectively act as a thermostat and locally conserve momentum, thus allowing for a correct description of long-range hydrodynamics (Ripoll et al. 2001).

The soft nature of the interactions in DPD allows for timesteps of few picoseconds, which are orders of magnitude larger than those in coarse-grained molecular dynamics simulations based on Lennard–Jones interactions (Goetz and Lipowsky 1998) or the Martini force field (Marrink et al. 2007). As a result, the DPD approach has proven to be very useful for investigations of soft materials that are usually characterized by slow kinetics. These include spinodal decomposition (Groot and Warren 1997, Hore and Laradji 2010), transport of polymer solutions (Jiang et al. 2007, Millan and Laradji 2009), micro-phase separation of block copolymers (Groot et al. 1999), nanocomposite materials (Hore and Laradji 2007, 2008), living polymers (Prathyusha et al. 2013, Thakur et al. 2010), and lipid membranes (Laradji and Kumar 2004, Venturoli et al. 2005, Yamamoto et al. 2002).

Different levels of coarse-graining exist within the DPD approach. For example, lipids can be treated as a single head-group bead connected to a single hydrophobic chain (Laradji and Kumar 2004, Shillcock and Lipowsky 2002) (Fig. 1(C)) or, more microscopically, as a head-group composed of few hydrophilic beads connected to two hydrophobic chains (Shillcock and Lipowsky 2002, Venturoli et al. 2005) (Fig. 1(D)). Regardless of the details of a single lipid configuration, proper choices of interaction parameters lead to self-assembled lipid membranes in solvent. Although it is difficult to develop DPD parameters that are system specific (Eriksson et al. 2009, Murtola et al. 2004), parameters can be estimated by reproducing physical quantities such as bending modulus, thickness, and compressibility of the bilayer. However, DPD has been mainly useful for understanding generic trends that are seen in different systems when parameters such as temperature or membrane compositions are changed.

Kranenburg et al. (2004, 2005) performed DPD simulations of DMPC membranes with a relatively fine coarse-graining level, as depicted in Fig. 1(D), and predicted a phase behavior (see Fig. 4) that is consistent with the one derived from experiments (Nagle and Tristram-Nagle 2000). In particular, they found that: At low temperatures, the bilayer is in the gel $L_{\beta'}$-phase, in which the lipids exhibit both translational and chain order, and are tilted with respect to the bilayer normal; at high temperatures, the bilayer is in the fluid L_α-phase, and lacks both translational and chain order; at intermediate temperatures, there occurs a rippled structure, akin to the $P_{\beta'}$-phase, in which the bilayer exhibits laterally alternating gel and fluid regions (Sun et al. 1996).

The abovementioned DPD model for DMPC was used by Venturoli et al. (2005) to study the effect of hydrophobic mismatch on the tilt of transmembrane proteins with respect to the bilayer normal, and on the local perturbations of the bilayer thickness in the vicinity of proteins. As Fig. 5 (A) depicts, it was found that the degree of protein-tilt depends on hydrophobic mismatch and also on the cross-sectional area of the protein. In particular, for proteins with small cross-sectional area, the hydrophobic mismatch leads to a tilt of the protein, as shown by the snapshots in Fig. 5(B). The predicted functional dependence of the tilt angle on hydrophobic mismatch was later confirmed by

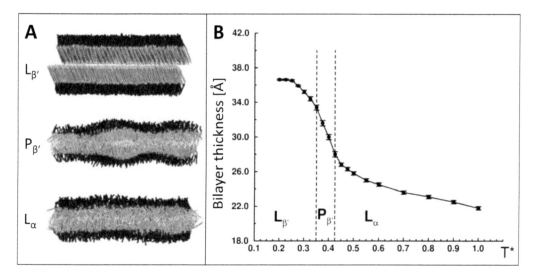

FIGURE 4 (A) Equilibrium snapshots of a simulated DMPC lipid bilayer in the gel $L_{\beta'}$-phase at low temperatures, a $P_{\beta'}$-like phase at intermediate temperatures, and the fluid L_{α}-phase at high temperatures. (B) The thickness of the lipid bilayer vs. temperature. From Venturoli et al. (2005).

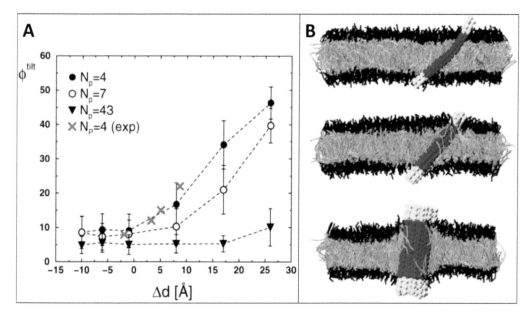

FIGURE 5 (A) Protein tilt angle vs. hydrophobic mismatch $\Delta d = d_p - d_L^{(0)}$, where d_p and $d_L^{(0)}$ are the hydrophobic thicknesses of the protein and the pure DMPC bilayer, respectively, for three values of protein size, N_p. Proteins are simulated as a bundle of N_P amphiphilic chains. The red crosses are from fluorescence spectroscopy experiments for the M13 natural peptide (Koehorst et al. 2004). (B) Configuration snapshots that correspond, from top to bottom, to $N_p = 4$, 7 and 43, respectively. The hydrophobic and hydrophilic beads of the protein are in blue and yellow, respectively. The hydrophobic and hydrophilic beads of the lipids are in green and black, respectively. From Venturoli et al. (2005). (See color insert for the color version of this figure)

experiments (Holt et al. 2009, Venturoli et al. 2005) for both the case of α-helical synthetic peptides WALP23 and KALP23, and the natural peptide M13. Both the artificial and the natural peptides have a small cross-sectional area. Instead, in the case of proteins with large cross-sectional area, hydrophobic mismatch tends to induce a local perturbation of the bilayer thickness, rather than protein tilt, as illustrated by the snapshot of Fig. 5(D). These results were consistent with the outcome of experimental measurements of the orientational tilt angle of two *Escherichia coli* outer membrane proteins, OmpA and FhuA, with different cross-sectional areas (Holt et al. 2009, Ramakrishnan et al. 2005).

Motivated by the ongoing debate regarding the cholesterol-enriched biomembrane domains, so-called rafts (Nicolson 2014, Simons and Sampaio 2011), the DPD approach was also used to investigate the kinetics of phase separation in multi-component planar lipid membranes and lipid vesicles (Laradji and Kumar 2004, 2005, Sornbundit et al. 2014, Yamamoto and Hyodo 2003, Yamamoto et al. 2002). For example, Laradji and Kumar (2005, 2006) showed that domain growth during the phase separation of multicomponent lipid membranes is due to their coalescence following their Brownian motion. During the early stages of these kinetics, the average size of the domains grows as $t^{1/3}$, as shown in Fig. 6 (E), irrespective of the amount of excess area on the membrane. When the excess area is available on the lipid membrane, domains buckle once the excess energy associated with their boundary exceeds their curvature energy, i.e., when the domain size exceeds κ/λ, where κ and λ are the bending modulus and domain line tension, respectively. This leads to faster late-time kinetics characterized by a $t^{4/9}$-growth law, as demonstrated by Fig. 6(E). Laradji and Kumar (2006) also showed that the presence of transbilayer asymmetry in the lipid distribution of the membrane causes a dramatic slowing down of the kinetics, as depicted by the blue curve of

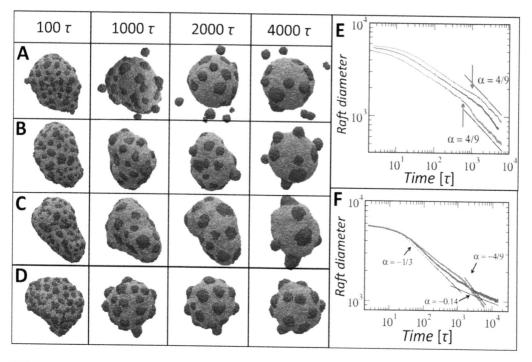

FIGURE 6 Time sequence of snapshots of a multi-component lipid vesicles undergoing phase separation for: (A) a vesicle with high line tension, (B) a vesicle with intermediate line tension, (C) a vesicle with low line tension, and (D) a vesicle with low line tension and asymmetric transbilayer lipid distribution. (E) Average domain size vs. time for the case of A (red), B (blue) and C (green). (F) Average domain size vs. time for the case of snapshots series in C (red) and D (blue). From Laradji and Kumar (2004, 2005, 2006). (See color insert for the color version of this figure)

Fig. 6(F). This study is particularly important, since the plasma membrane of eukaryotic cells is characterized by an actively maintained asymmetric lipid distribution, with the exoplasmic leaflet containing more saturated lipids, such as sphingomyelin and cholesterol, than the cytosolic leaflet. As a result, lipid rafts on the exoplasmic leaflet should be larger than those on the cytosolic leaflet. Taking into account that lipid domains on both sides of the membrane are in register, then the lipid rafts must be buckled outward, as shown by the late-times snapshots in Fig. 6(D). The late-times growth of these domains, in the presence of transbilayer asymmetry, was shown to be exceedingly slow, and may therefore explain the finite size of lipid rafts on the plasma membrane (Laradji and Kumar 2006).

IV Implicit Solvent Modeling

In simulations of lipid membranes, the solvent occupies a substantial fraction of the system's volume. As a result, most of the computational time in explicit solvent simulations is spent in the calculation of forces between solvent beads and, to a lesser extent, in the integration of their equations of motion. It is therefore useful to develop models of lipid membranes in which the solvent's degrees of freedom are effectively integrated out. The absence of the solvent implies that interactions between the lipid beads have to be modified in order to self-assemble stable membranes. This is achieved via attractive interactions between the lipid-tail beads.

Several implicit solvent models for self-assembled lipid membranes have been developed in the past (Brannigan et al. 2004, Cooke and Deserno 2005, Drouffe et al. 1991, Farago 2003, Morikawa and Saito 1994, Noguchi and Takasu 2001, Revalee et al. 2008, Whitehead et al. 2001). These include the pioneering model of Drouffe et al. (1991) in which the bilayer is treated as a monolayer of self-assembled anisotropic hard-spheres, shown in Fig. 1(A). The cross section of each particle represents a lateral patch of the lipid bilayer, and the diameter of each particle corresponds to the thickness of the bilayer. Each hard sphere has three positional degrees of freedom and an orientational vector representing the local normal direction to the bilayer. A hard sphere is subdivided into three segments: an equatorial segment representing the hydrophobic core of a lipid bilayer patch, and top and bottom spherical caps representing the hydrophilic top and bottom parts of the bilayer patch, respectively. Self-assembly is achieved via pairwise interactions, as well as multi-body interactions that depend on local density. Molecular dynamics simulations of this model predict both gel and fluid phases and shape fluctuations dominated by bending modes, in qualitative agreement with experiments. However, the predicted values of the bending modulus are much weaker than the experimental values. This model was later extended by Kohyama (2009) and Yuan et al. (2010) who modeled the coarse-grained particles as points with both positional and orientational degrees of freedom, thereby overcoming the use of the computationally inefficient multi-body nonadditive interactions in the model of Drouffe et al. (1991). The model by Yuan et al. (2010) was recently used to investigate blebbing and vesiculation induced by cytoskeleton defects in red blood cells (Li and Lykotrafitis 2015).

Noguchi and Takasu (2001) proposed an improved coarse-grained implicit solvent model in which lipid molecules are treated as rigid rods, each composed of one hydrophilic bead and two hydrophobic beads. The beads interact via a two-body repulsive interaction and an effective hydrophobic interaction in the form of a multi-body potential that is a function of the local density of hydrophobic beads. Farago (2003) refined this model by using two-body generalized Lennard–Jones interactions only, and avoiding the computationally cumbersome density-dependent multi-body interactions in the model of Noguchi and Takasu (2001).

Few years later, Brannigan et al. (2004) proposed a model in which a lipid molecule has a rigid sphero-cylindrical shape with a director vector along its orientation as depicted by Fig. 1(B). The lipid particles interact via excluded volume interactions, attractive forces between the hydrophobic ends, and orientation-dependent forces that tend to align the lipids to each other. Monte–Carlo simulations of this model predicted a self-assembly of the particles into gel and fluid membranes at low and high temperatures, respectively, and a fluctuations spectrum that, in the fluid phase, is dominated by bending modes.

The studies mentioned above demonstrate that simple implicit solvent particle-based models can reproduce thermodynamically stable gel and fluid membranes, with structural and elastic properties that are in qualitative agreement with those from experiments. A limitation of the implicit solvent models described above is that, although they allow for self-assembly without explicit solvent, they are too coarse-grained and do not account for the internal degrees of freedom of the lipids and their conformational changes. Cooke and Deserno (2005) developed the first implicit solvent particle-based model that partially accounts for the lipid's internal degrees of freedom. This is achieved by coarse-graining the lipids into semi-flexible chains of point beads. Non-bonded beads interact via generalized Lennard–Jones potentials, while bonded beads interact via the finitely extendible nonlinear elastic (FENE) potential (Kremer and Grest 1990). The bending rigidity of the lipids is maintained via a two-body harmonic potential between the lipid-chain end-beads. Self-assembly is achieved by making the interaction between two tail beads more attractive than that between two head beads, or between a tail and head beads. This model has been used to investigate, among others, membrane-induced protein aggregation (Reynwar et al. 2007), formation of pores by antimicrobial peptides (Kabelka and Vácha 2015), and membrane–nanoparticles interactions (Aydin and Dutt 2016, Ruiz-Herrero et al. 2012).

To improve computational efficiency, and be able to sample longer time scales compared to what previous implicit solvent models of lipid membranes enabled, Laradji's group developed an even more efficient implicit particle-based model of semi-flexible chains of point beads (Revalee et al. 2008), which interact via soft two-body interactions instead of Lenard–Jones interactions. Furthermore, for bonded beads, this model uses harmonic interactions instead of the FENE potential. This model (Laradji et al. 2016, Revalee et al. 2008) reproduces stable self-assembled membranes over a wide range of interactions and number of hydrophobic beads per lipid, as shown in Fig.7(A). Fig.7(B) shows that the fluctuations of the membrane shape are dominated by bending modes, and Fig.7(C) shows that the values of the bending modulus are in line with experimental values for phospholipid membranes.

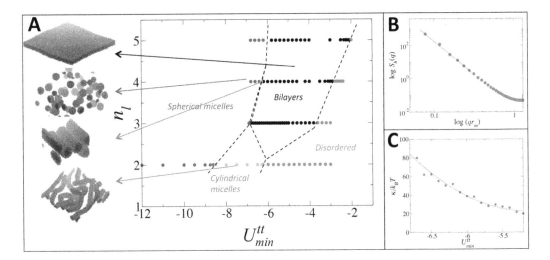

FIGURE 7 (A) Phase diagram of formation self-assembled membranes in terms of the interaction between tail beads, U_{min}^{tt} and the number of beads per lipid N_l; on the left are shown the snapshots of the bilayer, cylindrical, and spherical phases. The results refer to the model by of Revalee et al. (2008). (B) Bilayer height structure factor vs. wavevector showing the q^{-4}-behavior for small q, implying that long wavelength shape fluctuations of the bilayer are dominated by bending modes. (C) Bilayer bending modulus vs. strength of interaction between tail beads. From Laradji et al. (2016).

This model was used in several studies including the phase behavior of SLBs (Poursoroush et al. 2017), the effect of finite size of nanoscale liposomes on their phase behavior (Spangler et al. 2012), the combined effect of cortical cytoskeleton and transmembrane proteins on membrane compositional heterogeneities, such as lipid rafts (Sikder et al. 2014), cytoskeleton-induced membrane blebbing, and vesiculation (Spangler et al. 2011), partial wrapping and endocytosis of nanoparticles (Spangler et al. 2016), and membrane-induced self-assembly of isotropic (Spangler et al. 2018) and anisotropic nanoparticles (Olinger et al. 2016). In the following subsections, we describe the results of four illustrative studies based on the implicit solvent model of Revalee et al. (2008).

IV.A Phase Behavior of Supported Lipid Bilayers

The use of SLBs as biomimetic systems began already in the 1980s (Brian and McConnell 1984, McConnell et al. 1986, Tamm and McConnell 1985). SLBs preparation is relatively straightforward and these systems can be studied by surface-sensitive techniques, such as atomic force microscopy and neutron reflection. Since the 1990s, SLBs have been used (Alessandrini and Facci 2014) for investigating the in-plane morphology of lipid membranes, their thermodynamics, transbilayer asymmetry, and lipid transverse diffusion alias flip-flop (Bayerl and Bloom 1990, Crane et al. 2005, Gerelli et al. 2012, Gerelli et al. 2013, Johnson et al. 1991, Kim et al. 2001).

Experimental studies have shown that the proximity of a solid substrate induces bilayer changes that do not occur in free-standing bilayer systems (Bayerl and Bloom 1990, Crane et al. 2005, Johnson et al. 1991, Kim et al. 2001). For instance, it was found that the rate of lipid flip-flop in SLBs is higher than that in free-standing bilayers (Anglin et al. 2010, Brown and Conboy 2013, Marquardt et al. 2017). In a review-chapter by Sperotto and Ferrarini (2017), both experimental and computational values of lipid flip-flop are listed. Furthermore, controlled-temperature atomic force microscopy experiments showed that the temperature at which Langmuir phospholipid monolayers are transferred to the substrate affects the phase behavior of the resulting SLB (Ramkaran and Badia 2014).

A recent study by Poursoroush et al. (2017), based on molecular dynamics simulations of SLBs using the implicit solvent coarse-grained model of Revalee et al. (2008), showed that the substrate induces an asymmetric lipid distribution across the bilayer, with a higher lipid density of the proximal leaflet than the distal leaflet, as depicted in Fig. 8(C). The translational order parameter of the proximal leaflet exhibits an abrupt transition at the melting point from the gel phase to a fluid phase for all considered values of adhesion strength (see Fig. 8(D)). In contrast, and as shown in Fig. 8(D), although the change of the translational order parameter of the distal leaflet is also abrupt for low adhesion strength, for strong adhesion strength the change of the order parameter of the distal leaflet is only gradual. Inspection of the equilibrium configurations of the lipid bilayer, at finite bilayer-substrate interactions (see Fig. 8(B)), shows that the proximal leaflet is homogenous at both high and low temperatures. However, while the distal leaflet is homogeneous at high temperatures, it is heterogeneous at low temperatures, with coexisting gel and fluid domains. The structural inhomogeneities at low temperatures in the distal leaflet are due to the fact that its average density is lower than that of the gel phase, as demonstrated in Fig. 8(C), which prevents the distal leaflet form being entirely in the gel phase, in agreement with experiments (Ramkaran and Badia 2014). The observed asymmetry in the lipid distribution in SLBs is due to a high rate of substrate-induced flip-flop. This is in accord with recent 1H NMR measurements of DPPC unilamellar vesicles (Marquardt et al. 2017), which shows that the rate of flip-flop in supported DPPC bilayers is several orders of magnitude higher than that of DPPC unilamellar vesicles at the same temperature.

IV.B Anomalous Melting Behavior of Small Liposomes

As indicated earlier, one-component phospholipid membranes undergo a phase transition from a gel-phase to a fluid-phase around or at a system-specific temperature called melting temperature

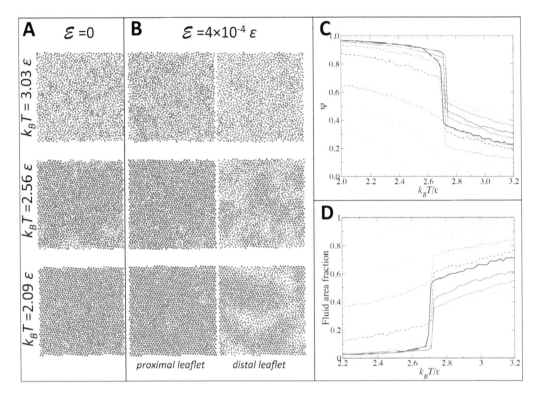

FIGURE 8 (A) Sequence of equilibrium snapshots at different temperatures of a non-supported lipid bilayer ($E = 0$). (B) Sequence of equilibrium snapshots at different temperatures of an SLB at a finite bilayer–substrate interaction, $E = 4 \times 10^{-4}\varepsilon$. In (B), both proximal and distal leaflets are shown. Blue and red dots correspond to chains in the gel and fluid state, respectively. (C) Lipid density of the proximal leaflet (solid lines) and distal leaflet (dotted lines) for different bilayer–substrate interactions. Red corresponds to $E = 1 \times 10^{-4}\varepsilon$, blue corresponds to $E = 2 \times 10^{-4}\varepsilon$, green corresponds to $E = 4 \times 10^{-4}\varepsilon$, orange corresponds to $E = 6 \times 10^{-4}\varepsilon$, and cyan corresponds to $E = 8 \times 10^{-4}\varepsilon$. The black curve corresponds to a non-supported lipid bilayer. (D) Translational order parameter of both leaflets. Same color coding as in (C) is used. From Poursoroush et al. (2017). (See color insert for the color version of this figure)

(Mabrey and Sturtevant 1976). Several experiments have shown that the melting temperature of phospholipid liposomes decreases with decreasing liposomes diameter (Biltonen and Lichtenberg 1993, Brumm et al. 1996, Nagano et al. 1995).

To understand the effect of liposomes finite size on their phase changes, Spangler et al. (2012) performed systematic molecular dynamics simulations of one-component liposomes, with diameter ranging between 24 nm and 91 nm, using the implicit solvent model of Revalee et al. (2008). The results of these simulations indicate that the specific heat of liposomes with diameter larger than about 40 nm is the same as that of planar unsupported lipid membranes, as shown in Fig. 9(D). For smaller liposomes, however, the specific heat has two distinct peaks, as shown in the inset of Fig. 9(D). This anomalous melting behavior of small liposomes was inferred from the chain order susceptibility, shown in Fig. 9(E) and Fig. 9(F), for the inner and outer leaflets, respectively. From this figure, one can see that the low temperature peak of the specific heat coincides with the peak of the susceptibility of the inner leaflet of the liposome, while the high temperature peak of the specific heat coincides with the peak of the susceptibility of the outer leaflet of the liposome. This indicates that the anomalous melting behavior of small liposomes is due to a decoupling of the melting transitions of the two leaflets.

Relatively large liposomes are faceted at temperatures below the melting point, with the edges of the facets coinciding with the boundaries of the gel domains, as shown in Fig. 9(A). Small liposomes, however, are faceted at temperatures below the melting point of the inner leaflet, as

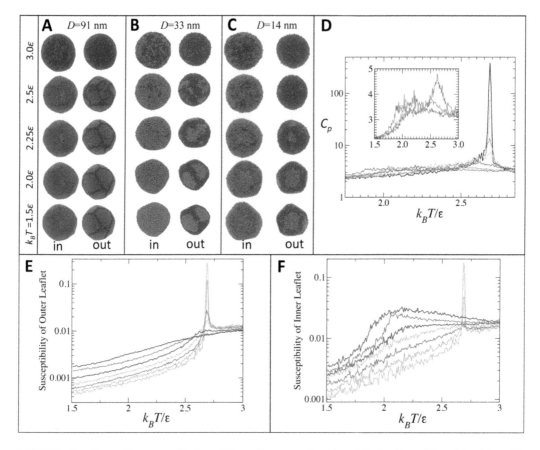

FIGURE 9 Snapshots of liposomes with three different diameters, $D =91$ nm (A), $D =33$ nm (B), and $D =14$ nm (C), at different temperatures. For each diameter, both the inner (In) and outer (Out) leaflets are shown. (D): Specific heat vs. temperature, where the color of the shown curves corresponds to liposome diameters $D = 128$ nm (black), 91 nm (green), 64 nm (magenta), 45 nm (orange), 33 nm (blue), 23 nm (red), and 14 nm (violet). Inset shows zoomed-in specific heats for liposomes with diameters 33 nm (blue), 23 nm (red), and 14 nm (violet). (E) and (F): Susceptibility vs. temperature for the outer and inner leaflet, respectively. Color code as described in (D). From Spangler et al. (2012). (See color insert for the color version of this figure)

shown in Fig. 9(C). At temperatures between the melting temperatures of the two leaflets of small liposomes, the lipids of the outer leaflet lack translational order while they exhibit chain order. Therefore, in this range of temperatures, the lipids of the outer leaflet are in a state akin to the LO phase observed in multicomponent lipid membranes containing saturated lipids and cholesterol (Ipsen et al. 1987).

IV.C Cytoskeleton-Induced Blebbing of Lipid Membranes

Cells can undergo morphological changes in the form of spherical exoplasmic protrusion, known as blebs, during processes, such as cytokinesis, apoptosis, and cell migration, that are mediated by the activity of the cortical cytoskeleton (Barros et al. 2003, Burton and Taylor 1997, Charras et al. 2008, Föller et al. 2008, Mercer and Helenius 2008, Mills et al. 1998, Paluch and Raz 2013, Paluch et al. 2006). Experiments have shown that blebs inhibition is correlated with an increased activity of myosin-II and the ensuing contraction of the cortical cytoskeleton (Paluch et al. 2005). Blebs can also be caused by laser-induced localized damage of the cortex (Tinivez et al. 2009), actin depolymerisation (Paluch et al. 2005), or rapid micropipette suction (Merkel et al. 2000). Blebs are devoid of actomysoin, and grow into spherical shapes up to 2 µm in diameter. While blebbing is mostly associated with nucleated cells, the suicidal death of red blood cells (eryopotosis) is also characterized by blebbing of their plasma membrane (Föller et al. 2008). Furthermore, during the late stages of the life of a red blood cell, a large fraction of its plasma membrane is shed into small vesicles that are devoid of spectrin cytoskeleton. The fact that the shed vesicles are small, about 100 nm in diameter, suggests that their precursor blebs have a size of about the cytoskeleton corral size (Sheetz et al. 2006).

Only few theoretical and computational studies based on continuum elasticity models have been performed to investigate blebbing (Sens and Gov 2007, Strychalski and Guy 2013, Tozluoglu et al. 2013, Woolley et al. 2014, Young and Mitran 2010). Spangler et al. (2011) used the coarse-grained implicit solvent model of Revalee et al. (2008) to investigate blebbing induced by the interplay between the elasticity of the cytoskeleton and that of the lipid bilayer. According to this model, the inner side of the lipid vesicle is apposed to a semi-flexible polymeric meshwork, representing the cortical cytoskeleton. The meshwork is tessellated by triangles with vertices linked to the lipid bilayer and with sides corresponding to semi-flexible chains, as shown by the snapshots in Fig. 10. This therefore models the spectrin meshwork of red blood cells. The phase diagram of the system, depicted in Fig. 10(A), in terms of the relaxed cytoskeleton area per corral, $a_{CSK}^{(0)}$, and the mismatch parameter, s, between the cytoskeleton relaxed area and the vesicles area, shows that increasing tension of the cytoskeleton leads to a morphological first-order phase transition from

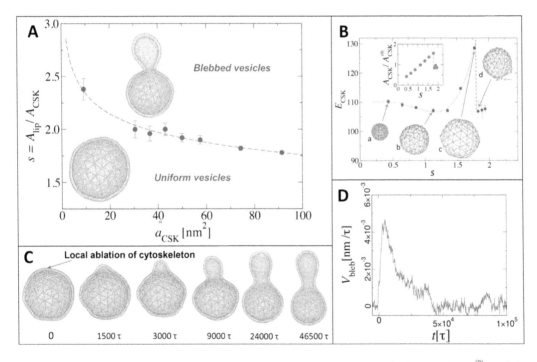

FIGURE 10 (A) Blebbing-formation phase diagram in terms of relaxed corral area (horizontal axis), $a_{CSK}^{(0)}$ and the mismatch parameter, $s = A_{lip}/A_{CSK}^{(0)}$ (vertical axis), where A_{lip} is the area of the lipid vesicle and $A_{CSK}^{(0)} = N_{cor}a_{CSK}^{(0)}$ is the relaxed area of the cytoskeleton (N_{cor} is the number of cytoskeleton corrals). $s > 1$ therefore corresponds to a stretched cytoskeleton. Red and yellow in the configurations represent the lipid membrane and the cytoskeleton, respectively. (B) Cytoskeleton elastic energy versus the mismatch parameter, s, along with configurations of the cytoskeleton. The orange arrow on snapshot (d) indicates where the bleb is located. The inset of (B) shows that, in the uniform phase, the area of the cytoskeleton increases linearly with increasing s, since in this phase the whole vesicle is apposed to the cytoskeleton. However, in the blebbing phase, the cytoskeleton is relaxed, hence $A_{CSK} \approx A_{CSK}^{(0)}$. (C) Time sequence of configurations of a vesicle undergoing blebbing following a localized ablation (blue arrow) at $t = 0$. (D) Front velocity of the bleb growth shown in (C). From Spangler et al. (2011). (See color insert for the color version of this figure)

uniform vesicles, i.e., without protrusions, to blebbed vesicles. In agreement with experiments (Tinivez et al. 2009), the blebbed vesicles are highly deformed with one part apposed to the cytoskeleton and another part in the form of a bleb that is devoid from the cytoskeleton, as shown by the top snapshot in Fig. 10(A). Fig. 10(B) shows that the minimum of the cytoskeleton elastic energy, E_{CSK}, is at about $s_{min} \approx 1$ or at $s \gtrsim 1.8$. At $s_{min} \approx 1$, the vesicle is spherical with a tensionless cytoskeleton (snapshot (b) in Fig. 10(B)). For smaller values of s, the cytoskeleton is compressed (snapshot (a) in Fig. 10(B)) since, in this case, the vesicle's area is smaller than that of the relaxed cytoskeleton. As s approaches 1.8 from below (snapshot (c) in Fig. 10(B)), the cytoskeleton is stretched since it is apposed to a vesicle with an area that is larger than the rest area of the cytoskeleton. The increased tension of the cytoskeleton for $1 \lesssim s \lesssim 1.8$ is compensated by the low curvature energy of the almost spherical vesicle. For $s \gtrsim 1.8$, the cytoskeleton would have to be extensively stretched in order to uniformly conform to the vesicle. Instead, it retracts to a relaxed configuration very similar to that at $s \approx 1$ with the excess lipids forming a bleb (snapshot (d) in Fig. 10(B)). The sharp discontinuity in the cytoskeleton elastic energy indicates that the blebbing transition is first order, in accord with Sens and Gov (2007).

Motivated by an experiment by Tinivez et al. (2009), Spangler et al. (2011) also investigated the kinetics of blebbing as a result of a localized ablation of the cytoskeleton. Here, the localized ablation is mimicked by the detachment of one anchor from the lipid membrane and the removal of its six links at $t = 0$ (blue arrow) in an equilibrated uniform vesicle with $a_{CSK}^{(0)} = 43.1\,nm^2$ and $s = 1.9$. Note that for this set of parameters, the cytoskeleton is initially under tension. The time-sequence of snapshots in Fig. 10(C), shows that, initially, the cytoskeleton undergoes compression while a small cap forms at the location of the ablation. The cap then grows into a bleb, with a speed depicted in Fig. 10(D), while the cytoskeleton retracts to the main part of the vesicle, which then equilibrates at about $40\,000\tau$. In the abovementioned study of cytoskeleton-induced blebbing (Spangler et al. 2011), the solvent is accounted for implicitly, and thus volume constraint and cytosol flow are ignored. While these effects are important, the study by Spangler et al. (2011) indicates that either contraction of the cytoskeleton or a localized disturbance of the cytoskeleton induces the formation of blebs. The incorporation of an explicit solvent in the simulations to account for both volume constraint and cytosol flow is numerically very costly but would be desirable.

IV.D Partial Wrapping and Endocytosis of Nanoparticles

The coarse-grained implicit solvent model of Revalee et al. (2008) was also adopted by Spangler et al. (2016) to study interactions of hydrophilic spherical nanoparticles with tensionless lipid membranes. The series of snapshots in Fig. 11(A) show that increasing the adhesion strength leads to an increased wrapping of the nanoparticle by the membrane, and that for strong enough adhesion strength, the nanoparticle is endocytosed. The phase diagram of adhesion and endocytosis of spherical nanoparticles, depicted in Fig. 11(C), shows three main phases: The unbound phase at low adhesion strength, followed by the wrapping phase at intermediate adhesion strength, and then the endocytosis phase at high adhesion strength. The transition adhesion strengths, between the different phases, decrease with increasing the nanoparticle diameter. In particular, larger nanoparticles are endocytosed at lower adhesion strength than small nanoparticles, in agreement with experiments (Le Bihan et al. 2009).

Earlier theoretical calculations based on the Helfrich–Canham Hamiltonian for tensionless membranes interacting with spherical nanoparticles, or microparticles, predicted that spherical particles cannot be partially wrapped by the membrane, i.e., the particles are either unbound or completely wrapped (Deserno and Bickel 2003, Deserno and Gelbart 2002). This is in contrast to the simulation results of Spangler et al. (2016) who found that the degree of wrapping continuously increases toward full wrapping as the adhesion strength increases, as shown in Fig. 11(D). However, the theory based on the Helfrich–Canham Hamiltonian assumes infinitesimally thin lipid membranes, with contact interactions with the nanoparticles (Deserno and Bickel 2003, Deserno and Gelbart 2002). Spangler et al. (2016) proposed a theoretical treatment of the Helfrich–Canham Hamiltonian, in which the interaction between the nanoparticle and the membrane is not treated as

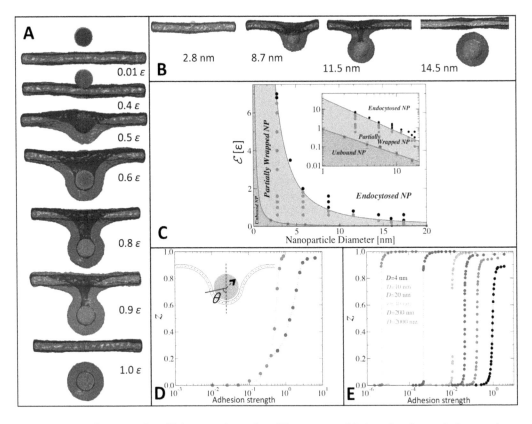

FIGURE 11 (A) Sequence of equilibrium snapshots of an 8.7-nm nanoparticle (green) and a tensionless membrane (brown) for different adhesion strengths. (B) Snapshots of nanoparticles with diameter ranging between 2.8 nm (left) and 14.5 nm (right), for the same adhesion strength corresponding to 0.6ε. (C) Wrapping-endocytosis phase diagram in terms of nanoparticle diameter and adhesion strength. Black symbols (in white region) represent the state of endocytosed nanoparticles. Red symbols (in cyan region) correspond to nanoparticles that are partially wrapped by the bilayer. Nanoparticles are unbound from the bilayer in the green region. The inset shows the same data in log–log plot. The slope of the solid line is −1.3, and that of the dashed line is −1.8. (D) Degree of wrapping, defined by $z = (1 - \cos\theta)/2$, where θ is the wrapping angle, vs. adhesion strength for nanoparticles of diameter 2.8 nm (blue) and 7.8 nm (red). (E) Degree of wrapping vs. adhesion strength as obtained from a continuum theory that takes into account the thickness of the membrane and the finite range of the nanoparticle. From Spangler et al. (2016). (See color insert for the color version of this figure)

a contact interaction and the finite thickness of the membrane is taken into account. As depicted in Fig. 11(E), they showed that the degree of wrapping for small nanoparticles indeed increases continuously with the adhesion strength, in agreement with their simulations (Spangler et al. 2016). For large nanoparticles, however, a discontinuity in the degree of wrapping emerges, in agreement with the previous theory of Deserno et al. (Deserno and Bickel 2003, Deserno and Gelbart 2002).

V Conclusion

During the last few decades, both explicit and implicit solvent particle-based mesoscopic models have been developed with the aim of investigating biomimetic membranes at time and length scales that were out of reach by AMD simulations. Explicit solvent models are suitable for studying both equilibrium and non-equilibrium properties of lipid bilayer systems. Implicit solvent models can also be adopted for studying non-equilibrium properties, provided that hydrodynamic interactions are suppressed in the systems under investigation, i.e., in the case where hydrodynamic modes do not play a major role. Nevertheless, a limitation of both explicit and implicit solvent models is their inability to correctly predict temperature-dependent properties of lipid membranes. This is due to the fact that, because of the fluctuations in networks of hydrogen bonds in water, the self-assembly of lipids is driven by entropy. As a result, the effective interactions between lipid beads include an entropic component and should therefore be temperature dependent. This dependence is however usually neglected in molecular dynamics simulations of coarse-grained models.

Mesoscale particle-based model studies, such as the ones described in this chapter, may be useful to understand if and how biomembrane function results from the collective properties of biomembrane components. Although biomimetic membranes constitute a simplified representation of biomembranes, their physical- and physico-chemical behavior is nevertheless complex. The study of these by one type of modeling and computational method can only help to unravel few aspects of their complexity. Also, one should keep in mind that equilibrium and non-equilibrium phenomena in biomembranes occur over a wide range of time and length scales, and that biomembrane function is the outcome of molecular interactions at the cellular and multicellular levels. Therefore, a multiscale computational approach, combined with experiments, would be beneficial for understanding which role complexity plays for the function of biomembranes. For instance, such an approach may be needed to understand whether biomembrane function crucially depends on the presence of the many different lipid types that compose biomembranes, or rather on the collective behavior of few of these types. From the biotechnological point of view, the resulting knowledge is a necessary step for understanding the cause of diseases and therefore for developing novel therapeutics and diagnostic tools (Al-Jamal and Kostarelos 2011, Lee et al. 2013).

ABBREVIATIONS

AMD, atomistic molecular dynamics; DAPC, diarachidonoyl phosphatidylcholine; DMPC, dimyristoyl phosphatidylcholine; DOPC, dioleoyl phosphatidylcholine; DPD, dissipative particles dynamics; DPPC, dipalmitoyl phosphatidylcholine; DUPC, dilinoleyl phosphatidylcholine; FENE, finitely extendible nonlinear elastic (potential); LD, liquid-disordered; LO, liquid-ordered; SLBs, supported lipid bilayers.

REFERENCES

Alessandrini, A. and P. Facci. 2014. Phase transitions in supported lipid bilayers studied by AFM. *Soft Matter* 10: 7145–64.

Al-Jamal, W.T. and K. Kostarelos. 2011. Liposomes: From a clinically established drug delivery system to a nanoparticle platform for theranostic nanomedicine. *Acc. Chem. Res.* 44: 1094–104.

Anglin, T.C., M.P. Cooper, H. Li, et al. 2010. Free energy and entropy of activation for phospholipids flip-flop in planar supported lipid bilayers. *J. Phys. Chem. B* 114: 1903–14.

Arnarez, C., J.J. UUsitalo, M.F. Masman, et al. 2015. Dry Martini, a coarse-gained force field for lipid membrane simulations with implicit solvent. *J. Chem. Theory Comp.* 11: 260–75.

Aydin, F. and M. Dutt. 2016. Surface reconfiguration of binary lipid vesicles via electrostatically induced nanoparticle adsorption. *J. Phys. Chem.* 120: 6646–56.

Baoukina, S., S.J. Marrink, and D.P. Tieleman. 2012. Molecular structure of membrane tethers. *Biophys. J.* 102: 1866–71.

Baoukina, S. and D.P. Tieleman. 2010. Direct simulation of protein-mediated vesicle fusion: Lung surfactant Protein B. *Biophys. J.* 99: 2134–42.

Barros, L.F., T. Kanaseki, R. Sabirov, et al. 2003. Apoptotic and necrotic blebs in epithelial cells display similar neck diameters but different kinase dependency. *Cell Death Diff* 10: 687–97.

Bayerl, T. and M. Bloom. 1990. Physical properties of single phospholipid bilayers adsorbed to micro glass beads. A new vesicular model system studied by H-nuclear magnetic resonance. *Biophys. J.* 58: 357–62.

Berkowitz M.L., S.L. Bostick, and S. Pandit. 2006. Aqueous solutions next to phospholipid membrane surfaces: Insights from simulations. *Chem. Rev.* 4: 1527–39.

Biltonen, R.L. and D. Lichtenberg. 1993. The use of differential scanning calorimetry as a tool to characterize liposome preparations. *Chem. Phys. Lipids* 64: 129–42.

Brannigan, G., L.C.L. Lin, and F.L.G. Brown. 2006. Implicit solvent simulation models for biomembranes. *J. Biophys. Lett.* 35: 104–24.

Brannigan, G., A.C. Tamboli, and F.L.G.Brown. 2004. The role of molecular shape in bilayer elasticity and phase behavior. *J. Chem. Phys.* 121: 3259–71.

Brian, A.A. and H.M. McConnell. 1984. Allogeneic stimulation of cyto-toxic t-cells by supported planar membranes. *Proc. Natl. Acad. Sci. USA* 81: 6159–63.

Brown, K.L. and J.C. Conboy. 2013. Lipid flip-flop in binary membranes composed of phosphatidylserine and phosphatidylcholine. *J. Phys. Chem. B* 117: 15041–50.

Brumm, T., K. Jorgensen, O.G. Mouritsen, et al. 1996. The effect of increasing membrane curvature on the phase transition and mixing behavior of a dimyristoyl-sn-glycero-3-phosphatidylcholine/distearoyl-sn-glycero-3-phosphatidylcholine lipid mixture as studied by Fourier transform infrared spectroscopy and differential scanning calorimetry. *Biophys. J.* 70: 1373–79.

Burton, K. and D.L. Taylor. 1997. Traction forces of cytokinesis measured with optically modified elastic substrata. *Nature (London)* 385: 450–54.

Canham, P.B. 1970. The minimum energy of bending as a possible explanation of the biconcave shape of the human red blood cell. *J. Theor. Biol.* 26: 61–81.

Charras, G.T., M. Coughlin, T.J. Michison, et al. 2008. Life and times of a cellular bleb. *Biophys. J.* 94: 1836–53.

Cooke, I.R. and M. Deserno. 2005. Solvent-free model for self-assembling fluid bilayer membranes: Stabilization of the fluid phase based on broad attractive tail potentials. *J. Chem. Phys.* 123: 224710.

Crane, J.M., V. Kiessling, and L.K. Tamm. 2005. Measuring lipid asymmetry in planar supported bilayers by fluorescence interference contrast microscopy. *Langmuir* 21: 1377–88.

Curtis, E.M. and C.K. Hall. 2013. Molecular dynamics simulations of DPPC bilayers using LIME, a new coarse-grained model. *J. Phys. Chem. B* 117: 5019–30.

Da Rocha, E.L., G.F. Caramorib, and C.R. Rambo. 2013. Nanoparticle translocation through a lipid bilayer tuned by surface chemistry. *Phys. Chem. Chem. Phys.* 15: 2282–90.

Davis, R.S., P.B.S. Kumar, M.M. Sperotto, et al. 2013. Predictions of phase separation in three-component lipid membranes by the MARTINI force field. *J. Phys. Chem. B* 117: 4072–80.

Deserno, M. and T. Bickel. 2003. Wrapping of a spherical colloid by a fluid membrane. *Europhys. Lett.* 62: 767–73.

Deserno, M. and W.M. Gelbart. 2002. Adhesion and wrapping in colloid-vesicle complexes. *J. Phys. Chem. B* 106: 5543–52.

Doniach, S. 1978. Thermodynamic fluctuations in phospholipid bilayers. *J. Chem. Phys.* 68: 4912–16.

Drouffe, J.-M., A.C. Maggs, and S. Leibler. 1991. Computer simulations of self-assembled membranes. *Science* 254: 1353–56.

Eriksson, A., M. Nilsson Jacobi, J. Nyström, et al. 2009. A method for estimating the interactions in dissipative particle dynamics from particle trajectories. *J. Phys.: Cond. Matt.* 21: 095401.

Español, E. and P. Warren. 1995. Statistical-mechanics of dissipative particle dynamics. *Europhys. Lett.* 30: 191–96.

Fahy, E., S. Subramaniam, R.C. Murphy, et al. 2009. Update of the LIPID MAPS comprehensive classification system for lipids. *J. Lipid Res.* 50: S9-S14.

Farago, O. 2003. Water-free computer model for fluid bilayer membranes. *J. Chem. Phys.* 119: 596–605.

Föller, M., S.M. Huber, and F. Lang. 2008. Erythrocyte programmed death. *IUBMB Life* 60: 661–8.

Gerelli, Y., L. Porcar, and G. Fragneto. 2012. Lipid rearrangement in DSPC/DMPC bilayers: A neutron reflectometry study. *Langmuir* 28: 15922–28.

Gerelli, Y., L. Porcar, L. Lombardi, et al. 2013. Lipid Exchange and Flip-Flop in Solid Supported Bilayers. *Langmuir* 29: 12762–9.

Goetz, R., G. Gompper, and R. Lipowsky. 1999. Mobility and elasticity of self-assembled membranes. *J. Phys. Rev. Lett.* 82: 221–4.

Goetz, R. and R. Lipowsky. 1998. Computer simulations of bilayer membranes: Self-assembly and interfacial tension. *J. Chem. Phys.* 108: 7397–409.

Gompper, G. and D. Kroll. 1994. Phase diagram of fluid vesicles. *Phys. Rev. Lett.* 73: 2139–42.

Groot, R.D., T.J. Madden, and D.J. Tildesley. 1999. On the role of hydrodynamic interactions in block copolymer microphase separation. *J. Chem. Phys.* 110: 9739–49.

Groot, R.D. and P.B. Warren. 1997. Dissipative particle dynamics: Bridging the gap between atomistic and mesoscopic simulation. *J. Chem. Phys.* 107: 4423–35.

Gupta, R. and B. Rai. 2017. Effect of size and surface charge of gold nanoparticles on their skin permeability: A molecular dynamics study. *Sci. Rep.* 7: 45292.

Hamm, M. and M. M. Kozlov. 2000. Elastic energy of tilt and bending of fluid membranes. *Eur. Phys. J. E* 3: 323–35.

Helfrich, W. 1973. Elastic properties of lipid bilayers: Theory and possible experiments. *Z. Naturforsch.* 28C: 693–703.

Holt, A., R.B.M. Koehorst, T. Rutters-Meijneke, et al. 2009. Tilt and rotation angles of a transmembrane model peptide as studied by fluorescence spectroscopy. *Biophys. J.* 97: 2258–66.

Hoogerbrugge, P.J. and J.M.V.A. Koelman. 1992. Simulating microscopic hydrodynamic phenomena with dissipative particle dynamics. *Europhys. Lett.* 19: 155–60.

Hore, M.J.A. and M. Laradji. 2007. Microphase separation induced by interfacial segregation of isotropic, spherical nanoparticles. *J. Chem. Phys.* 126: 224903.

Hore, M.J.A. and M. Laradji. 2008. Prospects of nanorods as an emulsifying agent of immiscible blends. *J. Chem. Phys.* 128: 054901.

Hore, M.J.A. and M. Laradji. 2010. Dissipative particle dynamics simulation of the interplay between spinodal decomposition and wetting in thin film binary fluids. *J. Chem. Phys.* 132: 024908.

Ipsen, J.H., G. Kalstr'om, O.G. Mouritsen, et al. 1987. Phase equilibira in the phosphatidylcholine-cholesterol system. *Biochim. Biophys. Acta* 905: 162–72.

Izvekov, S. and G.A. Voth. 2009. Solvent-free lipid bilayer model using multiscale coarse-graining. *J. Phys. Chem. B* 113: 4443–55.

Javanainen, M., H. Martinez-Seara, and I. Vattulainen. 2017. Excessive aggregation of membrane proteins in the Martini model. *PloS One* 12: e0187936.

Jiang, W., J. Huang, Y. Wang, et al. 2007. Hydrodynamic interaction in polymer solutions simulated with dissipative particle dynamics. *J. Chem. Phys.* 126: 044901.

Johnson, S.J., T.M. Bayerl, D.C. McDermott, et al. 1991. Structure of an adsorbed dimyristoylphosphatidylcholine bilayer measured with specular reflection of neutrons. *Biophys. J.* 59: 289–94.

Kabelka, I. and R. Vácha. 2015. Optimal conditions for opening of membrane pore by amphiphilic peptides. *J. Chem. Phys.* 143: 243115.

Kim, J., G. Kim, and P.S. Cremer. 2001. Investigations of water structure at the solid/liquid interface in the presence of supported lipid bilayers by vibrational sum frequency spectroscopy. *Langmuir* 17: 7255–60.

Koehorst, R.B.M., R.B. Spruijt, F.J. Vergeldt, et al. 2004. Lipid bilayer topology of the transmembrane - helix of M13 major coat protein and bilayer polarity profile by site-directed fluorescence spectroscopy. *Biophys. J.* 87: 1445–55.

Kohyama, T. 2009. Simulations of flexible membranes using a coarse-grained particle-based model with spontaneous curvature variables. *Physica A* 388: 3334–44.

Konig, S. and E. Sackmann. 1996. Molecular and collective dynamics of lipid bilayers. *Curr. Opin. Colloid Interface Sci.* 1: 78–82.

Kozlovsky, Y. and M.M. Kozlov. 2002. Stalk model of membrane fusion: Solution of energy crisis. *Biophys. J.* 82: 882–95.

Kranenburg, M., C. Laforge, and B. Smit. 2004. Mesoscopic simulations of phase transitions in lipid bilayers. *Phys. Chem. Chem. Phys.* 6: 4531–34.

Kranenburg M. and B. Smit. 2005. Phase behavior of model lipid bilayers. *J. Phys. Chem. B* 109: 6553–63.

Kranenburg, M., M. Venturoli, and B. Smit. 2003. Molecular simulations of mesoscopic bilayer phases. *Phys. Rev. E* 67: 060901.

Kremer, K. and G.S. Grest. 1990. Dynamics of entangled linear polymer melts: A molecular-dynamics simulation. *J. Chem. Phys.* 92: 5057–86.

Kumar, P.B.S., G. Gompper, and R. Lipowsky. 2001. Budding dynamics of multicomponent membranes. *Phys. Rev. Lett.* 86: 3911–14.

Laradji, M. and P.B.S. Kumar. 2004. Dynamics of domain growth in self-assembled fluid vesicles. *Phys. Rev. Lett.* 93: 198105.

Laradji, M. and P.B.S. Kumar. 2005. Domain growth, budding, and fission in phase-separating self-assembled fluid bilayers. *J. Chem. Phys.* 123: 224902.

Laradji, M. and P.B.S. Kumar. 2006. Anomalously slow domain growth in fluid membranes with asymmetric transbilayer lipid distribution. *Phys. Rev. E* 73: 040901.

Laradji, M., P.B.S. Kumar, and E.J. Spangler. 2016. Exploring large-scale phenomena in composite membranes through an efficient implicit-solvent model. *J. Phys. D: Appl. Phys.* 49: 293001.

Le Bihan, O., P. Bonnafous, and L. Marak, et al. 2009. Cryo-electron tomography of nanoparticle transmigration into liposome. *J. Struct. Biol.* 168: 419–25.

Lee, H. and R.G. Larson. 2008. Coarse-grained molecular dynamics studies of the concentration and size dependence of fifth- and seventh-Generation PAMAM dendrimers on pore formation in DMPC bilayer. *J. Phys. Chem. B* 112: 7778–84.

Lee, H. and R.G. Larson. 2009. Multiscale modeling of dendrimers and their interactions with bilayers and polyelectrolytes. *Molecules* 14: 423–38.

Lee, Y.K., H. Lee, and J.-M. Nam. 2013. Lipid-nanostructure hybrids and their applications in nanobiotechnology. *NPG Asia Materials* 5: e48.

Leventhal, I. and S.L. Veatch. 2016. The continuing mystery of lipid rafts. *J. Mol. Biol.* 428: 4749–64.

Li, H. and G. Lykotrafitis. 2015. Vesiculation of healthy and defective red blood cells. *Phys. Rev. E* 92: 012715.

Lin, J.Q., Y.G. Zheng, H.W. Zhang, et al. 2011. A simulation study on nanoscale holes generated by gold nanoparticles on negative lipid bilayers. *Langmuir* 27: 8323–32.

Lin, L.C.L., and F.L.H. Brown. 2004. Brownian Dynamics in Fourier Space: Membrane Simulations over Long Length and Time Scales. *Phys. Rev. Lett.* 93: 256001.

Lopez, C.A., A.J. Rzepiela, A.H. de Vries, et al. 2009. Martini coarse-grained force fields: Extension to carbohydrates. *J. Chem. Theor. Comput.* 5: 3195–210.

Louhivuori, M., H.J. Risselada, E. van der Giessen, et al. 2010. Release of content through mechano-sensitive gates in pressurized liposomes. *Proc. Natl. Acad. Sci.* 107: 19856–60.

Lyubartsev, A.P., A. Mirzoev, L. Chen, et al. 2009. Systematic coarse-graining molecular models by the Newton inversion method. *Faraday Discuss* 144: 43–56.

Mabrey, S. and J.M. Sturtevant. 1976. Investigation of phase transitions of lipids and lipid mixtures by sensitivity differential scanning calorimetry. *Proc. Nat. Acad. Sci.* 73: 3862–66.

Marquardt, D., F.A. Heberle, T. Miti, et al. 2017. H NMR shows slow phospholipid flip-flop in gel and fluid bilayers. *Langmuir* 33: 3731–41.

Marrink, S.J., M. Berkowitz, and H.J.C. Berendsen. 1993. Molecular dynamics simulation of a membrane/water interface: The ordering of water and its relation to the hydration force. *Langmuir* 9: 3122–31.

Marrink, S.J., A.H. de Vries, and A.E. Mark. 2004. Coarse-grained model for semiquantitative lipid simulations. *J. Phys. Chem. B* 108: 750–60.

Marrink, S.J., H.J. Risselada, S. Yefimov, et al. 2007. The MARTINI force field: Coarse grained model for biomolecular simulations. *J. Phys. Chem. B* 111: 7812–24.

Marrink, S.J. and D.P. Tieleman 2013. Perspective on the Martini model. *Chem. Soc. Rev.* 42: 6801–22.

May, S. 2002. Structure and energy of fusion stalks: The role of membrane edges. *Biophys. J.* 83: 2969–80.

McConnell, H.M., T.A. Watts, R.M. Weis, et al. 1986. Supported planar membranes in studies of cell-cell recognition in the immune system. *Biochim. Biophys. Acta* 864: 95–106.

McGreevy, R.L. and L. Pusztai. 1988. Reverse Monte Carlo simulations: A new technique for the determination of disordered structures. *Mol. Simul.* 1: 359–67.

Mercer, J. and A. Helenius. 2008. Vaccinia virus uses macropinocytosis and apoptotic mimicry to enter host cells. *Science* 320: 531–35.

Merkel, R., R. Simson, D.A. Simson, et al. 2000. A micormechanic study of cell polarity and plasma membrane cell body coupling in Dictyostelium. *Biophys. J.* 79: 707–19.

Millan, J.A. and M. Laradji. 2009. Cross-stream migration of driven polymer solutions in nanoscale channels: A numerical study with generalized dissipative particle dynamics. *Macromolecules* 42: 803–10.

Mills, J.C., N.L. Stone, J. Erhardt, et al. 1998. Apoptotic membrane blebbing is regulated by myosin light chain phosphorylation. *J. Cell Biol.* 140: 627–36.

Monticelli, L., S.K. Kandasamy, X. Periole, et al. 2008. The MARTINI coarse-grained force field: Extension to Proteins. *J. Chem. Theory and Comput.* 4: 819–34.

Moore, T.C., C.R. Iacovella, and C. McCabe. 2014. Derivation of coarse-grained potentials via multistate iterative Boltzmann inversion. *J. Chem. Phys.* 140: 224104.

Morikawa, R. and Y. Saito. 1994. Hard rod and frustum model of two-dimensional vesicles. *J. Phys. II* 4: 145–61.

Mouritsen, O.G., A. Boothroyd, R. Harris, et al. 1983. Computer simulation of the main gel-fluid phase transition of lipid bilayers. *J. Chem. Phys.* 79: 2027–41.

Murtola, T., E. Falck, M. Patra, et al. 2004. Coarse-grained model for phospholipid/cholesterol bilayer. *J. Chem. Phys.* 121: 9156–65.

Nagano, H., T. Nakanishi, H. Yao, et al. 1995. Effect of vesicle size on the heat capacity anomaly at the gel to liquid-crystalline phase transition in unilamellar vesicles of dimyristoylphosphatidylcholine. *Phys. Rev. E* 52: 4244–50.

Nagle, J.F. 1973. Theory of biomembrane phase transitions. *J. Chem. Phys.* 58: 252–64.

Nagle, J.F. and S. Tristram-Nagle. 2000. Structure of lipid bilayers. *Biochim. Biophys. Acta* 1496: 159–95.

Nicolson, G.L. 2014. The fluid-mosaic model of membrane structure: Still relevant to understanding the structure, function and dynamics of biological membranes after more than 40 years. *Biochim. Biophys. Acta* 1838: 1451–66.

Noguchi, H. and G. Gompper. 2005. Shape transitions of fluid vesicles and red blood cells in capillary flows. *Proc. Natl. Acad. Sci. USA* 102: 14150–64.

Noguchi, H. and M. Takasu. 2001. Self-assembly of amphiphiles into vesicles: A Brownian dynamics simulation. *Phys. Rev. E* 64: 041913.

Olinger, A.D., E.J. Spangler, P.B.S. Kumar, et al. 2016. Membrane-mediated aggregation of anisotropically curved nanoparticles. *Faraday Discuss* 186: 265–75.

Oroskar, P.A., C.J. Jameson, and S. Murad. 2016. Simulated permeation and characterization of PEGylated gold nanoparticles in a lipid bilayer system. *Langmuir* 32: 7541–55.

Paluch, E., M. Piel, J. Prost, et al. 2005. Cortical actomyosin breakage triggers shape oscillations in cells and cell fragments. *Biophys. J.* 89: 724–33.

Paluch, E., C. Sykes, J. Prost, et al. 2006. Dynamic modes of the cortical actomyosin gel during cell locomotion and division. *Trends Cell Biol* 16: 5–10.

Paluch, E.K. and E. Raz. 2013. The role and regulation of blebs in cell migration. *Curr. Op. in Cell Biol.* 25: 582–90.

Panizon, E., D. Bochicchio, L. Monticelli, et al. 2015. MARTINI coarse-grained models of polyethylene and polypropylene. *J. Phys. Chem. B* 119: 8209–16.

Parton, D.L., J.W. Kliongelhoefer, and M.S. Sansom. 2011. Aggregation of model membrane proteins, modulated by hydrophobic mismatch, membrane curvature, and protein class. *Biophys. J.* 101: 691–99.

Periole, Z., T. Huber, S.J. Marrink, et al. 2007. G Protein-coupled receptors self-assemble in dynamics simulations of model bilayers. *J. Am. Chem. Soc.* 129: 10126–32.

Pink, D.A., T.J. Green, and D. Chapman. 1980. Raman scattering in bilayers of saturated phosphatidyl-cholines. Experiment and theory. *Biochemistry* 19: 349–56.

Poursoroush, A., M.M. Sperotto, and M. Laradji. 2017. Phase behavior of supported lipid bilayers: A systematic study by coarse-grained molecular dynamics simulations. *J. Chem. Phys.* 146: 154902.

Prathyusha, K.R., A.P. Deshpande, M. Laradji, et al. 2013. Shear-thinning and isotropic-lamellar-columnar transition in a model for living polymers. *Soft Matter* 9: 9983–90.

Ramakrishnan, M., J. Qu, C. L. Pocanschi, et al. 2005. Orientation of -barrel proteins OmpA and FhuA in lipid membranes. Chain length dependence from infrared dichroism. *Biochemistry* 44: 3515–23.

Ramakrishnan, N., P.B.S. Kumar, and R. Radhakrishnan. 2014. Mesoscale computational studies of membrane bilayer remodeling by curvature-inducing proteins. *Phys. Rep.* 543: 1–60.

Ramkaran, M. and A. Badia. 2014. Gel-to-fluid phase transformations in solid-supported phospholipid bilayers assembled by the Langmuir-Blodgett technique: Effect of the Langmuir monolayer phase state and molecular density. *J. Phys. Chem. B* 118: 9708–21.

Reister-Gottfried, E., S.M. Leitenberger, and U. Seifert. 2007. Hybrid simulations of lateral diffusion in fluctuating membranes. *Phys. Rev. E* 75: 011908.

Revalee, J.D., M. Laradji, and P.B.S. Kumar. 2008. Implicit-solvent mesoscale model based on soft-core potentials for self-assembled lipid membranes. *J. Chem. Phys.* 128: 035102.

Reynwar, B.J., G. Ilya, V.A. Harmandaris, et al. 2007. Aggregation and vesiculation of membrane proteins by curvature-mediated interactions. *Nature* 447: 461–64.

Ripoll, M., M.H. Ernst, and P. Español. 2001. Large scale and mesoscopic hydrodynamics for dissipative particle dynamics. *J. Chem. Phys.* 115: 7271–84.

Risselada, H.J., C. Kutzner, and H. Grubmüller. 2011. Caught in the act: Visualization of SNARE-mediated fusion events in molecular detail. *Chem. Bio. Chem.* 12: 1049–55.

Risselada, H.J. and S.J. Marrink. 2008. The molecular face of lipid rafts in model membranes. *Proc. Natl. Acad. Sci.* 105: 17367–72.

Rossi, G., P.F. Fuchs, J. Barnoud, et al. 2012. A coarse-grained MARTINI model of polyethylene glycol and of polyoxyethylene alkyl ether surfactants. *J. Phys. Chem. B* 116: 14353–62.

Ruiz-Herrero, T., E. Velasco, and M. F. Hagan. 2012. Mechanisms of budding of nanoscale particles through lipid bilayers. *J. Phys. Chem. B* 116: 9595–603.

Seifert, U. 1997. Configurations of fluid membranes and vesicles. *Adv. Phys.* 46: 13–137.

Sens, P. and N. Gov. 2007. Force balance and membrane shedding at the red-blood-cell surface. *Phys. Rev. Lett.* 98: 018102.

Sheetz, M.P., J.E. Sable, and H.-G. Döbereiner. 2006. Continuous membrane-cytoskeleton adhesion requires continuous accommodation to lipid and cytoskeleton dynamics. *Annu. Rev. Biophys. Biomol. Struct.* 35: 417–34.

Shelley, J.C., M.Y. SHelley, R.C. Reeder, et al. 2001. A coarse-grain model for phospholipid simulation. *J. Phys. Chem. B* 105: 4464–70.

Shillcock, J.C. and R. Lipowsky. 2002. Equilibrium structure and lateral stress distribution of amphiphilic bilayers from dissipative particle dynamics simulations. *J. Chem. Phys.* 117: 5048–61.

Shinoda, W., R. DeVane, and M.L Klein. 2010. Zwitterionic lipid assemblies: Molecular Dynamics studies of monolayers, bilayers, and vesicles using a new coarse grain force field. *J. Phys. Chem. B* 114: 6836–49.

Sikder, M.K.U., K.A. Stone, P.B.S. Kumar, et al. 2014. Combined effect of cortical cytoskeleton and transmembrane proteins on domain formation in biomembranes. *J. Chem. Phys.* 141: 054902.

Simonelli, F., D. Bochicchio, R. Ferrando, et al. 2015. Monolayer-protected anionic Au nanoparticles walk into lipid membranes step by step. *J. Phys. Chem. Lett.* 6: 3175–79.

Simons, K. and J.L. Sampaio. 2011. Membrane organization and lipid rafts. *Cold Spring Harb. Perspect. Biol.* 3: a004697.

Sintes, T. and A. Baumgärtner. 1997. Protein attraction in membranes induced by lipid fluctuations. *Biophys. J.* 73: 2251–59.

Smondyrev, A.M. and M.L. Berkowitz. 1999. Structure of dipalmitoylphosphatidylcholine/cholesterol bilayer at low and high cholesterol concentrations: Molecular dynamics simulation. *Biophys. J.* 77: 2075–89.

Sodt, A.J. and T. Head-Gordon. 2010. An implicit solvent coarse-grained lipid model with correct stress profile. *J. Chem. Phys.* 132: 205103.

Sornbundit, K., C. Modchang, W. Triampo, et al. 2014. Kinetics of domain registration in multi-component lipid bilayer membranes. *Soft Matter* 10: 7306–15.

Spangler, E.J., C.W. Harvey, J.D. Revalee, et al. 2011. Computer simulation of cytoskeleton-induced blebbing in lipid membranes. *Phys. Rev. E* 84: 051906.

Spangler, E.J., P.B.S. Kumar, and M. Laradji. 2012. Anomalous freezing behavior of nanoscale liposomes. *Soft Matter* 8: 10896–904.

Spangler, E.J., P.B.S. Kumar, and M. Laradji. 2018. Stability of membrane-induced self-assemblies of spherical nanoparticles (under review).

Spangler, E.J., S. Upreti, and M. Laradji. 2016. Partial wrapping and spontaneous endocytosis of spherical nanoparticles by tensionless lipid membranes. *J. Chem. Phys.* 144: 044901.

Sperotto, M.M., and A. Ferrarini. 2017. Spontaneous lipid flip-flop in membranes: A still unsettled picture from experiments and simulations. In *The Biophysics of Cell Membranes*, eds. R. Epand, and J.-M. Ruysschaert, 29–60.Singapore: Springer.

Sreeja, K.K., J.H. Ipsen, and P.B.S. Kumar. 2015. Monte Carlo simulations of fluid vesicles. *J. Phys. Condens. Matter* 27: 273104.

Stark, A. C, C.T. Andrews, and A.H. Elcock. 2013. Toward optimized potential functions for protein–protein interactions in aqueous solutions: Osmotic second Virial coefficient calculations using the Martini coarse-grained force field. *J Chem. Theory Comput.* 9: 4176–85.

Strychalski, W. and R.D. Guy. 2013. A computational model of bleb formation. *Math. Med. Biol.* 30: 115–30.

Sun, W.J., S. Tristram-Nagle, R.M. Suter, et al. 1996. Structure of the ripple phase in lecithin bilayers. *Proc. Nat. Acad. Sci. USA* 93: 7008–12.

Tamm, L.K. and H.M. McConnell. 1985. Supported phospholipid bilayers. *Biophys. J.* 47: 105–13.

Taniguchi, T. 1996. Shape deformation and phase separation dynamics of two-component vesicles. *Phys. Rev. Lett.* 76: 4444–7.

Terzi, M.M. and M. Deserno. 2017. Novel tilt-curvature coupling in lipid membranes. *J. Chem. Phys.* 147: 084702.

Thakur, S., K.R. Prathyusha, A.P. Deshpande, et al. 2010. Shear induced ordering in branched living polymer solutions. *Soft Matter* 6: 489–92.

Tinivez, J.-Y., U. Shulze, G. Salbreux, et al. 2009. Role of cortical tension in bleb growth. *Proc. Nat. Acad. Sci. USA* 106: 18581–6.

Tozluoglu, M., A.L. Tournier, R.P. Jenkins, et al. 2013. Matrix geometry determines optimal cancer cell migration strategy and modulates response to interventions. *Nat. Cell Biol.* 15: 751–62.

Uusitalo, J.J, H.I. Ingólfsson, P. Akhshi, et al. 2015. Martini coarse-grained force field: Extension to DNA. *J. Chem. Theory Comput.* 11: 3932–45.

van Meer, G. and A.I.P.M. de Kroon. 2011. Lipid map of the mammalian cell. *J. Cell Science* 124: 5–8.

van Meer, G., D.R. Voelker, and G.W. Feigenson. 2008. Membrane lipids: Where they are and how they behave. *Nat. Rev. Mol. Cell Biol.* 9: 112–24.

Veatch, S.L. and S.L. Keller. 2002. Organization in lipid membranes containing cholesterol. *Phys. Rev. Lett.* 89: 268101.

Veatch, S.L. and S.L. Keller. 2003. Separation of liquid phases in giant vesicles of ternary mixtures of phospholipids and cholesterol. *Biophys. J.* 85: 3074–83.

Venturoli, M., B. Smit, and M.M. Sperotto. 2005. Simulation studies of protein-induced bilayer deformations, and lipid-induced protein tilting, on a mesoscopic model for lipid bilayers with embedded proteins. *Biophys. J.* 88: 1778–98.

Venturoli, M., M.M. Sperotto, M. Kranenburg, et al. 2006. Mesoscopic models of biological membranes. *Physics Reports* 437: 1–54.

Wang, Z.-J. and M. Deserno. 2010. A systematically coarse-grained solvent-free model for quantitative phospholipid bilayer simulations. *J. Phys. Chem. B* 114: 11207–20.

Watson, M.C., E.G. Brandt, P.M. Welch, et al. 2012. Determining biomembrane bending rigidities from simulations of modest size. *Phys. Rev. Lett.* 109: 028102.

Whitehead, L., C.M. Edge, and J.W. Essex. 2001. Molecular dynamics simulation of the hydrocarbon region of a biomembrane using a reduced representation model. *J. Comput. Chem.* 22: 1622–33.

Woolley, T.E., E.A. Gaffney, S.L. Walters, et al. 2014. Three mechanical models for blebbing and multi-blebbing. *IMA J. App. Math.* 79: 636–60.

Yamamoto, S. and S.-A. Hyodo. 2003. Budding and fission dynamics of two-component vesicles. *J. Chem. Phys.* 118: 7937–43.

Yamamoto, S., Y. Maruyama, and S.A. Hyodo. 2002. Dissipative particle dynamics study of spontaneous vesicle formation of amphiphilic molecules. *J. Chem. Phys.* 116: 5842–49.

Yanagisawa, M., M. Imai, T. Masui, et al. 2007. Growth Dynamics of Domains in Ternary Fluid Vesicles. *Biophys. J.* 92: 115–25.

Yesylevskyy, S.O., L.V. Schäfer., D. Sengupta, et al. 2010. Polarizable Water Model for the Coarse-Grained MARTINI Force Field. *PLoS Comp. Biol.* 6: e1000810.

Young, J. and S. Mitran. 2010. A numerical study of cellular blebbing: A volume-conserving, fluid-structure interaction model of the entire cell. *J. Biomech.* 43: 210–20.

Yuan, H., C. Huang, J. Li, et al. 2010. One-particle-thick, solvent-free, coarse-grained model for biological and biomimetic fluid membranes. *Phys. Rev. E* 82: 011905.

3

Continuum Elastic Description of Processes in Membranes

Alexander J. Sodt
*Eunice Kennedy Shriver National Institutes of Child Health and Human Development,
National Institutes of Health*

Continuum models are defined by their lack of particles. Particle models are typically understood in terms of Newton's laws: differential equations that indicate how the particles move subject to forces. With a continuum model, the molecular details of the system are discarded. Molecular scale modelers are often interested in questions that intersect with the long lengthscales accessible to continuum modeling. For example: How does a protein bend a membrane? Yet the lack of atomic detail, combined with our training in the physical laws that rely on force sites, can make continuum modeling particularly opaque. This chapter provides a general guide, to the models (§1) and the variables (§2), as used by software implementations, that define them.

Ignoring molecular details makes perfect sense to model the walls of my office. In fact, my strong preference is that the engineers of my wall did not use all-atom simulation. To do so would severely limit the lengthscale necessary to model its structure, make the effective material constants subject to the limitations of forcefield accuracy, and possibly mislead the engineer into thinking that molecular motions had undue significance. What matters is the load that the wall will bear.

Do the molecular details of a bilayer, like a membrane in a cell, matter? This depends on the question asked of a model of the membrane. Whether the molecular details matter will only be answered when models are put to the test and fail. At the end of this chapter, a few recent studies are summarized that push continuum modeling to its limits, see Section 3.

Membranes are fascinating candidates for continuum modeling in part because of the large array of chemically distinct lipids that constitute them, and the wonderful shapes that result, in part, from this variation. For modeling large systems (for example, 100s of nanometers), an effective choice is to reduce the thin membrane (approximately 5 nanometers thick) to a two-dimensional sheet. At any point on the membrane, the local deformation can be described by just a few parameters: First is a deformation in thickness that is energetically equivalent at first order to a change in lateral area, and its modulus is the area compressibility modulus, here denoted K_A:

$$\overline{E}_A = \frac{K_A}{2}\left(\frac{A - A_0}{A_0}\right)^2 = \frac{K_A}{2}\varepsilon_A^2 \tag{1}$$

where ε_A is the area strain. Second is the sum of the two principal curvatures $c_1 + c_2$. Its modulus is the curvature or bending modulus, k_c:

$$\overline{E}_c = \frac{k_c}{2}\left(c_1 + c_2 - c_0\right)^2 \tag{2}$$

Lipid and membrane 'spontaneous curvature', denoted c_0, varies strongly with lipid composition and has functional consequences. Third is the product of the two principal curvatures, $c_1 c_2$, often denoted K. Its modulus is the Gaussian curvature modulus, k_G. How these parameters vary with lipid composition and the shapes that result are of principal interest of this chapter: they are the critical elements of describing a biological process with a continuum model membrane. With these concepts in mind, this chapter begins with an extended discussion of how these constants are represented and used in modeling.

1 Continuum Modeling Paradigms

Modeling lipid membranes crosses biological, chemical, physical, mathematical, and engineering disciplines. Different objectives and perspectives have led to a number of techniques capable of describing membranes accurately at various lengthscales. The following perspective is by no means comprehensive but generally categorizes these approaches:

• **Membrane perspective** – Energetic model focused primarily on *curvature* to explain observations of biological membranes. This, the principal model used by lipid biophysical studies, is the Helfrich/Canham/Evans (Canham 1970; Helfrich 1973; Evans 1974) model. It is based on experimental observables of surface deformations (e.g., total curvature and area) and assigns them moduli (k_c and K_A). *Gaussian* curvature, the product of the two principal curvatures, is handled explicitly.

• **Engineering perspective** – Energetic model focused on efficient computation of the deformations of *thin plates*. Classically applied to *homogeneous* materials, spontaneous curvature can be introduced using a lateral pressure profile applied to the planar state. Inhomogeneity along the membrane normal can be modeled with layered material properties. In this chapter, engineering models are applied to model processes in Sections 2.4.1, 3.0.1, and 3.0.3. To be consistent with lateral fluidity, Gaussian curvature energetics enter only through lateral stress (see Section 1.1.3). The work of William Klug stands out as an effective application of this perspective to describe the response of membrane shapes (Feng and Klug 2006; Lee et al. 2008; Ma and Klug 2008; Ursell et al. 2009; Kahraman et al. 2014, 2016).

• **Molecular perspective** – Extending basic model treatments using information on the *internal degrees of freedom* of the constituent lipids at the level of chemical moieties and nanoscopic lipid lateral structure. An example is lipid 'tilt,' a continuous field overlaid on the surface that models the orientation of lipids relative to the normal. This chapter concludes with a short discussion of the potential impact characterizing these internal variables will have on lipid biophysics.

An excellent reason for a lipid biophysicist to explore beyond the bounds of Helfrich theory is the enormous body of analysis tools and experience targeting the behavior of thick and thin plates from the engineering community. An engineering tactic is to model an inhomogeneous material with varying elastic constants that reflect the structure. For example, analysis of the relation between K_A and k_c indicates that the lipid bilayer is stiff near the surface but softer in the acyl chain region (Campelo et al. 2008). Furthermore, an inhomogeneous distribution of lateral stresses is generally accepted to lead to spontaneous curvature. For example, a flat leaflet has a static force acting to contract the surface to minimize hydrophobic/water exposure balanced by an expansive force acting to expand the disordered tails. This engineering approach is used in two examples below that push the limits of continuum modeling (Sections 3.0.1 and 2.4.2).

An important distinction between models separates those that explicitly include a description of membrane structure along the normal. The membrane perspective falls into the surface category, typically ignoring thickness variations, while the engineering model is capable of including this dimension. An immediate connection between the two models is found in the effort of the engineering community to reduce the dimensionality of solid materials by modeling 'thin plates.' The methodology for how to handle deformations of elastic plates dates back centuries, to at least the work of Germain (1821) Kirchhoff (1850), and Love (1888). An intriguing difference between lipid bilayers and homogeneous thin plates immediately arises: the bilayer is not a homogeneous material, its thickness is only twice that of its constituent molecules, whose chemistry is highly asymmetric. Moreover, it has in-plane fluidity. Because of this fluidity, a lipid surface will adapt itself to the lateral shape of its container. As shown below in Section 1.1.2, this leads to constraints that simplify the elastic material coefficients.

A particularly important quantity related to spontaneous curvature is the degree of curvature stress, the forces leading to the origin of spontaneous curvature. The two leaflets of a bilayer have intrinsic forces acting to curve them; typically these forces are counteracted by the opposite leaflet (this is especially clear in the case of a symmetric bilayer). This *per-leaflet* curvature stress can be quantified by the derivative of the leaflet free energy with respect to curvature, evaluated at zero curvature. In the case of the Helfrich model, this quantity has the value $-k_c c_0$, but in general it will simply be denoted $\overline{F}'(0) \equiv \frac{\partial \overline{F}}{\partial c}|_{c=0}$ where the free energy per unit area $\overline{F}$ is written for notational convenience as a function of only curvature.

The apparent dependence of observables on *Gaussian* curvature is particularly vexing when interpreting the bilayer as a material. Due to in-plane fluidity, the two principal curvature directions are not expected to couple. This dynamic fluidity raises the possibility for unexpected internal degrees of freedom to be significant; for example, if the constituent lipids re-orient according to local curvature in one direction, the average stress in the *other* direction of curvature may change. These possibilities would not be anticipated to be significant for simple materials in engineering.

The membrane and engineering perspectives lack to varying degrees of how molecular information, if found to be important, should be introduced. For example, as oriented molecules, lipids can tilt (Hamm and Kozlov 2000). From the membrane perspective, this is not a fundamental property of a surface, rather, tilt is overlaid on the surface to describe an internal molecular degree of freedom. From the engineering perspective, lipid tilt has been modeled using the lateral-normal *shear* modulus (Campelo et al. 2008) although in this case shear is not a molecular property like tilt. Once tilt is accepted as a useful variable, it becomes less clear when to stop. Although a flat bilayer might be isotropic on average, an individual lipid is not. Internal variables like lipid orientation can thus be coupled to fluctuations in curvature (that provide an anisotropic environment), influencing second-order properties like k_G (Terzi and Deserno 2017). That is, simply because the Helfrich theory presents such nice simplicity in terms of the differential geometry of curved surfaces is no reason to assume this is the limit of model effectiveness, or that the underlying 'constants' might not vary systematically with a variable (like tilt) that could be characterized (Wang & Deserno, 2016).

1.1 Surface (2D) and Material (3D) Modeling of Membranes

1.1.1 Helfrich Energy of a Surface Model

A typical analysis of membrane bending might employ the *Monge gauge*, in which the surface is represented as the height $h(x, y)$ as a function of lateral coordinates x and y. This representation is convenient for analyzing small fluctuations of planar surfaces in Fourier components where, within the assumptions of small deformations, the mode energetics are separable. Cartesian coordinates are not always convenient to parameterize arbitrary surfaces that may be very curved, such as a vesicle. Thus, it is necessary to evaluate the curvature of a surface with the so-called generalized coordinates, u and v that must be mapped to cartesian space. Now, the surface is defined as $\mathbf{r}(u, v)$, a cartesian 3-vector.

The bending energy of a surface model is a function of the two principal curvatures of the surface. These can be computed using the concept of the first and second fundamental forms used to represent the surface when using arbitrary coordinates u and v; see Kreyszig (1991). The first fundamental form describes the *metric* of the surface, which we may consider a generalization of the 'units' when transferring between $\{u, v\}$ and tangible spatial units. These units change with position on the surface, similarly to how longitudinal displacements are different at the pole and equator. The computation of the form also illustrates the expression of the tangent plane of the surface. Consider a surface defined by $\mathbf{r}(u, v)$, where u and v are general surface coordinates. Represent the tangent plane of the surface at $\{u_0, v_0\}$ as:

$$\mathbf{r}_{\text{tangent}}(u, v) = \mathbf{r}(u_0, v_0) + \mathbf{r}_u(u_0, v_0)u + \mathbf{r}_v(u_0, v_0)v \tag{3}$$

where $\mathbf{r}_u(u, v)$ is the derivative of the surface with respect to u, accomplished by differentiating each vector component. The two-by-two matrix 'metric tensor' is defined by

$$\hat{g}_{uv} = \mathbf{r}_u \cdot \mathbf{r}_v \tag{4}$$

The elements of $\hat{g}_{uv}$ are typically represented by Roman letters:

$$E \equiv \hat{g}_{uu} = \mathbf{r}_u \cdot \mathbf{r}_u \tag{5}$$

$$F \equiv \hat{g}_{uv} = \hat{g}_{vu} = \mathbf{r}_u \cdot \mathbf{r}_v \tag{6}$$

$$G \equiv \hat{g}_{vv} = \mathbf{r}_v \cdot \mathbf{r}_v \tag{7}$$

The determinant of $\hat{g}$ will be denoted as g, $g = EF - G^2$.

Consider the area of a triangle at $\{u_0, v_0\}$ with side along the u (v) direction of length δ_u (δ_v). Compute the area as

$$A = \frac{1}{2}\delta_u\delta_v|\mathbf{r}_u(u_0, v_0) \otimes \mathbf{r}_v(u_0, v_0)|, \tag{8}$$

where $\otimes$ denotes the cross product and the surrounding bars denote the vector magnitude. This is equal to one half $\delta_u\delta_v$ times the determinant of the metric tensor $g = |EG - F^2|$. The term F indicates the nonorthogonality of vectors u and v when represented in real space ($r_u \cdot r_v \neq 0$) and thus subtracts from the area. The squared length of any displacement vector s in the vicinity of $\{u_0, v_0\}$, when represented by the generalized coordinates, is given by the inner product:

$$|\mathbf{s}|^2 = \mathbf{s}^T \cdot \hat{g} \cdot \mathbf{s} \tag{9}$$

The tangent plane is a first-order property, indicating the changing position with displacements in u and v. Curvature is a second-order property: a curved surface has a *tangent plane that changes with displacements in u and v.* Just as the metric depends on the direction expressed in u and v, curvature does as well. Curvature is typically computed using the mechanism of the *second fundamental form* with another three scalar values that are functions of u and v:

$$L \equiv \mathbf{r}_{uu} \cdot \mathbf{n}$$

$$M \equiv \mathbf{r}_{uv} \cdot \mathbf{n} = \mathbf{r}_{vu} \cdot \mathbf{n} \tag{10}$$

$$N \equiv \mathbf{r}_{vv} \cdot \mathbf{n}$$

with

$$\hat{b} = \begin{bmatrix} L & M \\ M & N \end{bmatrix} \tag{11}$$

and the determinant b computed as:

$$b = LN - M^2. \tag{12}$$

Here $\mathbf{r}_{uu}$ is the vector of second derivatives of the (for example) cartesian x, y, and z coordinates with respect to u. The second fundamental form is measured with respect to the normal: The variation of the surface *in plane* does not reflect curvature. For example, the surface $\mathbf{r}(u, v) = \{u^2, v^2, 0\}$, with $\mathbf{n} = \{0, 0, 1\}$, is flat even though coordinates vary at second order.

To compute the Helfrich energy, the sum and product of the principal curvatures are required. These are computed as:

$$c_1 + c_2 = \mathrm{Tr}\{\hat{g}_{uv}^{-1} \cdot \hat{b}\} \tag{13}$$

$$c_1 c_2 = \frac{b}{g} \tag{14}$$

The values c_1 and c_2 can be computed by diagonalizing $\hat{g}_{uv}^{-1} \cdot \hat{b}$.

Roughly speaking, analogous to a curve in one dimension, the values of curvature are equivalent to second derivatives of the surface projected into the normal (the matrix $\hat{b}$). However, we want these values in the spatial representation; thus, we must transform them appropriately (the matrix $\hat{g}$). The matrix $\hat{g}$ is 'in the denominator' (inverted) because unlike lengths and areas (Eq. (9)), derivatives have the spatial coordinate in the denominator. For example, looking in one dimension at a curve of height with respect to position, $h(x)$, the curvature $\frac{d^2 h}{dx^2}$ has the spatial measure of x in the denominator, and thus requires the inverse of the metric. For more complex operations, the covariant/contravariant formalism is typically employed for bookkeeping this distinction.

In Section 2.2, a practical method for evaluating Eqs. 13 and 14 on an arbitrary surface, the *subdivision limit surface* algorithm is discussed.

1.1.2 Mechanical Energy of a Three-Dimensional Model

The standard engineering material model can be extended to a bilayer. In this model, each continuum element of the material has a resting coordinate $\mathbf{r}(x, y, z)$. Deformations of the material are tracked with the deformation field $\mathbf{u}(x, y, z)$ that specifies the displacement of the element from its minimum energy resting position. A constant value of the field is simply a lateral translation of the material. Thus, the spatial gradient of the deformation field indicates distortions of the material.

Consider two elements of the material, with resting positions 1 and 2 located at $\mathbf{r}_1 = \{x, y, z\}$ and $\mathbf{r}_2 = \{x + \delta, y, z\}$, respectively. The energetic model stated below will be equivalent to attaching a spring between sites 1 and 2. The resting length of the spring is defined to be δ. The deformation field acts to move the points to new positions at $\mathbf{r}_1'$ and $\mathbf{r}_2'$.

$$\mathbf{r}_1' = \{x, y, z\} + \mathbf{u}(x, y, z) \tag{15}$$

$$\mathbf{r}_2' = \{x + \delta, y, z\} + \mathbf{u}(x + \delta, y, z) \tag{16}$$

To first order in δ, the distance between the two sites is now $\delta + \nabla \mathbf{u} \cdot (\delta \hat{x}) + \mathcal{O}(\delta^2)$. That is, the deformation (per length of material) is determined by the *gradient* of the field $\mathbf{u}$. The notation $u_{\alpha\beta}(\mathbf{r})$ is used to describe the derivative of the α component of the field in the β direction. The diagonal quantities, e.g., $u_{xx}(\mathbf{r})$ describe compression and expansion along an axis, a typical case when loading a beam. The field u can vary perpendicularly to the displacement it describes, for example, creating a shear.

The energy is quadratic in this deformation gradient, relative to the resting state. The general expression for the quadratic expansion of the energy per unit volume at a point is:

$$\overline{E} = \frac{1}{2}\lambda_{\alpha\beta,\gamma\delta}u_{\alpha\beta}u_{\gamma\delta} \tag{17}$$

which can be integrated over the position-dependent $\mathbf{u}(\mathbf{r})$ field to obtain the total energy. The λ values are the continuum equivalent of finite spring constants.

Consider a slab continuum model of a *single leaflet* of a membrane of thickness h, defined between $-\frac{h}{2}$ and $\frac{h}{2}$ whose properties vary only with z. Apply the volume-preserving area deformation:

$$\mathbf{u}A(\mathbf{r}) = \varepsilon_A\lambda\{\frac{x}{2},\frac{y}{2},-z\} \tag{18}$$

The gradient tensor is simply $u_{xx} = \frac{\varepsilon_A}{2}$, $u_{yy} = \frac{\varepsilon_A}{2}$ and $u_{zz} = -\varepsilon_A$. The area of the patch is deformed as $(1 + \frac{\varepsilon_A}{2})^2 = 1 + \varepsilon_A$ or $A = A_0 + A_0\varepsilon_A$. Comparing with Eq. (1) yields the formula for computing leaflet compressibility K_A^l from the local elastic coefficients:

$$K_A^l = \int_{-\frac{h}{2}}^{\frac{h}{2}} \overline{K}_A(z)dz \tag{19}$$

where the *per unit thickness* local area compressibility modulus $\overline{K}_A(z)$ is defined as:

$$\overline{K}_A(z) = \frac{1}{2}\left[\frac{1}{2}\lambda_{xx,xx}(z) + \frac{1}{2}\lambda_{yy,yy}(z) - \lambda_{xx,zz}(z) - \lambda_{yy,zz}(z) + \lambda_{zz,zz}(z)\right] \tag{20}$$

The use of the volume-preserving deformation assumes that the material deforms at constant volume, which may not be precisely true; see for example, Deserno (2015) for a discussion of Poisson's ratio. The terms in Eq. (20) are determined from the coefficients of $u_{\alpha\beta}$ present in the gradient tensor. The quantity $\overline{K}_A(z)$ is not an experimental property; it is simply a convenient way to parameterize the continuum model that is consistent with the concept of lateral membrane fluidity and which is convenient when assuming that the material locally preserves volume as it deforms. *Shear* can be modeled with off-diagonal constants such as $\lambda_{xz,xz}$ (Campelo et al. 2008).

A limited set of deformation fields u(r) such as $u_A(\mathbf{r})$ preserve the volume of the membrane pointwise at limited order; the divergence of $\mathbf{u}(\mathbf{r})$ must be zero.

Lateral fluidity of the bilayer can be modeled with elastic constants that fully couple the lateral dimensions (here taken to be x and y). Consider a deformation that changes the lateral dimensions of a container while leaving its thickness untouched:

$$\mathbf{u}_{\text{lateral}}(\mathbf{r}) = \lambda\{x, -y, 0\} \tag{21}$$

To first order this deformation preserves volume and area everywhere, while expanding the x dimension and contracting in y. The elastic constant k_{xy} for the deformation is:

$$k_{xy} = \int \lambda_{xx,xx} - 2\lambda_{xx,yy} + \lambda_{yy,yy}dz = 0 \tag{22}$$

With x and y equivalent by symmetry, it is zero if $\lambda_{xx,xx} = \lambda_{yy,yy} = \lambda_{xx,yy}$.

It is important when working with the 3D continuum model to make the distinction between bilayer and single leaflet properties. A useful model of a bilayer is that of two leaflets which slide across each other. This relatively weak coupling between leaflets is difficult to model using a 'whole' material model of the bilayer. The separation of the leaflet deformations is accomplished using separate neutral surfaces z_0 for each leaflet – the pivot around which the leaflets bend at constant area.

A curvature deformation can also be constructed that preserves volume at first order:

$$\mathbf{u}_c(\mathbf{r}; c) = c\left\{x(z - z_0), 0, \frac{x^2 - (z - z_0)^2}{2}\right\} \tag{23}$$

Consider the $u_x(\mathbf{r})$ term, $x(z - z_0)$. The x dependence of the field acts to make the spacing between particles wider with magnitude $z - z_0$. The effect of this field is to make slabs with $z > z_0$ wider and shorter and slabs with $z < z_0$ narrower and taller. This is equivalent, at first order, to *curvature c* (Szleifer et al. 1990; Sodt and Pastor 2013) with direction of principal curvature along $\hat{x}$; thus, the parameter 'c' is included as a parameter in its input. The 'pivot' of curvature is z_0 and is defined separately for each leaflet.

For clarity, the notation for the gradient tensor $u_{\alpha\beta}(\mathbf{r})$ will be reduced to a simplified short hand because, in the following, it will only be contracted with the terms present in $\overline{K}_A(z)$. The gradient tensor for this field is:

$$u_c(\mathbf{r}; c) = u_{\alpha\beta}(\mathbf{r}; c) = c\{\{z - z_0, 0, -x\}, \{0, 0, 0\}, \{x, 0, -(z - z_0)\} \tag{24}$$

where only terms $u_{xx}(z; c) = c(z - z_0)$ and $u_{zz}(z; c) = -c(z - z_0)$ contribute to the energy because, for example $\lambda_{xz,xz}(z) = \lambda_{zx,zx}(z)$ (the deformation does not create shear). For the flat system, the dependence on x can thus be disregarded. The energy of the deformation can be computed by inserting $u_c(z; c)$ into Eq. (17), yielding:

$$\Delta \overline{E}_{\text{ideal}}(c) = \int \frac{1}{2} c^2 (z - z_0)^2 \overline{K}_A(z) dz \tag{25}$$

where the bar over E indicates this is the energy per lateral unit area subject to curvature c. The second derivative of the energy with respect to curvature is equal to the single-leaflet bending modulus:

$$k_c^m = \int (z - z_0)^2 \overline{K}_A(z) dz \tag{26}$$

where z_0, the position of the neutral surface, is chosen to minimize the energy of the deformation:

$$\frac{\partial \Delta \overline{E}_{\text{ideal}}}{\partial z_0} = c^2 \int_{-\frac{h}{2}}^{\frac{h}{2}} (z - z_0) \overline{K}_A(z) dz = 0 \tag{27}$$

The continuum model description has the flexibility to set the resting shape of the material arbitrarily. Consider, for example, a leaflet with material constants $\overline{K}_A(z)$ that, when flat, has been slightly deformed from its resting shape by a field $c_0 \mathbf{u}^{\text{intrinsic}}(\mathbf{r}; c_0)$ where c_0 is small. The superscript text 'intrinsic' has been inserted to make distinct a trial deformation of the system from that which is intrinsically 'pre-deformed.' The energy of the material, when subject to curvature deformation $\mathbf{u}(\mathbf{r}; c)$, is:

$$E = \frac{1}{2}[u_c(z;c) - u^{\text{intrinsic}}{}_c(z;c_0)]\overline{K}_A(z)[u_c(z;c) - u^{\text{intrinsic}}{}_c(z;c_0)] \tag{28}$$

where implicitly the deformation gradient tensor $u_c(z;c)$ is contracted with relevant non-zero terms of $K_A(z)$. The minimum energy c is found at c_0, the so-called 'spontaneous curvature.' Under the above approximations, the quantity $K_A(z)u^{\text{intrinsic}}{}_c(z;c_0)$ is equivalent to the *stress tensor*, $\sigma(z)$. Furthermore, consider the following mechanism for probing the property of a continuum model of a flat slab. We seek to probe a static property of the slab (which will result from the stress tensor) using a trial deformation, $\mathbf{u}(\mathbf{r})$. We apply the trial deformation and look at the change in energy at first order in the deformation magnitude, λ:

$$\overline{E} = \overline{E}_0 + \int \underbrace{u_c(z;\lambda)}_{\text{trial deformation}} \times \underbrace{\overline{K}_A(z)u_c^{\text{intrinsic}}(z;c_0)}_{\text{system property}} dz \tag{29}$$

The derivative with respect to the deformation magnitude (λ) is:

$$\frac{\mathrm{d}\overline{E}}{\mathrm{d}\lambda} = \int [z - z_0]\sigma(z)dz = -c_0 \int [z - z_0]^2 K_A(z)dz = -k_c c_0 \tag{30}$$

1.1.3 Curvature Stress Computed from Three Different Models

Below we demonstrate that the equivalent derivative can be computed from simulation using the continuum formalism, but that mistakes of misinterpretation can be made when extended to *second-order* quantities.

The derivative of the free energy with respect to the curvature deformation is computed as:

$$\frac{\mathrm{d}F}{\mathrm{d}\lambda} = \Big\langle \sum_i \underbrace{\mathbf{u}_c(\mathbf{r}_i)}_{\text{trial deformation}} \cdot \underbrace{\mathbf{f}_i}_{\text{static property}} \Big\rangle. \tag{31}$$

where $\mathbf{r}_i$ and $\mathbf{f}_i$ are the position of and force on particle i, respectively. See Schofield and Henderson (1982, section 4) for the complete derivation of the general expression. The deformation $\mathbf{u}_c(\mathbf{r})$ preserves volume at first order so there is no contribution from kinetic energy. Unfortunately a few technical details make Eq. (31) impractical. First, the bilayer system is condensed. It has infinite periodic repeats in each direction. This creates an 'edge' effect where the net force on the box may be non-zero due to the force of neighboring periodic images. Second, molecular simulations are often executed with the bonds to hydrogens constrained and thus these vibrations have zero kinetic energy. These constrained bonds are not guaranteed to be isotropically averaged and thus must be included as an effective force. A consequence is that the kinetic energy is not isotropically distributed and its change with the magnitude of the deformation must be included.

The solution to these problems is to *project* the forces between sites of the molecular model onto a continuum model. A histogram of the anisotropic particle kinetic energy is built. Unfortunately, the projection of the molecular forces onto the continuum model is *not unique*. That is, there are many $\sigma_{\alpha\beta}$ that represent the effect of the force pair interacting with the deformation:

$$\int \sigma_{\alpha\beta}(\mathbf{r})u_{\alpha\beta}(\mathbf{r})dz = \mathbf{f}_i \cdot \mathbf{u}(\mathbf{r}_i) + \mathbf{f}_j \cdot \mathbf{u}(\mathbf{r}_j) \tag{32}$$

for any continuously differentiable material deformation $\mathbf{u}(\mathbf{r})$. The quantity $\sigma_0(z)$ becomes an intermediate quantity with a few convenient practical purposes. It correctly accounts for the anisotropic kinetic energy density that arises from constraints and (with proper selection of the path between particles) yields zero stress in the aqueous region of molecular simulations. The widely used Harasima (1958) and Irving and Kirkwood (1950) contours properly account for zero stress in the aqueous layer of bilayer simulations. The question arises as to which properties can be computed from $\sigma_{\alpha\beta}$.

It is frequently claimed that the Gaussian curvature modulus can be determined from the second moment of the stress profile, as extracted from a *molecular simulation*. Roughly, the derivation follows the logic that Gaussian curvature results from simultaneously applying a deformation yielding, for example, positive curvature in the $\hat{x}$ direction and negative curvature of the same magnitude in the $\hat{y}$ direction:

$$\mathbf{u}_{(-)}(\mathbf{r}; c) = \mathbf{u}_c^x(\mathbf{r}; c) + \mathbf{u}_c^y(\mathbf{r}; -c), \tag{33}$$

where $\mathbf{u}_c^x(\mathbf{r})$ is the deformation defined above that creates curvature along $\hat{x}$ and that with superscript y creates curvature along $\hat{y}$. To first order in c, the energy change is zero if the system is isotropic in x and y; the energetics of the x and y dimensions cancel. A new variable $K = -c^2$ could certainly be defined to recast the expression as a first derivative. In this case, the expression for the continuum model is:

$$\frac{\mathrm{d}E}{\mathrm{d}K} = \int \underbrace{(z - z_0)^2}_{\text{trial deformation}} \underbrace{\sigma(z)}_{\text{static property}} \mathrm{d}z \text{ (continuum model)} \tag{34}$$

In the continuum model with λ chosen to represent lateral fluidity, an explicit assumption is made for how the lateral pressure profile in x changes with curvature along orthogonal direction y, leading to Eq. (34). This is not necessarily the case with a general *molecular* model, where internal degrees of freedom like tilt, conformation, and lateral orientation would couple stresses in x and y. This would invalidate the use of a first-order expression to evaluate K, as in Eq. (34). See for example, in the derivation of Eq. (34) in Szleifer et al. (1990), the assumption

$$\frac{\partial\sigma(z)}{\partial(c_x + c_y)} \neq 0 \tag{35}$$

$$\frac{\partial\sigma(z)}{\partial(c_x - c_y)} = 0 \text{ (continuum assumption)} \tag{36}$$

appears to make the Gaussian curvature modulus available from stress profile information, whereas the bending modulus is not available. The assumption of Eq. (36) is not built into molecular models, rendering the application of Eq. (34) incorrect in general.

The presence of a molecular mechanism for curvature stress (hydrogen bonding in sphingomyelin and DOPE (Venable et al. 2014; Sodt et al. 2016)) that yields oriented lipid interactions (and thus stresses) suggests the origin of a violation of Eq. (36). In Section 2.1.2 below, the well-known importance of k_G for vesicle creation is shown.

---

2 Variables of a Continuum Model

While the molecular dynamics energy is defined in terms of the atomic coordinates, the energy functional in a continuum model is defined 'pointwise' on the membrane surface or throughout its material. The finite number of atomic coordinates can be propagated by forces computed by derivatives of the energy; however, the infinite continuum of points on the surface cannot. Rather, a finite set of variables is used to control the positions of the material everywhere.

2.1 Highly Symmetric Cases

In special cases, a continuum model can be controlled simply by a limited number of continuous parameters and analyzed analytically. This is often the case when the Helfrich Hamiltonian is used as a tool to interpret experiments with simple geometry. In some process, including hypothetically, membrane fusion, the most likely transition path is assumed to lie along a path of axial symmetry (for example, a spherical vesicle merging with another spherical vesicle has an approximate symmetry axis connecting the two). In Ryham et al. (2016), the membrane is parameterized by a finite-element one-dimensional curve, which defines a surface by rotating about the symmetry axis. For surfaces of rotation, the two directions of principal curvature are: one, the projection of the symmetry axis into the tangent plane and, two, the direction along the path of rotation. This allows the authors to compute the curvature of two-dimensional surfaces using only a one-dimensional curve. A similar mechanism was used in the analysis of the function of the gramicidin ion channel, using continuum modeling (Sodt et al. 2017). In this work, both the surface and material models were compared using the same convenience of defining a higher dimensional material by rotation. These two processes, membrane fusion and channel function, involve pushing the limits of the continuum model, and are discussed in depth below (Section 3).

2.1.1 Inverse Hexagonal Swelling

In inverted hexagonal phase experiments proposed by Gruner et al. (1986) to measure spontaneous curvature, osmotic pressure is applied to hexagonally arrayed single-leaflet lipid tubes, arranged with their headgroups (and water) on the cylindrical interior. Lipid acyl chains meet between cylinders. Removing water from the system by applying osmotic pressure changes the cylinder volume ($\pi R^2 h$). For most lipid compositions and accessible tube radii, this deformation occurs at constant surface area ($2\pi Rh$). Expressed in terms of the radial strain ($\varepsilon_R = \frac{R-R_0}{R_0}$), the energetics of the area per lipid are:

$$E_{\text{area}} = \frac{K_A}{2} A_0 \left(\frac{A - A_0}{A_0} \right)^2 = \frac{K_A}{2} A_0 \varepsilon_R^2 \tag{37}$$

while curvature strain is

$$E_{\text{curvature}} = \frac{k_c}{2} A_0 \left(\frac{1}{R} - c_0 \right)^2 = \frac{k_c}{2R_0^2} A_0 \varepsilon_R^2 \tag{38}$$

Thus, K_A can be compared to $\frac{k_c}{R_0^2}$ to determine the relative energetics when the tube is near its spontaneous curvature. For a typical hexagonal phase leaflet, the ratio $\frac{k_c}{K_A/R_0^2}$ is on the order of 25, that is, the area deformation is 25 times stiffer. In this regime, the relationship between pressure and radius primarily reflects bending energetics. Eq. (7) of Chen and Rand (1997) is:

$$\Pi R^2 = 2k_c^m \left(\frac{1}{R} - \frac{1}{R_0} \right) \tag{39}$$

where geometries are measured at the pivotal plane rather than the bilayer midplane. Eq. (39) relates the osmotic pressure Π to the tube radius R via mechanical parameters k_c and R_0. Note that the product $k_c^m R_0^{-1} \approx k_c^m c_0$, the quantification of curvature stress of a planar leaflet in the

Helfrich model discussed in Section 1.1.3. The inverse hexagonal phase technique is the principal experimental method for determining spontaneous curvature.

Of particular relevance here is that the inverse hexagonal phase cannot be formed for arbitrary lipid species, as stability of the phase requires a sufficiently negative c_0. A typical experiment is thus to add small amounts of the target lipid into a dioleoylphosphatidylethanolamine (DOPE) matrix and record the perturbation on c_0. This raises the question of what *molecular* variables change in different lipid matrices. All-atom molecular dynamics simulations, using $\overline{F}'(0)$ to quantify curvature stress, have identified nonadditive mixing trends with liquid-ordered mixtures, as well as with sphingomyelin (Sodt et al. 2016). This is discussed further in Section 4.

2.1.2 Process: Curvature Energy Required to Create a Vesicle

For a vesicle with symmetric lipid composition, the spontaneous curvature is zero, the inner and outer leaflets will have approximately opposite curvature, and thus, the constituent leaflets will have canceling c_0. The curvature energy is:

$$E_{vesicle,symmetric} = \frac{k_c}{2} \int_{surface} A \left(\frac{2}{R}\right)^2 = 8\pi\left(k_c + \frac{1}{2}k_G\right) \tag{40}$$

This quantity is independent of vesicle size (see Eq. (27) of Helfrich (1973)). This continuum model has provided the conceptual basis for the thermodynamics of vesicle creation, establishing an energetic baseline. In mixtures, lipids with spontaneous curvature will tend to populate their preferred leaflet (Berden et al. 1975; Mattjus et al. 2002) and an energetic preference for a particular size develops. This size dependence is consistent with the observation of increased shape uniformity in vesicles of bilayer mixtures Berden et al. (1975). Note however, the importance of the poorly known quantity k_G!

2.2 Evaluation of Curvature on a Splined Surface

A surface can be parameterized by a *box spline*, using a finite order polynomial of the general coordinates u and v. An example of a simple surface so-defined would be a triangular segment of a plane:

$$\mathbf{r} = \mathbf{r}_1(1 - u - v) + \mathbf{r}_2 u + \mathbf{r}_2 v \text{ (plane)} \tag{41}$$

where u and v are greater than zero and their sum is less than one. Higher order polynomial expansions yield curved surfaces. These triangular patches can be fit together to yield a continuous surface.

A beautiful algorithm for yielding continuous surfaces is the subdivision limit surface algorithm (Cirak et al. 2000; Feng and Klug 2006). The input to the algorithm is a triangular mesh of a surface with arbitrary topology. A 'rule' is repeatedly applied that subdivides the triangles to smaller and smaller triangles (see Fig. 1). The rule specifies the positions of all the vertices from one iteration to the next; the vertex positions are averages of the neighboring points on the mesh. As the mesh gets finer, the vertex positions converge to their limit; the collection of points defines a continuous limit surface. Importantly, *the positions of the limit surface are defined as a spline of the original mesh.*

The limit spline yields $\mathbf{r}(u, v)$ for which the curvature can be computed at any point on the triangle; see section A.1 of Cirak et al. (2000). A fixed integration scheme can then be applied to evaluate the curvature energy at a set of points, for example, a single point scheme using the middle $(u = \frac{1}{3}, v = \frac{1}{3})$. Derivatives of the energy with respect to vertex coordinates are

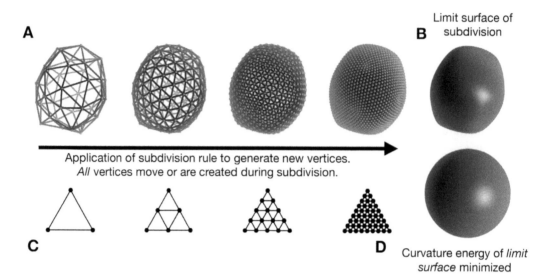

FIGURE 1 A depiction of the limit surface algorithm. (A) Subdivision of an initial coarse mesh yields a finer and finer mesh. (B) The limit surface of the subdivision algorithm. (C) Illustration of how triangles are subdivided. (D) The Helfrich energy of the surface is minimized yielding an approximate sphere.

feasible because the surface is a linear function of the control vertices (Eq. (41)). In Fig. 1, the curvature energy of the limit surface is minimized using the gradient of the quadrature-integrated energy.

2.3 Coarse Evaluation of Curvature at Vertices and Edges of a Triangular Mesh

A membrane representation equivalent to Ramakrishnan et al. (2010) (and see Jülicher 1996) has been used recently to evaluate membrane reshaping in autophagy (Bahrami et al. 2017) and by endophilin (Tourdot et al. 2015). Curvature is computed from the change in the membrane normal relative to the original tangent plane (in a particular direction), or equivalently, the change in the tangent plane (r_u, r_v) relative to the original normal direction (see Eqs. 10, 11 and 13). Thus, the angle made between the normal of two triangles of a mesh indicates curvature on a coarse scale, the scale of the triangles themselves. In Ramakrishnan et al. (2010), the so-called 'shape operator' of the surface is reconstructed, using information about curvature determined along the edges between triangles. This yields sufficient information to evaluate total and Gaussian curvature. Being a continuous collection of flat triangles, this model does not have curvature or tangent plane continuity. However, evaluation of the curvature energy can be accomplished efficiently.

2.4 Finite Element and Finite Difference Modeling

Two standard techniques for evaluating the curvature of a surface or thin plate are using finite element modeling (FEM) and finite-difference modeling. In finite element modeling, a triangle or tetrahedron (for example) represents the material over which it is continuously defined, with the vertices (and possibly additional internal coordinates) of the elements used as control points. Constraints between elements yield properties that improve the fidelity to a higher resolution model, for example, enforcing tangent plane continuity. The so-called *shape functions* interpolate the material between nodes. Finite difference modeling, in contrast, does not present a continuous description of the material, but rather evaluates properties only at a finite set of points. Explicitly or implicitly, both of these methods enable evaluation of changes in the deformation field $\mathbf{u}(\mathbf{r})$.

2.4.1 Process: Protein–Protein Bilayer Deformation-Mediated Coupling

The article 'Bilayer-thickness-mediated interactions between integral membrane proteins' (Kahraman et al. 2016) offers a particularly illuminating treatment of the membrane deformation between proteins. The authors have determined the membrane deformation around a set of protein-like 'shapes,' including how the deformations couple between proteins in the same bilayer. They solve for the minimum continuum model energy (including K_A, k_c) analytically, using both FEM and finite difference approaches.

In their model of interactions between cylindrical inclusions, which are 0.6 nm thicker than the surrounding bilayer, the authors find approximately 5 $k_B T$ of 'binding energy' between the inclusions, with the interaction decaying over several nanometers, the characteristic lengthscale of bilayer deformations ($\sqrt{\frac{k_c}{K_A}}$; Huang 1986). Complex shapes and larger proteins have dramatically larger interaction energies.

In their finite element approach (see, for example, Kahraman et al. 2016; Fig. 3), the authors use interpolating shape functions to compute leaflet curvature at three points on each triangular face of the membrane, using these points for numerical integration. In their finite difference approach, the authors use the leaflet height in neighborhoods of finitely separated points to compute high-order derivatives (Fig. 4 of Kahraman et al. 2016), using the Monge gauge for evaluation. The authors are able to compute FEM and finite difference energetics to be nearly indistinguishable (Fig. 8 of Kahraman et al. 2016).

2.4.2 Process: Gramicidin A Dissociation

The seminal work of Huang (1986) established the utility of K_A and k_c for modeling a lipid-regulated protein process, here the association and dissociation of gramicidin monomer units to form a dimer channel. In this process, single-leaflet gramicidin A monomers move laterally until they encounter a monomer on the *opposite* leaflet. Hydrogen bonding between monomers leads to formation of a transmembrane dimer channel which now imposes its hydrophobic thickness on the *bilayer* locally, inducing a membrane deformation. A composite continuum-all-atom representation is shown in Fig. 2. The deformation acts to push the units apart, increasing the rate of dissociation. Because this deformation force is proportional to the energetics of the deformation,

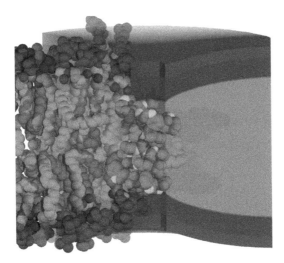

FIGURE 2 An illustration of a material continuum deformation around a gramicidin channel. The hydrophobic length of the channel 'pinches' the surrounding leaflets. (See color insert for the color version of this figure)

the kinetics can be used to probe membrane stiffness. Olaf Andersen and co-workers have used gramicidin A as a tool to not only study the mechanical properties of lipids but also the effect of additives like curcumin (Ingolfsson et al. 2007), docosahexaenoic acid (Bruno et al. 2007), capsaicin (Lundbaek 2005), and more.

A caveat looms over the interpretation of the deformation energy around a protein: The strength of the leaflet deformation depends critically on the *boundary condition* for how the bilayer meets the protein (Helfrich and Jakobsson 1990; Lundbaek and Andersen 1999). This has been a point of considerable interest for modeling gramicidin A, a classic tool for analysis of lipid mechanical properties directly in the context of their effect on protein function. Letting the boundary be free versus constraining it (for example, to meet the protein with zero slope) can result in a factor of four difference in energetics (Sodt et al. 2017). The protein–lipid boundary of gramicidin, including its molecular form and how it couples to curvature, is discussed below in Section 3.3.

3 Pushing the Limits of a Continuum Model

When the continuum surface or its energy varies on the scale of molecular features, the validity of the model itself must certainly be questioned. However, determining model failure can be difficult, as this regime often sits at the edge of what experiments and molecular simulations can test. The examples below illustrate a mixture of success and failure at aggressive application of continuum modeling.

3.1 Process: Alpha Helix Curvature Induction

In Ryham et al. (2016), the authors use the material model to predict the curvature induction ability of an amphipathic helix via the 'wedge' mechanism, in which the helix pushes aside lipid material near the headgroups, creating curvature stress. A critical element of the paper is their determination of the λ values (for example, Eqs. 17 and 20) that are consistent with a fluid bilayer. The authors introduce a cylindrical deformation boundary field $\mathbf{u}(z)$ at $x = 0$ (in the undeformed coordinate system) and then solve for the $\mathbf{u}(\mathbf{r})$ that minimizes the energy. The result is a material that deflects down away from the helix, creating space, equivalent to curvature along the deformed seam. The authors observe extremely high *positive* spontaneous curvature, with magnitude greater than that of PE lipids. This result has provided quantitative data for interpreting the role of alpha helices in membrane reshaping processes.

The curvature stress that results from the embedded alpha helical wedge should also determine $\overline{F}'(0)$ and thus correspond with an equivalent molecular simulation. In Sodt and Pastor (2014), the material model is both reproduced and compared to all-atom molecular simulations, using $\overline{F}'(0)$ as an intermediate. The effect in molecular simulation is found to be twice as high as in the continuum model. This observation is extended in Perrin et al. (2015), which includes correlation of the depth of helix insertion with curvature.

3.2 Process: Membrane Fusion

Ryham et al. (2016) use an axisymmetric continuum model to describe a complete path through the process of membrane fusion. At its most basic, membrane fusion describes the process of two vesicles joining to become one. The pathway must describe the initial repulsion between opposing membranes, the merging of the hydrophobic regions of the membrane to form and enlarge what is called a *hemi-fusion diaphragm* (two vesicles joined to share a circular portion of a bilayer, the diaphragm), the poration of this diaphragm, and finally, expansion of the fusion pore.

To accurately describe the energetic barriers for this process (and thus the minimum energy coordinate), the precise balance of energetic contributions must be determined. In regions where the forces are strong, small errors in material constants as well as terms beyond harmonic will disrupt this balance.

It is inevitable in a model of topological change there will be a *break* in the continuity of the surface energy because the surface itself breaks. The authors retain the model constants for the intact surface but now use an empirical potential to handle the hydrophobic effect, including attraction between nearby exposed hydrophobic regions. The conclusion of the study is that the energy barriers are relatively moderate, with fissure expansion – fluctuation of a hydrophobic pocket in a leaflet – during the initial merging of leaflets the most energetically expensive.

The study is unable to comment on the influence of k_G on the energetics of fusion, a key component. While the authors conclude that the pathway is independent of k_G, the energy change, and thus the likelihood of fusion itself, is highly sensitive.

3.3 Process: Lipid Redistribution around Gramicidin

In a pair of companion articles, Beaven et al. (2017) and Sodt et al. (2017), experiment, molecular simulation, and continuum modeling are applied to predict the redistribution of two lipids around the small channel gramicidin A. The driving force for lipid redistribution is hydrophobic mismatch. One lipid with tails sixteen carbons long is nearly a perfect match for the channel, while the other lipid, at 24 carbon units, is much longer. The variation of the surface thickness depends strongly on curvature energetics and the details of the boundary where lipids and protein meet. This in turn determines the degree of redistribution based on mismatch.

The material (3D) and surface (2D) models both have the minimal energetic ingredients to solve the problem: k_c to account for curvature near the protein, K_A to account for the thickness deformation, and an explicit accounting of the boundary (in the case of the surface model, this is represented by the *slope* of the surface as it meets the protein). However, the material model is able to account for deformations beyond this treatment. The authors used all-atom simulations to determine the effective shape of the channel, providing the boundary condition to the material model. Excellent agreement of the curvature stress, available from the all-atom and material models, partially validates the energetics of the material model. Rearrangements, approximately 58% lipids near the channel, were indistinguishable from the experimental result to statistical precision (Beaven et al. 2017; Sodt et al. 2017).

4 Molecular Modifications to Continuum Modeling

The material constants of the lipid bilayer are critical for understanding the energetics of biological processes that involve membrane reshaping. However, we currently lack sufficient understanding of how these constants change with lipid composition, particularly in relation to cholesterol and sphingolipids. In a giant vesicle experiment, the bending modulus of a mixture of the sphingolipid GM1 with POPC is reduced to 25% of that of POPC at less than 10 mole% GM1 (Fricke and Dimova 2016)!

Molecular simulations indicate that cholesterol and sphingolipids will have unexpected nonadditive curvature energetics, likely related to nanometer-scale partitioning (Sodt et al. 2016). Figure 3 shows how the curvature stress of an all-atom simulated mixture containing a sphingolipid varies nonlinearly with sphingolipid fraction, suggesting that interactions between sphingolipids are coupling to curvature.

The so-called polymer brush theory, used effectively by Evans (Rawicz et al. 2000) to describe the relation between k_c and k_G for simple lipids, is an excellent theoretical point of focus for interpreting the material parameters of complex mixtures. The polymer brush model includes the molecular feature of the lipid tails, together with an effective (or equivalently, continuum) measure of the tension exerted by the hydrophobic/hydrophilic leaflet interface. It provides a model link between the flexibility of lipid tails and the area compressibility modulus that is so central to all the material properties of the bilayer:

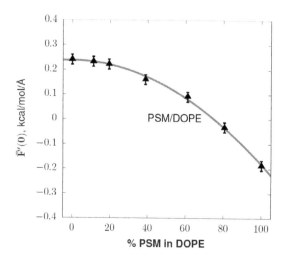

FIGURE 3 The value of $\bar{F}'(0)$ (see section on 'Curvature Stress') for mixtures of palmitoylsphingomyelin and dioleoylphosphatidylethanolamine (all-atom simulation) shows a nonadditive relationship.

$$K_A \approx \frac{18 n_s k_B T x^3}{a_c} \text{ (polymer brush)} \tag{42}$$

where x, the chain extension, varies up to 1 (full extension), a_c is the acyl chain area, and n_s is the number of independent polymer segments of the chain, reflecting flexibility. It is possible that an understanding of the impact of cholesterol, in ordered and disordered phases, may be found in such a treatment (Sodt et al. 2016). A complete theory for membrane reshaping may only be possible following the development of a quantitative model for how lipids redistribute and change the material properties of complex mixtures.

REFERENCES

A. H. Bahrami, M. G. Lin, X. Ren, J. H. Hurley, and G. Hummer. Scaffolding the cup-shaped double membrane in autophagy. *PLoS Comput. Biol.*, 13(10), 2017. ISSN 15537358. doi: 10.1371/journal.pcbi.1005817.

A. H. Beaven, A. M. Maer, A. J. Sodt, H. Rui, R. W. Pastor, O. S. Andersen, and W. Im. Gramicidin a channel formation induces local lipid redistribution I: experiment and simulation. *Biophys. J.*, 112 (6), 1185–1197, 2017. ISSN 15420086. doi: 10.1016/j.bpj.2017.01.028.

J. A. Berden, R. W. Barker, and G. K. Radda. NMR studies on phospholipid bilayers. Some factors affecting lipid distribution. *BBA—Biomembr.*, 375(2), 186–208, 1975. ISSN 00052736. doi: 10.1016/0005-2736(75)90188-1.

M. J. Bruno, R. E. Koeppe, and O. S. Andersen. Docosahexaenoic acid alters bilayer elastic properties. *Proc. Natl. Acad. Sci.*, 104(23), 9638–9643, 2007. ISSN 0027-8424. doi: 10.1073/pnas.0701015104.

F. Campelo, H. T. McMahon, and M. M. Kozlov. The hydrophobic insertion mechanism of membrane curvature generation by proteins. *Biophys. J.*, 95(5), 2325–2339, 2008. ISSN 15420086. doi: 10.1529/biophysj.108.133173.

P. B. Canham. The minimum energy of bending as a possible explanation of the biconcave shape of the human red blood cell. *J. Theor. Biol.*, 26(1), 61–81, 1970. ISSN 10958541. doi: 10.1016/S0022-5193(70)80032-7.

Z. Chen and R. P. Rand. The influence of cholesterol on phospholipid membrane curvature and bending elasticity. *Biophys. J.*, 73(1), 267–276, 1997. ISSN 00063495. doi: 10.1016/S0006-3495(97)78067-6.

F. Cirak, M. Ortiz, and P. Schroder. Subdivision surfaces: a new paradigm for thin-shell finite-element analysis. *Int. J. Numer. Methods Eng.*, 47(12), 2039–2072, 2000. ISSN 0029-5981. doi: 10.1002/ (SICI)1097-0207(20000430)47:12<2039::AID-NME872>3.0.CO;2-1.

M. Deserno. Fluid lipid membranes: from differential geometry to curvature stresses. *Chem. Phys. Lipids*, 185, 11–45, 2015. ISSN 18732941. doi: 10.1016/j.chemphyslip.2014.05.001.

E. A. Evans. Bending resistance and chemically induced moments in membrane bilayers. *Biophys. J.*, 14 (12), 923–931, 1974. ISSN 00063495. doi: 10.1016/S0006-3495(74)85959-X.

F. Feng and W. S. Klug. Finite element modeling of lipid bilayer membranes. *J. Comput. Phys.*, 220(1), 394–408, 2006. ISSN 00219991. doi: 10.1016/j.jcp.2006.05.023.

N. Fricke and R. Dimova. GM1 softens POPC membranes and induces the formation of micron-sized domains. *Biophys. J.*, 111(9), 1935–1945, 2016. ISSN 15420086. doi: 10.1016/j.bpj.2016.09.028.

S. Germain. *Recherches sur la th'eorie des surfaces elastiques*. Veuve Courcier, Paris, 1821.

S. M. Gruner, V. A. Parsegian, and R. P. Rand. Directly measured deformation energy of phospholipid H<inf>II</inf> hexagonal phases. *Faraday Discuss. Chem. Soc.*, 81, 29–37, 1986. ISSN 03017249. doi: 10.1039/DC9868100029.

M. Hamm and M. M. Kozlov. Elastic energy of tilt and bending of fluid membranes. *Eur. Phys. J. E*, 3(4), 323–335, 2000. ISSN 12928941. doi: 10.1007/s101890070003.

A. Harasima. Molecular theory of surface tension. *Adv. Chem. Phys.*, 1, 203–237, 1958.

W. Helfrich. Elastic properties of lipid bilayers elastic properties of lipid bilayers: theory and possible experiments. *Z. Naturforsch*, 28(6), 3–7, 1973. ISSN 0341-0471. URL www.researchgate.net/ publication/18891534_Helfrich_W_Elastic_properties_of_lipid_bilayers_theory_and_possible_ experiments_Z_Naturforsch_28C_693-703.

P. Helfrich and E. Jakobsson. Calculation of deformation energies and conformations in lipid membranes containing gramicidin channels. *Biophys. J.*, 57(5), 1075–1084, 1990. ISSN 00063495. doi: 10.1016/ S0006-3495(90)82625-4.

H. W. Huang. Deformation free energy of bilayer membrane and its effect on gramicidin channel lifetime. *Biophys. J.*, 50(6), 1061–1070, 1986. ISSN 00063495. doi: 10.1016/S0006-3495(86)83550-0.

H. I. Ingolfsson, R. E. Koeppe, and O. S. Andersen. Curcumin is a modulator of bilayer material properties. *Biochemistry*, 46(36), 10384–10391, 2007. ISSN 00062960. doi: 10.1021/bi701013n.

J. H. Irving and J. G. Kirkwood. The statistical mechanical theory of transport processes. IV. The equations of hydrodynamics. *J. Chem. Phys.*, 18(6), 817–829, 1950. ISSN 00219606. doi: 10.1063/ 1.1747782.

F. Jülicher. The morphology of vesicles of higher topological genus: conformal degeneracy and conformal modes. *J. Phys. II*, 60(12), 1797–1824, 1996. ISSN 1155-4312. doi: 10.1051/jp2:1996161. URL jp2. journaldephysique.org/articles/jp2/abs/1996/12/jp2v6p1797/jp2v6p1797.html%5Cnwww.edps ciences.org/10.1051/jp2:1996161.

O. Kahraman, W. S. Klug, and C. A. Haselwandter. Signatures of protein structure in the cooperative gating of mechanosensitive ion channels. *EPL*, 1070(4), 2014. ISSN 12864854. doi: 10.1209/0295- 5075/107/48004.

O. Kahraman, P. D. Koch, W. S. Klug, and C. A. Haselwandter. Bilayer-thickness-mediated interactions between integral membrane proteins. *Phys. Rev. E*, 930(4), 2016. ISSN 24700053. doi: 10.1103/ PhysRevE.93.042410.

G. Kirchhoff. Uber das gleichgewicht und die bewegung einer elastischen scheibe. *Journal fr die reine und angewandte Mathematik*, 40, 51–88, 1850.

E. Kreyszig. *Differential Geometry*. Dover Publications Inc., New York, 1991. ISBN 0-486-66721-9.

H. J. Lee, E. L. Peterson, R. Phillips, W. S. Klug, and P. A. Wiggins. Membrane shape as a reporter for applied forces. *Proc. Natl. Acad. Sci.*, 105(49), 19253–19257, 2008. ISSN 0027-8424. doi: 10.1073/ pnas.0806814105.

A. Love. On the small free vibrations and deformations of thin elastic shells. *Phil. Trans. Roy. Soc.*, 179, 491–549, 1888. URL http://rsta.royalsocietypublishing.org/content/179/491.abstract.

J. Lundbaek. Capsaicin regulates voltage-dependent sodium channelsby altering lipid bilayer elasticity. *Mol. Pharmacol.*, 68(3), 680–689, 2005. ISSN 0026-895X. doi: 10.1124/mol.105.013573.

J. A. Lundbæk and O. S. Andersen. Spring constants for channel-induced lipid bilayer deformations estimates using gramicidin channels. *Biophys. J.*, 76(2), 889–895, 1999. ISSN 00063495. doi: 10.1016/ S0006-3495(99)77252-8.

L. Ma and W. S. Klug. Viscous regularization and r-adaptive remeshing for finite element analysis of lipid membrane mechanics. *J. Comput. Phys.*, 227(11), 5816–5835,2008. ISSN 00219991. doi: 10.1016/j.jcp.2008.02.019.

P. Mattjus, B. Malewicz, J. T. Valiyaveettil, W. J. Baumann, R. Bittman, and R. E. Brown. Sphingomyelin modulates the transbilayer distribution of galactosylceramide in phospholipid membranes. *J. Biol. Chem.*, 277(22), 19476–19481, 2002. ISSN 00219258. doi: 10.1074/jbc.M201305200.

B. S. Perrin, A. J. Sodt, M. L. Cotten, and R. W. Pastor. The curvature induction of surface-bound antimicrobial peptides piscidin 1 and piscidin 3 varies with lipid chain length. *J. Membr. Biol.*, 248 (3), 455–467, 2015. ISSN 14321424. doi: 10.1007/s00232-014-9733-1.

N. Ramakrishnan, P. B. Sunil Kumar, and J. H. Ipsen. Monte Carlo simulations of fluid vesicles with in-plane orientational ordering. *Phys. Rev. E—Stat. Nonlinear, Soft Matter Phys.*, 81(4), 2010. ISSN 15393755. doi: 10.1103/PhysRevE.81.041922.

W. Rawicz, K. C. Olbrich, T. McIntosh, D. Needham, and E. A. Evans. Effect of chain length and unsaturation on elasticity of lipid bilayers. *Biophys. J.*, 79(1), 328–339, 2000. ISSN 00063495. doi: 10.1016/S0006-3495(00)76295-3.

R. J. Ryham, T. S. Klotz, L. Yao, and F. S. Cohen. Calculating transition energy barriers and characterizing activation states for steps of fusion. *Biophys. J.*, 110(5), 1110–1124,2016. ISSN 15420086. doi: 10.1016/j.bpj.2016.01.013.

P. Schofield and J. R. Henderson. Statistical mechanics of inhomogeneous fluids. *Proc. R. Soc. Lond. A*, 379(1776), 231–246, 1982. doi: 10.1098/rspa.1982.0015.

A. J. Sodt, A. H. Beaven, O. S. Andersen, W. Im, and R. W. Pastor. Gramicidin a channel formation induces local lipid redistribution II: a 3D continuum elastic model. *Biophys. J.*, 112(6), 1198–1213, 2017. ISSN 15420086. doi: 10.1016/j.bpj.2017.01.035.

A. J. Sodt and R. W. Pastor. Bending free energy from simulation: correspondence of planar and inverse hexagonal lipid phases. *Biophys. J.*, 104(10), 2202–2211, 2013. ISSN 00063495. doi: 10.1016/j.bpj.2013.03.048.

A. J. Sodt and R. W. Pastor. Molecular modeling of lipid membrane curvature induction by a peptide: more than simply shape. *Biophys. J.*, 106(9), 1958–1969, 2014. ISSN 15420086. doi: 10.1016/j.bpj.2014.02.037.

A. J. Sodt, R. M. Venable, E. Lyman, and R. W. Pastor. Nonadditive compositional curvature energetics of lipid bilayers. *Phys. Rev. Lett.*, 117(13), 2016. ISSN 10797114. doi: 10.1103/PhysRevLett.117.138104.

I. Szleifer, D. Kramer, A. Ben-Shaul, W. M. Gelbart, and S. A. Safran. Molecular theory of curvature elasticity in surfactant films. *J. Chem. Phys.*, 92(11), 6800, 1990. ISSN 00219606. doi: 10.1063/1.458267. URL http://scitation.aip.org/content/aip/journal/jcp/92/11/10.1063/1.458267%5Cnhttp://scitation.aip.org/deliver/fulltext/aip/journal/jcp/92/11/1.458267.pdf?itemId=/content/aip/journal/jcp/92/11/10.1063/1.458267&mimeType=pdf&containerItemId=content/aip/journal.

M. M. Terzi and M. Deserno. Novel tilt-curvature coupling in lipid membranes, 2017. ISSN 00219606.

R. W. Tourdot, N. Ramakrishnan, T. Baumgart, and R. Radhakrishnan. Application of a free-energy-landscape approach to study tension-dependent bilayer tubulation mediated by curvature-inducing proteins. *Phys. Rev. E—Stat. Nonlinear, Soft Matter Phys.*, 92(4), 2015. ISSN 15502376. doi: 10.1103/PhysRevE.92.042715.

T. S. Ursell, W. S. Klug, and R. Phillips. Morphology and interaction between lipid domains. *Proc. Natl. Acad. Sci. U. S. A.*, 106(32), 13301–13306, 2009. ISSN 0027-8424. doi: 10.1073/pnas.0903825106.

R. M. Venable, A. J. Sodt, B. Rogaski, H. Rui, E. Hatcher, A. D. MacKerell, R. W. Pastor, and J. B. Klauda. CHARMM all-atom additive force field for sphingomyelin: elucidation of hydrogen bonding and of positive curvature. *Biophys. J.*, 107(1), 134–145, 2014. ISSN 15420086. doi: 10.1016/j.bpj.2014.05.034.

X. Wang and M. Deserno. Determining the lipid tilt modulus by simulating membrane buckles. *J. Phys. Chem. B*, 120(26), 6061–6073, 2016. ISSN 15205207. doi: 10.1021/acs.jpcb.6b02016.

4

Water between Membranes: Structure and Dynamics

Sotiris Samatas[1], **Carles Calero**[1], **Fausto Martelli**[2], **and Giancarlo Franzese**[1,3]

I Introduction

It has been long recognized that the structure and function of biological membranes is greatly determined by the properties of hydration water (Berkowitz et al. 2006). Biological membranes are made of a large number of components, including membrane proteins, cholesterol, glycolipids, ionic channels, among others, but their framework is provided by phospholipid molecules forming a bilayer. The bilayer structure is a consequence of the interaction with water of the phospholipids, that are macromolecules made of a hydrophilic end (headgroup) and a hydrophobic side (chain or tailgroup). To reduce the free-energy cost of the interface, the polar heads expose to water, with their tails side by side, while the apolar hydrocarbon tails hide from water extending in the region between two layers of heads (hydrophobic effect).

But water is important for membranes not only because it induces the hydrophobic effect. On the one hand, interfacial water modulates the fluidity of the membrane, on the other hand, it mediates the interaction between membranes and between membranes and solutes (ions, proteins, DNA, etc.), regulating cell-membrane tasks as, for example, transport and signaling functions (Hamley 2007).

Phospholipid bilayers or monolayers of a single type of phospholipid are used as model systems to understand how basic biological membranes function and how they interact with the environment. For its important role in determining the properties of biological membranes, structure and dynamics of water at the interface with phospholipid bilayers have been extensively investigated both in experiments and in simulation studies.

Experiments have used Nuclear Magnetic Resonance (NMR) spectroscopy to study the translational dynamics of interfacial water, evidencing the different rates of lateral and normal diffusion and revealing the effect of lipid hydration on water dynamics (Volke et al. 1994, Wassall, 1996). The slow-down of water dynamics due to the interaction with the phospholipid membrane has also been observed with the help of molecular dynamics (MD) simulations (Berkowitz et al. 2006, Bhide and Berkowitz 2005).

NMR experiments and vibrational sum frequency generation spectroscopy have provided insight on the ordering and orientation of water molecules around phospholipid headgroups (Chen et al. 2010, König et al. 1994). These results are in agreement with the picture extracted from computer simulations of hydrated phospholipid membranes (Berkowitz et al. 2006, Pastor 1994).

Infrared spectroscopy measurements indicate the formation of strong hydrogen bonds with the phosphate and carbonyl groups of phospholipids, as well as an enhancement of the hydrogen bonds

---

[1] Secciò de Fisica Estadistica i Interdisciplinaria—Departament de Física de la Matèria Condensada & Institute of Nanoscience and Nanotechnology (IN2UB), Universitat de Barcelona, Martí Franqués 1, 08028 Barcelona.
[2] IBM Research, Hartree Centre, Daresbury WA4 4AD, United Kingdom.
[3] gfranzese@ub.edu.

between water molecules in the vicinity of phospholipid headgroups (Binder 2003, Chen et al. 2010). The rotational dynamics of water molecules is also dramatically affected by the presence of phospholipids and the hydration level of the membrane, as evidenced experimentally using a variety of techniques including ultrafast vibrational spectroscopy (Zhao et al. 2008), terahertz spectroscopy (Tielrooij et al. 2009), and neutron scattering (Trapp et al. 2010). MD simulations have complemented these studies by exploring the decay of water orientation correlation functions in phospholipid membranes with different hydration levels (Calero et al. 2016, Gruenbaum and Skinner 2011, Martelli, Ko, Borallo and Franzese 2018, Zhang and Berkowitz 2009).

These studies reveal that the structure and dynamics of hydration water in stacked phospholipid membranes are affected by both the interaction with phospholipids and by its level of hydration. In the following, we present the results of studies based on all-atom MD simulations which probe the structural and dynamical properties (both diffusion and rotational dynamics) of hydration water at stacked phospholipid membranes as a function of their hydration.

First, the focus is put on the influence of the membrane hydration level on the translational and rotational dynamical properties of confined water. Second, we perform a detailed analysis to determine the dependence of water dynamics on the local distance to the membrane and on temperature. We also relate the dynamical behavior of water confined between bilayers with the structure and dynamics of the hydrogen bond network formed with other water molecules and with selected groups of the lipid.

We finally discuss the surprising result showing that the water–membrane interface has a structural effect at ambient conditions that propagates further than the often-invoked 1 nm length scale. To this goal, we discuss the calculation of water intermediate range order at the interface adopting a parameter recently introduced by Martelli, Ko, Oğuz and Car (2018).

II Dynamics of Water Between Stacked Membranes at Different Hydration Levels

In the following, we focus on water between stacked membranes, an important part in several biological structures, including endoplasmic reticulum and Golgi apparatus, that processes proteins for their use in animal cells, or thylakoid compartments in chloroplasts (plant and algal cells) and cyanobacteria, both involved in the photosynthesis process. We will consider the water dynamics when the membrane hydration level changes, and will use as a model membrane made of dimyristoyl phosphatidylcholine (DMPC) phospholipid bilayers.

Among a wide variety of lipids, DMPC are phospholipids incorporating a choline as a headgroup and a tailgroup formed by two myristoyl chains (Figure 1). Choline-based phospholipids are ubiquitous in cell membranes and used in drug-targeting liposomes (Hamley 2007).

In MD simulations (Calero et al. 2016), by using periodic boundary conditions (Figure 2), it is possible to describe a system of perfectly stacked phospholipid bilayers with a homogeneous prescribed hydration level ω, including the low hydrated systems probed in experiment (Tielrooij et al. 2009, Trapp et al. 2010, Volkov et al. 2007, Zhao et al. 2008) and a fully hydrated membrane (with $\omega = 34$) (Nagle et al. 1996). The last case has been thoroughly studied using computer simulations (Bhide and Berkowitz 2005).

In particular, Calero et al. (2016) consider 128 DMPC lipids distributed in two leaflets in contact with hydration water, and perform MD simulations using the NAMD 2.9 (Phillips et al. 2005) package at a temperature of 303 K and an average pressure of 1 atm, setting the simulation time step to 1 fs, and describing the structure of phospholipids and their mutual interactions by the recently parameterized force field CHARMM36 (Klauda et al. 2010, Lim et al. 2012), which is able to reproduce the area per lipid in excellent agreement with experimental data. The water model employed in their simulations, consistent with the parametrization of CHARMM36, is the modified TIP3P (Jorgensen et al. 1983, MacKerell et al. 1998). They cut off the van der Waals interactions at 12Å with a smooth switching function starting at 10 Å and compute the long-

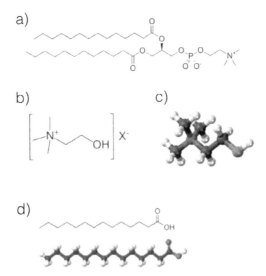

FIGURE 1 Dimyristoyl phosphatidylcholine (DMPC) phospholipid macromolecule. a) DMPC chemical structure. b) Headgroup: choline (quaternary ammonium salt) containing the N,N,N-trimethylethanolammonium cation and with an undefined counteranion X^-. c) Choline detailed structure as ball-and-stick, with carbon in black, hydrogen in white, oxygen in red, nitrogen in blue. d) Tailgroup: two myristoyl chains, in skeletal and ball-and-stick representations. (See color insert for the color version of this figure)

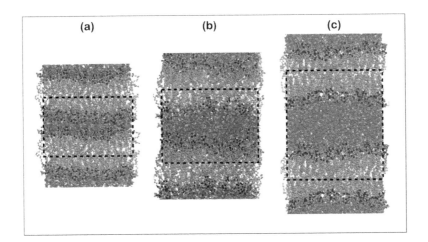

FIGURE 2 Snapshots of a DMPC bilayer with fixed hydration levels (a) $\omega = 7$, (b) 20, (c) 34. Other hydration levels considered in Calero et al. (2016) were $\omega = 4, 10$, and 15. Gray and red beads represent phospholipid tails and headgroups, respectively, orange beads the phosphorous atoms. Blue and white beads represent oxygen and hydrogen atoms of water. The dashed line indicates the size of the simulation box. (See color insert for the color version of this figure)

range electrostatic forces using the particle mesh Ewald method (Essmann et al. 1995) with a grid space of $\approx 1\text{Å}$, updating the electrostatic interactions every 2 fs. After energy minimization, they equilibrate the hydrated phospholipid bilayers for 10 ns followed by a production run of 50 ns in the NPT ensemble at 1 atm. They use a Langevin thermostat (Berendsen et al. 1984) with a damping coefficient of 0.1 ps^{-1} to control the temperature and a Nosé-Hoover Langevin barostat (Feller et al. 1995) with a piston oscillation time of 200 fs and a damping time of 100 fs to control the pressure.

II.A Translational Dynamics

The translational dynamics of water confined in stacked DMPC bilayers is characterized by the mean-square displacement of the center of mass of water molecules projected onto the plane of the membrane (MSD_p). The diffusion coefficients for confined water are obtained from the linear regime reached by the MSD at sufficiently long times through

$$D_p \equiv \lim_{t \to \infty} \frac{\langle |r_p(t) - r_p(0)|^2 \rangle}{4t}, \qquad (1)$$

where $r_p(t)$ is the projection of the center of mass of a water molecule on the plane of the membrane and the angular brackets $\langle ... \rangle$ indicate average over all water molecules and time origins (Figure 3).

Calero et al. (2016) find that water molecules are significantly slowed down when the hydration level of the membrane is reduced, in agreement with previous experimental and computational studies (Gruenbaum and Skinner 2011, Tielrooij et al. 2009, Wassall 1996, Zhang and Berkowitz 2009, Zhao et al. 2008). The diffusion coefficient increases monotonically with hydration (Figure 3), from $D_p = 0.13$ nm^2/ns for the lowest hydrated system ($\omega = 4$) to 3.4 nm^2/ns for the completely hydrated membrane ($\omega = 34$), in agreement with experimental results of similar systems (Rudakova et al. n.d., Wassall 1996). The agreement is not quantitative, because a comparison of the MD results with experimental results is problematic due to the difficulty in

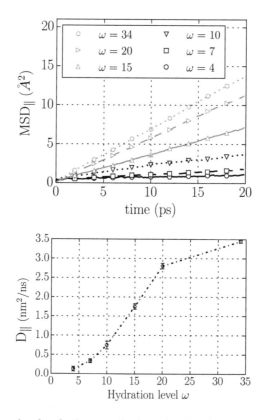

FIGURE 3 Translational dynamics of confined water molecules projected on the plane of the membrane for the different stacked phospholipid bilayers. (Upper panel) Mean-square displacement MSD_p on the plane of the membrane, as a function of time. (Bottom panel) Diffusion coefficient D_p of water molecules on the plane of the membrane for the different hydration levels considered. Adapted from Calero et al. (2016).

experiment to ensure the homogeneity of hydration and the perfect alignment of the membranes in measuring D_p. Nevertheless, the MD simulations reproduce qualitatively the large drop of the diffusion coefficient D_p for low hydrated membranes with respect to bulk water and its dependence with hydration, as seen in (Wassall 1996), where the diffusion of water confined in the lamellar phase of egg phosphatidylcholine was investigated as a function of the hydration of the membrane. For weakly hydrated systems, the authors of the experiments observed that the water diffusion coefficient had an important reduction (of approximately a factor 10), which was attributed to the interaction with the membrane. In addition, the authors found a monotonic increase of the diffusion coefficient with the membrane's hydration. It is, therefore, clear that the MD approach is able, at least qualitatively, to reproduce the translational behavior of water between stacked membranes, suggesting the possibility to get insight into the detailed mechanisms of the slowing down at the molecular level.

2.2 Rotational Dynamics

The rotational dynamics of the water molecules confined in stacked phospholipid bilayers is characterized by calculating from the trajectories of the MD simulations the rotational dipolar correlation function,

$$C_{sim}^{rot}(t) \equiv \langle \hat{\mu}(t) \cdot \hat{\mu}(0) \rangle, \tag{2}$$

where $\hat{\mu}(t)$ is the direction of the water dipole vector at time t and $\langle \ldots \rangle$ denotes ensemble average over all water molecules and time origins (Figure 4). This quantity is related to recent terahertz

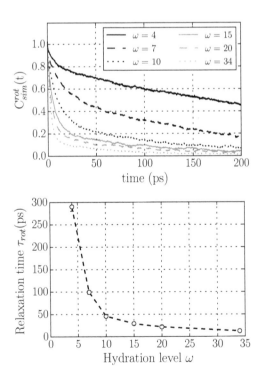

FIGURE 4 Rotational dynamics of water molecules confined in stacked phospholipid bilayers at different levels of hydration ω. (Upper panel) Rotational dipolar correlation function C_{sim}^{rot} of water molecules for $\omega = 4, 7, 10, 15, 20, 34$ (from top to bottom). (Lower panel) Rotational relaxation time τ_{rot} calculated from Eq. (3) as a function of the hydration level. Adapted from Calero (2016).

dielectric relaxation measurements used to probe the reorientation dynamics of water (Tielrooij et al. 2009).

To quantify the relaxation of the correlation functions $C_{sim}^{rot}(t)$, Calero et al. (Calero et al. 2016) define the relaxation time

$$\tau_{rot} \equiv \int_0^\infty C_{sim}^{rot}(t)dt, \tag{3}$$

which is independent of any assumptions on the functional form of the correlation function. Calero et al. find that τ_{rot} decreases from (290 ± 10) ps at low hydration $\omega = 4$, to (12.4 ± 0.3) ps at full hydration $\omega = 34$ (Figure 4), with a monotonic slowing-down of the rotational dynamics of water for decreasing membrane hydration. This result is consistent with experiments (Tielrooij et al. 2009, Zhao et al. 2008) and previous computational works (Gruenbaum and Skinner 2011, Zhang and Berkowitz 2009).

The decay of the rotational correlation function $C_{sim}^{rot}(t)$ represented in Figure 4 occurs through an initial rapid decrease of $C_{sim}^{rot}(t)$ followed by a much slower decay process. This dependence suggests the existence of different time-scales relevant in the rotational dynamics of water molecules. In particular, Tielrooij et al. (2009) assumed the existence of three different water species in regards to their reorientation dynamics: (i) *bulk-like water* molecules, whose reorientation dynamics resemble that of bulk water, with characteristic times τ_{bulk} of a few picoseconds; (ii) *fast water* molecules, which reorient significantly faster than bulk water with $\tau_{fast} \approx$ fraction of

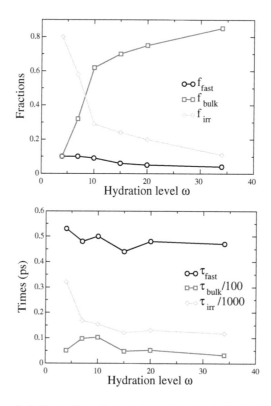

FIGURE 5 Partition into fast, bulk-like and irrotational water molecules as in Eq. (6). (Upper panel) Fractions f_{fast}, f_{bulk}, and f_{irr} with the condition in Eq. (5). (Lower panel) Characteristic times for each of the three species, τ_{fast}, τ_{bulk}, and τ_{irr}, respectively. For sake of clarity, times are rescaled as indicated in the legend. The characteristic times have no regular behavior for increasing hydration. Adapted from Calero (2016).

picosecond, and (iii) *irrotational water* molecules, which might relax with characteristic times $\tau_{irr} \gg 10$ ps. With this assumption, one can identify the fractions f_{fast}, f_{bulk} and f_{irr} of the three species as a function of ω by fitting $C_{sim}^{rot}(t, \omega)$ to a sum of pure exponentially decaying terms:

$$C_{fit}(t, \omega) = f_{fast}(\omega)e^{-t/\tau_{fast}} + f_{bulk}(\omega)e^{-t/\tau_{bulk}} + f_{irr}(\omega)e^{-t/\tau_{irr}}, \qquad (4)$$

with

$$f_{fast}(\omega) + f_{bulk}(\omega) + f_{irr}(\omega) = 1 \qquad (5)$$

for each ω.

Using this fitting procedure, Calero et al. (2016) analyze their MD results for water molecules for different hydration levels of the bilayer (Figure 5). Their results qualitatively account for the experimental behavior reported in Tielrooij et al. (2009). However, such a fitting procedure is not robust – there are five fitting parameters for each ω – the parameters τ_{fast}, τ_{bulk}, and τ_{irr} do not show any regular behavior as function of ω. This strongly suggests that the hypothesis made by Tielrooij et al. of the existence of such distinctive types of water might not be complete and that a more thorough analysis is needed, as described in the next section.

III Dependence of Water Dynamics on Distance to Membrane

III.A Water Structure at Stacked Phospholipid Bilayers

III.A.1 Definition of Distance to the Membrane

To understand the dependence of the dynamics of water molecules on membrane hydration, a detailed investigation of the water–membrane interface is necessary. To this end, a suitable definition of a distance to the membrane is required. In the relevant length-scales of the problem (defined, for example, by the size of the water molecule), the water–membrane interface is not flat, but exhibits spatial inhomogeneities of $\approx$ 1nm. In addition, these inhomogeneities are dynamical since the phospholipid membrane is in a two-dimensional fluid phase. Nevertheless, its dynamics is much slower than water dynamics, with characteristic timescales significantly longer than the relevant timescales of water dynamics, as evidenced by the disparity of diffusion coefficients: while the typical diffusion coefficient of water is of the order of 1 nm²/ns, the diffusion coefficient of phospholipids is of $\approx$ 0.001 nm²/ns (Yang et al. 2014). Additionally, the interface is soft and water molecules can penetrate into the membrane (Fitter et al. 1999, Lopez et al. 2004, Pandit et al. 2003).

In order to describe the interface and accommodate such features of the membrane, Berkowitz et al. (2006) and Pandit et al. (2003) devised a local definition of the distance to the membrane. Briefly, for each frame, a two-dimensional Voronoi tessellation of the plane of the membrane (the XY-plane) is performed using as centers of the cells the phosphorous and nitrogen atoms of the phospholipid heads. Each water molecule is assigned a Voronoi cell by its location in the XY-plane and has a distance, ξ, to the membrane given by the difference between the Z-coordinates of the water molecule and its corresponding Voronoi cell.

III.A.2 Water Density Profile for Different Hydration Levels

By adopting the local distance ξ to the membrane, it is possible to study the structure of the water–membrane interface, represented by the density profile as a function of ξ (Figure 6). The

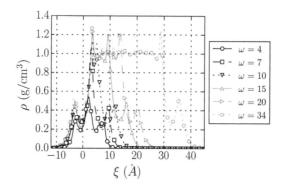

FIGURE 6 Water density profile for stacked DMPC phospholipid bilayers at different hydration levels: $\omega = 4$ (dots), $\omega = 7$ (squares), $\omega = 10$ (down triangles), $\omega = 15$ (up triangles), $\omega = 20$ (side triangles), and $\omega = 34$ (hexagons). Symbols are placed to identify the corresponding lines.

water density profile for stacked membranes at different hydration level emphasizes structural changes in the way water molecules locate themselves at the interface. From inspection of the water density profile for the fully hydrated membrane (with hydration level = 34), we can clearly distinguish three main regions: an inner (interior) layer ($\xi < 0$), the first hydration layer ($0 < \xi < 5\text{Å}$), and outer layers ($\xi > 5\text{Å}$) (Pandit et al. 2003).

For the other less hydrated cases, the same structure is preserved, although the exterior layer is thinner (in the cases with hydration $\omega = 15$, 20) or nonexistent (in the cases with hydration level $\omega = 4$, 7, 10). In addition, the density profiles in Figure 6 show that, as hydration increases, water molecules accumulate in a layering structure. Indeed, water molecules first fill the interior and first hydration layer before starting to accumulate in the outer region ($\xi > 5\text{Å}$). The interior and the first hydration layer become saturated when $7 < \omega < 10$, in agreement with X-ray scattering experiments (Kučerka et al. 2005).

For the least hydrated cases (hydration levels $\omega = 4$, 7), Calero et al. (2016) observe that, although not yet 'full,' the inner region and the first hydration layer are occupied. This implies that the inner water is a *structural* part of the membrane and remains even if the hydration is low. As will be discussed in the following, inner water can be seen as an essential component of the membrane that plays a structural role with hydrogen bonds that are forming bridges between lipids (Lopez et al. 2004, Pasenkiewicz-Gierula et al. 1997).

III.A.3 Structure: Hydrogen Bonds

The structure and dynamics of the hydrogen bond network regulates the dynamical and thermodynamical properties of liquid water (Bianco and Franzese 2014, de Los Santos and Franzese 2009, 2011, 2012, Franzese and de Los Santos 2009, Franzese et al. 2010, Franzese and Stanley 2002, Kumar et al. 2006, 2008, Laage and Hynes 2006, Mazza et al. 2011, 2012, 2009, Stanley et al. 2009, Stokely et al. 2010), as well as the properties of hydrated biological macromolecules (Bianco and Franzese 2015, Bianco, Franzese, Dellago and Coluzza 2017, Bianco et al. 2012, Bianco, Pagès-Gelabert, Coluzza and Franzese, 2017, Franzese et al. 2011, Vilanova et al. 2017). Here, we analyze in detail the case of the hydrogen bond network formed by water molecules between stacked membranes and we discuss its relation to the dynamics of the hydration water.

From the trajectories of all-atom MD simulations, Calero et al. (2016) evaluate the number of hydrogen bonds formed by water between membranes and its dependence with the hydration level. They adopt the widely employed geometric definition of the hydrogen bond: two molecules are hydrogen bonded if the distance between donor and acceptor oxygen atoms satisfies $d_{OO} < 3.5$ Å and the angle formed by the OH bond of the donor molecule with the OO direction is $\theta < 30^{o}$. In addition to hydrogen bonds formed between water molecules, they also consider hydrogen bonds created between water and phosphate or ester groups of the DMPC phospholipid (Bhide and Berkowitz 2005).

The average number $\langle n_{HB} \rangle$ of hydrogen bonds formed by water molecules depends on the hydration level of the membrane (Figure 7). The number of water–water hydrogen bonds decreases monotonically as the membrane hydration ω is reduced. In contrast, the fraction of hydrogen bonds formed by water with selected groups of the lipid increases for decreasing ω, amounting to almost half of all the hydrogen bonds at low hydration ($\omega = 4$).

Due to the importance of hydrogen bonds in water properties, the hydrogen bond profile provides a good measure of the structure of the water–membrane interface. The average number of hydrogen bonds $\langle n_{HB} \rangle$ formed by water molecules shows a complex dependence on distance ξ of water to the membrane with different hydration level (Figure 8). For the completely hydrated membrane (with $\omega = 34$), the number of water–water hydrogen bonds is ≈ 3.45 in the exterior part of the membrane, decreases in the first hydration layer, and becomes ≈ 2 in the interior part of the membrane. The number of water–lipid hydrogen bonds is, by necessity, zero at the exterior region of the interface, increases in the first hydration layer, and becomes ≈ 1 in the interior part of the membrane. The same profile is observed approximately for all hydration levels, except for the two least hydrated cases with $\omega = 4, 7$. In those cases, the number of water–water hydrogen

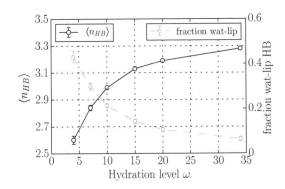

FIGURE 7 Hydrogen bonds formed by water molecules confined in stacked phospholipid membranes as a function of their hydration level. The average total number of hydrogen bonds (black circles) increases, while the average fraction of hydrogen bonds that water forms with lipid groups (gray squares) decreases, when ω increases. Adapted from Calero (2016).

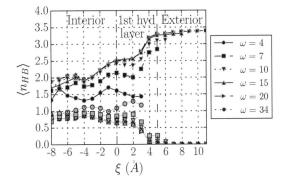

FIGURE 8 Average number of hydrogen bonds $\langle n_{HB} \rangle$ as a function of ξ for different hydrations: between water molecules (black) and between water molecules and selected groups of the phospholipid (gray). Adapted from Calero et al. (2016).

bonds is lower, whereas the number of water–lipid hydrogen bonds is higher than in cases where the interior and first hydration layers of the membrane are saturated with water molecules.

To further understand the structure of the water–membrane interface, we calculate here the probability distribution of the number of hydrogen bonds formed by water molecules located at different regions of the interface for the completely hydrated membrane ($\omega = 34$). We find that the probability distribution of hydrogen bonds is different in each region (Figure 9).

Water has preferentially two water–water hydrogen bonds in the inner part of the membrane, three in the first hydration shell, and four in the outer part. At the same time, a water molecule forms often one, and seldom two hydrogen bonds with the lipid in the inner part, while in the first

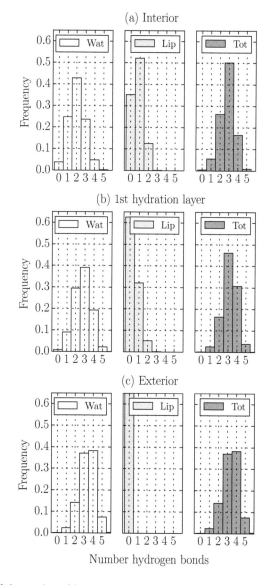

FIGURE 9 Distribution of the number of hydrogen bonds formed by a water molecule in a completely hydrated ($\omega = 34$) stacked membrane at different regions of the interface: (a) Interior ($\xi < 0$), (b) 1st hydration layer ($0 < \xi < 5\text{Å}$), and (c) Exterior ($\xi > 5\text{Å}$), showing the distribution of water–water (Wat), water–lipid (Lip) and total (Tot) hydrogen bonds. The probability distribution of hydrogen bonds is different in each region.

hydration shell, it forms one with significant probability and very rarely two, although most water does not bind to the phospholipid. However, as discussed by Calero et al. (2016), even at low hydration, there is no unbound water near the membrane.

This result is at variance with the hypothesis of *free*, fast water in weakly hydrated phospholipid bilayers that has been proposed to account for rapidly relaxing signals associated with the reorientation of water molecules in experiment (Tielrooij et al. 2009, Volkov et al. 2007). As discussed elsewhere (Calero and Franzese 2018), the experiments can be interpreted better by analyzing how the water dynamics changes as a function of distance to membrane, showing that it is possible to identify the existence of an interface between the first hydration shell, extremely slow and partially made of water bonded to the membrane, and the next shells, faster but still one order of magnitude slower than bulk water.

IV Effect of Temperature on the Hydration Water Dynamics

Next we study how hydration water dynamics changes with temperature T. We simulate a fully hydrated ($\omega = 34$) DMPC phospholipid membrane using the NAMD simulation package and study how the relaxation of the rotational dipolar correlation function $C_{sim}^{rot}(t)$, defined in Eq. (2), varies with T.

In particular, we carry out 10 independent simulations, with production runs of 20 ps each, in the NVT ensemble at different temperatures and at an average pressure P of 1 atm, following the same simulation protocol described in Calero et al. (2016).

To simplify our analysis, we analyze the dynamics of hydration water by grouping the molecules based on their average distance from the membrane at the beginning of each run, regardless of their following trajectories. This distance definition is different from the distance ξ presented in Section IIIA1, but, as we will show next, is able to quantify the effect of temperature changes within our 20 ps simulation and in the range of temperatures we consider here.

At the beginning of each run, we classify water molecules into three different groups: (1) those within a distance of up to 3 Å from any phospholipid atom, including those between the lipids; (2) those at larger distances up to 9 Å from the lipids, and (3) those at distances larger than 9 and up to 15 Å from the lipids (Figure 10). Because the classification is atom-based, there is a finite probability of having molecules that initially are between two regions. For sake of simplicity, these molecules are excluded from the analysis.

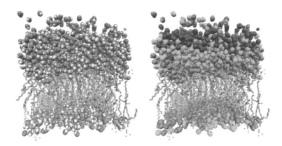

FIGURE 10 Snapshot of the hydrated DMPC phospholipid membrane, represented by the gray carbon backbone of the lipids. The system has periodic boundary conditions, reproducing water between stacked membranes. (Left panel) An initial configuration with water oxygens represented in red and water hydrogens in white. (Right panel) The same configuration where now oxygen and hydrogen atoms are colored based on their distance from the closest lipid atom: (1) red for those at distance up to 3 Å, (2) green for those at larger distances up to 9 Å, (3) blue for the rest in 9–15 Å range. Careful inspection reveals that there are water molecules with atoms with different colors. These molecules across regions are excluded from our analysis. (See color insert for the color version of this figure)

We simulate the systems for temperatures between 303 K and 288.6 K, at which the membrane remains in the liquid phase, finding a clear slowing-down of the rotational dipolar correlation function $C_{sim}^{rot}(t)$ as the temperature decreases. To quantify this effect, at each T, we fit $C_{sim}^{rot}(t)$ with the Kohlrausch–Williams–Watts stretched exponential function, typical of glassy systems (Fierro et al. 1999, Franzese, 2000, 2003, Franzese and Coniglio 1999, Franzese et al. 1999, Franzese and de Los Santos 2009, Franzese et al. 1998, Kumar et al. 2006),

$$C_{sim}^{rot}(t) = \exp\left[-\left(\frac{t}{\tau}\right)^{\beta}\right], \qquad (6)$$

where τ is the characteristic relaxation time and β is the stretched exponent (Figure 11, Table I). Note that β is always larger than 1/3, as expected from general considerations (Fierro et al. 1999, Franzese and de Los Santos 2009, Franzese et al. 1998) and that τ is comparable to the rotational correlation function of bulk water at 250 K (Kumar et al. 2006), that is ≈ 38 K less than the temperature of the hydrated membrane considered here.

We observe a much slower decay for the water molecules of the first group (up to 3 Å) compared to that of the molecules found in the second (3–9 Å) and third (9–15 Å) groups. The characteristic relaxation times obtained for each case increase for decreasing T (Table II). We find that the correlation time τ_1 estimated for $1/e$ at $T = 303$ K for group 1 of water molecules at the closest

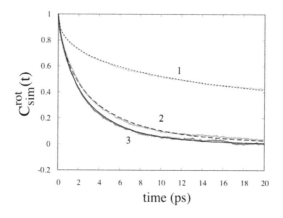

FIGURE 11   Rotational dipolar correlation function $C_{sim}^{rot}(t)$ as a function of time from MD simulations of water between stacked membranes at T=288.6 K and 1 atm. From top to bottom: Correlation of (1) water molecules within 3 Å average distance from the membrane, (2) at larger distances up to 9 Å, (3) at distances above 9 and up to 15 Å. Data from simulations are represented as colored points (blue for 1, green for 2, red for 3), fits with stretched exponential decay equation (6) with continuous lines (dotted for 1, dashed for 2, solid for 3). Fitting parameters are given in Table I. (See color insert for the color version of this figure)

TABLE I

Stretched exponential decay fitting parameters from Eq. (6) for the MD data of rotational dipolar correlation function of water between stacked membranes at T=288.6 K and 1 atm for molecules in groups at different average distances from the membrane: 1, 2, and 3, as defined in the text.

Parameter	β	τ (ps)
Group 1	0.440 ± 0.002	26.9 ± 0.1
Group 2	0.685 ± 0.003	3.05 ± 0.06
Group 3	0.767 ± 0.002	2.53 ± 0.02

TABLE II

Rotational relaxation time estimated as the time at which $C_{sim}^{rot}(t)$ reaches the value $1/e$ for different temperatures T 1 atm, for molecules in groups 1, 2, and 3, as defined in the text. For the lowest temperature $T = 288.6$ K, we use the times from Table I, that have approximately the same values as the estimate for $1/e$, but a smaller error.

T (K)	303.0	295.8	292.2	288.6
τ_1(ps)	15 ± 1	18 ± 1	20 ± 1	26.9 ± 0.1
τ_2(ps)	2.0 ± 0.5	2.5 ± 0.2	2.8 ± 0.2	3.05 ± 0.06
τ_3(ps)	1.5 ± 0.2	2.0 ± 0.1	2.2 ± 0.1	2.53 ± 0.02

distance from the membrane is approximately equal to the integral correlation time (Eq. (3)) at the same temperature and full hydration (12.4 ± 0.3) ps, and that τ_1 increases for decreasing T.

To make a better comparison of the relaxation times estimated here with those calculated in Section IIB, we calculate the integral rotational correlation time from Eq. (3) for the fitted functions of $C_{sim}^{rot}(t)$ for $T = 288.6$ K. We find: $\tau_{rot,1} = 70.1$ ps, $\tau_{rot,2} = 3.9$ ps, and $\tau_{rot,3} = 3.0$ ps for groups 1, 2, and 3 of water molecules, respectively. As expected, the integral correlation time for the group of water hydrating the membrane within 3 Å is much larger than the integral correlation time for $T = 303$ K at full hydration (12.4 ± 0.3) ps, consistent with the slowing down of the system for decreasing T.

We observe that our preliminary analysis of the rotational correlation times shows T-dependence also for water in groups 2 and 3, at distance larger than 3 Å from the membrane, but with only a minor difference between the two groups. These results seem to suggest that the effect on water of the soft interface is only minor at a distance larger than 3 Å from the membrane. In the next section, we will discuss if this is really the case.

V Extension of the Structural Effect of the Membrane on Interfacial Water

In order to investigate the effect of the membrane–water interactions to the structural properties of water, as in Martelli, Ko, Borallo and Franzese (2018), we employ a recently introduced sensitive local order metric (LOM) that maximizes the spatial overlap between a local snapshot and a given reference structure (Martelli, Ko, Oguz and Car 2018).

The LOM measures and grades the degree of order present in the neighborhood of an atomic or molecular site in a condensed medium. It is endowed with high resolving power and high flexibility (Martelli, Giovambattista, Torquato and Car, 2018, Martelli, Ko, Oğuz and Car 2018), and allows one to look for specific ordered domains defined by the location of selected atoms (e.g., water oxygens) in a given reference structure, for which it is found the best alignment out of a sufficiently large number of orientations picked at random and with uniform probability. Typically, the reference is taken to be the local structure of a perfect crystalline phase. From the LOM averaged over all sites in a snapshot, one obtains a score function S, representing an order parameter that tends to 0 for a completely disordered system, and is 1 for a system perfectly matching the reference structure.

As in Martelli, Ko, Borallo and Franzese (2018), we inspect the intermediate-range order (IRO) of water employing, as a reference structure, the configuration of the oxygens of cubic ice (Ic) at the level of the second shell of neighbors. Similar results hold when adopting the second shell of neighbors of hexagonal ice. The structure described by the 12 second neighbors in Ic (Figure 12) defines the Archimedean solid cuboctahedron (Cromwell 1999).

With this choice, we calculate the profile of the score S as a function of the distance from the membrane surface on the z-axis perpendicular to the surface. The membrane surface here is defined as the place where the density of lipid heads reaches a maximum along the z-axis perpendicular to

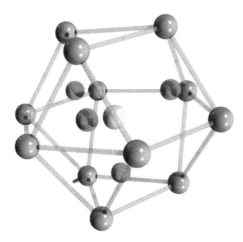

FIGURE 12 Pictorial representation of the second neighbor shell in cubic ice (or cuboctahedron). Gray spheres indicate the position of oxygen atoms, while the green lines emphasize the geometrical structure. The central oxygen is depicted by the yellow sphere, while the red spheres represent the first shell of neighboring oxygens. (See color insert for the color version of this figure)

the average surface plane and approximately coincides with the distance where the water density has the largest variation along z and an approximate value of half the bulk density (Figure 13).

The first observation is that the interface is smeared out over $\approx$ 1 nm as a consequence of the average over the membrane fluctuations, with the water density changing smoothly from 0 to the bulk value, 1 g/cm^3. The water density profile suggests no membrane-induced structure in the hydration water.

Nevertheless, we find that S approaches, but never reaches the value of S in bulk liquid water at the same thermodynamic conditions, indicating that, at a distance as far as $\simeq$ 2.5 nm from the surface, the structural properties of bulk water are not recovered yet (Figure 13).

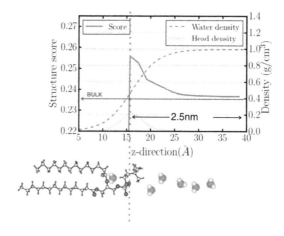

FIGURE 13 Upper panel: Structure of water and lipids along the z-axis perpendicular to the average surface plane of the membrane. The right vertical axis of the plot represents the density profile of lipid heads (in orange) and water (in gray) along z. The z of maximum heads density coincides approximately with the largest variation of water density and is marked by a vertical dotted line in red. The left vertical axis represents the average score S of water (in red) along z. S is zero inside the membrane and is higher than its bulk value (blue arrow) within at least 2.5 nm distance from the surface. Lower panel: Schematic representation of a configuration with a lipid head near the surface location (vertical dotted line in red) and water molecules mimicking their density profile as indicated in the upper panel. (See color insert for the color version of this figure)

On the other hand, Martelli, Ko, Borallo and Franzese (2018) show that dynamical bulk properties are recovered at much shorter distances. To characterize the translational dynamics of water at the interface over a 1 ns-long simulation, they calculate the average O—O distances from the first minimum of the O—O radial distribution function of bulk water, and for each water near the membrane they evaluate the 'standard displacement,' i.e., the distance traveled in units of O—O distances (Figure 14).

At the considered thermodynamic conditions, bulk water would travel a standard displacement between 8 or 10 in a run of 1 ns. This is consistent with the observed quantity for water between 1 nm and 2.5 nm from the membrane (Figure 14).

By approaching the membrane surface, instead, water slows down considerably and the standard displacement reduces to $\simeq 4$, similar to the displacement in deeply undercooled liquid water (Martelli, Ko, Oğuz and Car 2018). This value is quite small, considering that a standard displacement $\leqslant 1$ would correspond to a system in which translational degrees of freedom are frozen and water molecules would only rattle in their nearest neighbors cages. Such decrease occurs synergistically with a considerable increase of S (Figure 13), suggesting that the structuring in the IRO can be ascribed to the significant slowdown of translational degrees of freedom.

Based upon the observation that DMPC heads contains both P- and N-headgroups, with different charges, it is conceivable to imagine that P–OH and N–O interactions among lipids and water have different strengths and, ultimately, that the P- and the N-headgroups contribute differently to the structural properties of water close to the membrane surface. In order to quantify this effect, Martelli, Ko, Borallo and Franzese (2018) compute the dipole and the OH correlation functions. They fit the correlation functions with a double exponential function, obtaining two characteristic relaxation times, τ_1 and τ_2, with $\tau_2 \gg \tau_1$ (Figure 15).

The rotation of water considerably slows down as water molecules approach the membrane. In particular, the distances at which the change is more appreciable are the same as those for the translations and for the increase in the score function S. On approaching the lipid membrane, both τ_1 and τ_2 increase and, interestingly, the relaxation times for OH increase at a pace higher than the relaxation times for μ, indicating that the P–HO interaction is stronger than the N–O interaction.

The slowing down of the rotational dynamics decreases as water moves away from the interface. At distances 1.5 nm, the relaxation times are almost indistinguishable and, as one would expect in bulk liquid water, $\tau_1^{\mu} \simeq \tau_2^{\mu} \simeq \tau_1^{OH} \simeq \tau_2^{OH}$ (Figure 15).

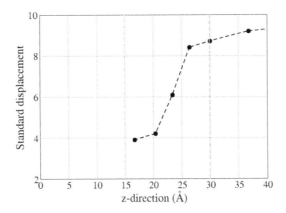

FIGURE 14 Standard displacement of water molecules near the membrane as a function of the distance z from the membrane. The standard displacement largely decreases where the variations in S with respect to bulk are appreciable (region between the two dotted vertical lines). The distance is calculated with respect to the center of the bilayer ($z = 0$) (Martelli, Ko, Borallo and Franzese 2018).

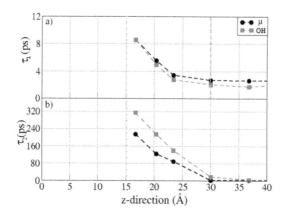

FIGURE 15 Relaxation times (a) τ_1 and (b) τ_2 for the dipole (black circles) and OH (red squares) vector correlation functions as a function of the distance from the membrane surface. All the relaxation times, with $\tau_2 \gg \tau_1$, rapidly increase where the variation in S with respect to bulk is appreciable (region between the two dotted vertical lines). The distance is calculated with respect to the center of the bilayer ($z = 0$). Adapted from Martelli and Frontiers (2018). (See color insert for the color version of this figure)

VI Conclusions

The dynamics and the structure of a membrane and the hydration water are related in a way that goes beyond the hydrophobic effect that allows the membrane to self-assemble. MD simulations of water confined between stacked phospholipid bilayers at different hydration levels reveal that, adopting an appropriate local distance of water from the membrane (Berkowitz et al. 2006, Pandit et al. 2003), water forms layers around the membrane (Calero et al. 2016). These layers are not observed on the macroscopic scale because the spatial fluctuations of the membrane smeared them out. Nevertheless, the local distance, and the analysis at different hydration levels, allow to identify different dynamics behavior of the layers clearly.

In particular, the closer is water to the membrane, the slower is its translational and rotational dynamics, being always slower than bulk water (Calero et al. 2016, Martelli, Ko, Borallo and Franzese 2018). These results are putting into question the possible existence of rotationally *fast* water molecules near the membrane, as proposed to rationalize recent experiments (Tielrooij et al. 2009). This possibility, however, cannot be ruled out completely because it could be associated to the presence of heterogeneities, such as those associated with water molecules with a single hydrogen bond to a lipid at low hydration (Volkov et al. 2007).

Following the structural and dynamical characterization of water near a membrane, it is possible to classify water as inner membrane, first hydration layer, and outer membrane water. The inner membrane water has a structural role forming bridges between lipids (Lopez et al. 2004, Pasenkiewicz-Gierula et al. 1997). This water has a dynamics up to two orders of magnitude slower than bulk water as a consequence of the robustness of water–lipid hydrogen bonds, which are more frequent the lower the hydration of the membrane.

In the first hydration layer, water–water hydrogen bonds have a dynamics one order of magnitude slower than bulk water. This is due to longer lifetime of the hydrogen bonds for lower hydration, a consequence of the slowdown of hydrogen bond-switching due to the decrease of water density (Calero et al. 2016).

These effects for the inner membrane water and the first hydration layer are emphasized when the temperature decreases. We show that water near the membrane has a glassy-like behavior when $T = 288.6$ K. In particular, the rotational correlation function of water within 3 Å from the membrane is comparable, in terms of correlation time, to that of bulk water at ≈ 30 K colder temperature, but with a characteristic stretched exponent, $\beta(T = 288.6$ K$) = 0.44$, that is much smaller than that of the bulk case, $\beta_{bulk}(T = 250$ K$) = 0.85$ (Kumar et al. 2006), suggesting a larger heterogeneity of relaxation modes in the membrane hydration water.

The water slowing down decreases rapidly for water at larger distance from the membrane. Although the dynamics, as well as density, of bulk water are recovered at a short distance, $\simeq 1.2$ nm from the surface of the membrane, the perturbation on the structural properties of liquid water propagates over a much larger distances, at least $\simeq 2.5$ nm (Martelli, Ko, Borallo and Franzese 2018).

This effect is emphasized by the adoption of a local structural parameter S that has a higher sensitivity, with respect to the correlation functions, to small effects arising from the interface. In particular, adopting a definition based on the second water hydration shell, and considering that water diameter and hydrogen bond length are quite similar, S is able to capture structural effects extending over nine water diameters. It is worthy to mention that short-range order parameters – e.g., the tetrahedrality – are unable to show such a results and are recovering bulk values at the same distances where the density and the dynamical properties do.

Finally, water molecules interacting with P in phospholipids are slightly more ordered than those interacting with N. Moreover, water molecules interacting with P-groups have closer second neighbors than those interacting with N-heads. These structural observations, together with the dynamical analysis of hydration water, could help us to understand the role of water at biomembrane interfaces (Calero and Franzese 2018).

ACKNOWLEDGMENTS

We are thankful to Fabio Leoni for helpful discussions. We acknowledge the support of Spanish MINECO grant FIS2015-66879-C2-2-P. F.M. acknowledges the support of the STFC Hartree Centre's Innovation Return on Research programme, funded by the Department for Business, Energy & Industrial Strategy. S.S. acknowledges the support of the A. G. Leventis Foundation. G.F. acknowledges the support of the ICREA Foundation.

REFERENCES

Berendsen, H. J. C., Postma, J. P. M., van Gunsteren, W. F., Nola, A. D. and Haak, J. R. (1984). Molecular dynamics with coupling to an external bath, *Jornal of Physical Chemistry* **81**: 3684.

Berkowitz, M. L., Bostick, D. L. and Pandit, S. (2006). Aqueous solutions next to phospholipid membrane surfaces: Insights from simulations, *Chemical Reviews* **106**(4): 1527–1539.

Bhide, S. Y. and Berkowitz, M. L. (2005). Structure and dynamics of water at the interface with phospholipid bilayers, *The Journal of Chemical Physics* **123**(22): 224702.

Bianco, V. and Franzese, G. (2014). Critical behavior of a water monolayer under hydrophobic confinement, Scientific Reports 4.

Bianco, V. and Franzese, G. (2015). Contribution of water to pressure and cold denaturation of proteins, *Physical Review Letters* **115**(10): 108101–.

Bianco, V., Franzese, G., Dellago, C. and Coluzza, I. (2017). Role of water in the selection of stable proteins at ambient and extreme thermodynamic conditions, *Physical Review X* **7**: 021047.

Bianco, V., Iskrov, S. and Franzese, G. (2012). Understanding the role of hydrogen bonds in water dynamics and protein stability, *Journal of Biological Physics* **38**(1): 27–48.

Bianco, V., Pagès-Gelabert, N., Coluzza, I. and Franzese, G. (2017). How the stability of a folded protein depends on interfacial water properties and residue-residue interactions, *Journal of Molecular Liquids* **245**(SupplementC): 129–139.

Binder, H. (2003). The molecular architecture of lipid membranes—New insights from hydration-tuning infrared linear dichroism spectroscopy, *Applied Spectroscopy Reviews* **38**(1): 15–69.

Calero, C. and Franzese, G. (2018). Membranes with different hydration levels: The membrane water–Hydration water interface. Submitted.

Calero, C., Stanley, H. E. and Franzese, G. (2016). Structural interpretation of the large slowdown of water dynamics at stacked phospholipid membranes for decreasing hydration level: All-atom molecular dynamics, *Materials* **9**(5): 319.

Chen, X., Hua, W., Huang, Z. and Allen, H. C. (2010). Interfacial water structure associated with phospholipid membranes studied by phase-sensitive vibrational sum frequency generation spectroscopy, *Journal of the American Chemical Society* **132**(32): 11336–11342.

Cromwell, P. R. (1999). *Polyhedra*, New York: Cambridge University Press.

de Los Santos, F. and Franzese, G. (2009). Influence of intramolecular couplings in a model for hydrogen-bonded liquids, *AIP Conference Proceedings* **1091**(1): 185–197.

de Los Santos, F. and Franzese, G. (2011). Understanding diffusion and density anomaly in a coarse-grained model for water confined between hydrophobic walls, *The Journal of Physical Chemistry B*, **115**(48).

de Los Santos, F. and Franzese, G. (2012). Relations between the diffusion anomaly and cooperative rearranging regions in a hydrophobically nanoconfined water monolayer, *Physical Review E* **85**(1): 010602.

Essmann, U., Perera, L., Berkowitz, M.L., Darden, T., Lee, H. and Pedersen, L.G. (1995). A smooth particle mesh Ewald method.*The Journal of Chemical Physics* **103**: 8577.

Feller, S.E., Zhang, Y., Pastor, R.W. and Brooks, B.R. (1995). Constant pressure molecular dynamics simulation: The Langevin piston method.*The Journal of Physical Chemistry* **103**: 4613.

Fierro, A., Franzese, G., de Candia, A. and Coniglio, A. (1999). Percolation transition and the onset of nonexponential relaxation in fully frustrated models, *Physical Review E* **59**(1): 60–66.

Fitter, J., Lechner, R. E. and Dencher, N. A. (1999). Interactions of hydration water and biological membranes studied by neutron scattering, *The Journal of Physical Chemistry B* **103**(38): 8036–8050.

Franzese, G. (2000). Potts fully frustrated model: Thermodynamics, percolation, and dynamics in two dimensions, *Physical Review E* **61**(6): 6383–6391.

Franzese, G. (2003). Fast relaxation time in a spin model with glassy behavior, *Fractals* **11**(supp01): 129–138. https://doi.org/10.1142/S0218348X03001793

Franzese, G., Bianco, V. and Iskrov, S. (2011). Water at interface with proteins, *Food Biophysics* 186–198. 10.1007/s11483-010-9198-4.

Franzese, G. and Coniglio, A. (1999). Precursor phenomena in frustrated systems, *Physical Review E* **59** (6): 6409–6412. http://link.aps.org/doi/10.1103/PhysRevE.59.6409

Franzese, G., Coniglio, A. Univ Roma Tre, Dipartimento Fis E Amaldi, V. V. N.. I.-. R. I. I. N. F. M. U. R. N. I. U. N. D. S. F. I.-. N. I. (1999). The potts frustrated model: Relations with glasses, *Ilosophical Magazine B-Physics Of Condensed Mat- Ter Statistical Mechanics Electronic Optical and Magnetic Properties* **79**(ISI:000083953300011): 1813.

Franzese, G. and de Los Santos, F. (2009). Dynamically slow processes in supercooled water confined between hydrophobic plates, *Journal of Physics: Condensed Matter* **21**(50): 504107.

Franzese, G., Fierro, A., De Candia, A. and Coniglio, A. (1998). Autocorrelation functions in 3d fully frustrated systems, *PHYSICA A* **257**(ISI:000075876500042): 379.

Franzese, G., Hernando-Mart´ınez, A., Kumar, P., Mazza, M. G., Stokely, K., Strekalova, E. G., de Los Santos, F. and Stanley, H. E. (2010). Phase transitions and dynamics of bulk and interfacial water, *Journal of Physics: Condensed Matter* **22**(28): 284103.

Franzese, G. and Stanley, H. E. (2002). Liquid-liquid critical point in a hamiltonian model for water: Analytic solution, *Journal of Physics-Condensed Matter* **14**(9): 2201–2209.

Gruenbaum, S. and Skinner, J. (2011). Vibrational spectroscopy of water in hydrated lipid multibilayers. i. infrared spectra and ultrafast pump-probe observables, *The Journal of Chemical Physics* **135**(7): 075101.

Hamley, I. W. (2007). *Introduction to Soft Matter*, West Sussex, England: John Wiley and Sons.

Jorgensen, W. L., Chandrasekhar, J., Madura, J. D., Impey, R. W. and Klein, M. L. (1983). Comparison of simple potential functions for simulating liquid water, *Journal of Chemical Physics* **79**(2): 926–935.

Klauda, J. B., Venable, R. M., Freites, J. A., O'Connor, J. W., Tobias, D. J., Mondragon-Ramirez, C., Vorobyov, I., MacKerell, A. D. and Pastor, R. W. (2010). Update of the CHARMM all-atom additive force field for lipids: Validation on six lipid types, *Journal of Physical Chemistry B* **114**(23): 7830–7843.

König, S., Sackmann, E., Richter, D., Zorn, R., Carlile, C. and Bayerl, T. (1994). Molecular dynamics of water in oriented dppc multilayers studied by quasielastic neutron scattering and deuterium-nuclear magnetic resonance relaxation, *The Journal of Chemical Physics* **100**(4): 3307–3316.

Kučerka, N., Liu, Y., Chu, N., Petrache, H. I., Tristram-Nagle, S. and Nagle, J. F. (2005). Structure of fully hydrated fluid phase dmpc and dlpc lipid bilayers using x-ray scattering from oriented multilamellar arrays and from unilamellar vesicles, *Biophysical Journal* **88**(4): 2626–2637.

Kumar, P., Franzese, G., Buldyrev, S. V. and Stanley, H. E. (2006). Molecular dynamics study of orientational cooperativity in water, *Physical Review E* **73**(4): 041505.

Kumar, P., Franzese, G., Buldyrev, S. V. and Stanley, H. E. (2008). *Dynamics of Water at Low Temperatures and Implications for Biomolecule, Vol. 752 of Lecture Notes in Physics*, Berlin Heidelberg: Springer, 3–22.

Laage, D. and Hynes, J. T. (2006). A molecular jump mechanism of water reorientation, *Science* **311** (5762): 832–835.

Lim, J. B., Rogaski, B. and Klauda, J. B. (2012). Update of the cholesterol force field parameters in CHARMM, *Journal of Physical Chemistry B* **116**(1): 203–210.

Lopez, C. F., Nielsen, S. O., Klein, M. L. and Moore, P. B. (2004). Hydrogen bonding structure and dynamics of water at the dimyristoylphosphatidylcholine lipid bilayer surface from a molecular dynamics simulation, *The Journal of Physical Chemistry B* **108**(21): 6603–6610.

MacKerell, A. D., Bashford, D., Bellott, M., R. L. Dunbrack, Evanseck, J. D., Field, M. J., Fischer, S., Gao, J., Guo, H., Ha, S., Joseph-McCarthy, D., Kuchnir, L., Kuczera, K., Lau, F. T., Mattos, C., Michnick, S., Ngo, T., Nguyen, D. T., Prodhom, B., Reiher, W. E., Roux, B., Schlenkrich, M., Smith, J. C., Stote, R., Straub, J., Watanabe, M., Wiórkiewicz-Kuczera, J., Yin, D. and Karplus, M. (1998). All-atom empirical potential for molecular modeling and dynamics studies of proteins, *The Journal of Physical Chemistry B* **102**(18): 3586–3616.

Martelli, F., Giovambattista, N., Torquato, S. and Car, R. (2018). Searching for crystal-ice domains in amorphous ices, *Physical Review Materials* **2**: 075601.

Martelli, F., Ko, H.-Y., Borallo, C. C. and Franzese, G. (2018). Structural properties of water confined by phospholipid membranes, *Frontiers of Physics* **13**(1): 136801.

Martelli, F., Ko, H.-Y., Oğuz, E. C. and Car, R. (2018). Local-order metric for condensed-phase environments, *Physical Review B* **97**: 064105.

Mazza, M. G., Stokely, K., Pagnotta, S. E., Bruni, F., Stanley, H. E. and Franzese, G. (2011). More than one dynamic crossover in protein hydration water, *Proceedings of the National Academy of Sciences* **108**(50): 19873–19878.

Mazza, M. G., Stokely, K., Stanley, H. E. and Franzese, G. (2012). Effect of pressure on the anomalous response functions of a confined water monolayer at low temperature, *The Journal of Chemical Physics* **137**(20): 204502–204513.

Mazza, M. G., Stokely, K., Strekalova, E. G., Stanley, H. E. and Franzese, G. (2009). Cluster monte carlo and numerical mean field analysis for the water liquid-liquid phase transition, *Computer Physics Communications* **180**(4): 497–502.

Nagle, J. F., Zhang, R., Tristram-Nagle, S., Sun, W., Petrache, H. I. and Suter, R. M. (1996). X-ray structure determination of fully hydrated l alpha phase dipalmitoylphosphatidylcholine bilayers., *Biophysical Journal* **70**(3): 1419–1431.

Pandit, S. A., Bostick, D. and Berkowitz, M. L. (2003). An algorithm to describe molecular scale rugged surfaces and its application to the study of a water/lipid bilayer interface, *The Journal of Chemical Physics* **119**(4): 2199–2205.

Pasenkiewicz-Gierula, M., Takaoka, Y., Miyagawa, H., Kitamura, K. and Kusumi, A. (1997). Hydrogen bonding of water to phosphatidylcholine in the membrane as studied by a molecular dynamics simulation: Location, geometry, and lipidlipid bridging via hydrogen-bonded water, *The Journal of Physical Chemistry A* **101**(20): 3677–3691.

Pastor, R. W. (1994). Molecular dynamics and monte carlo simulations of lipid bilayers, *Current Opinion in Structural Biology* **4**(4): 486–492.

Phillips, J. C., Braun, R., Wang, W., Gumbart, J., Tajkhorshid, E., Villa, E., Chipot, C., Skeel, R. D., Kalé, L. and Schulten, K. (2005). Scalable molecular dynamics with NAMD, *Journal of Computational Chemistry* **26**(16): 1781–1802.

Rudakova, M., Filippov, A. and Skirda, V. (n.d.). Water diffusivity in model biological membranes, *Applied Magnetic Resonance* **27**: 519–526.

Stanley, H. E., Kumar, P., Han, S., Mazza, M. G., Stokely, K., Buldrev, S. V., Franzese, G., Mallamace, F. and Xu, L. (2009). Heterogeinties in confined water and protein hydration water, *Journal of Physics: Condensed Matter* **21**: 504105.

Stokely, K., Mazza, M. G., Stanley, H. E. and Franzese, G. (2010). Effect of hydrogen bond cooperativity on the behavior of water, *Proceedings of the National Academy of Sciences* **107**(4): 1301–1306.

Tielrooij, K. J., Paparo, D., Piatkowski, L., Bakker, H. J. and Bonn, M. (2009). Dielectric relaxation dynamics of water in model membranes probed by terahertz spectroscopy, *Biophysical Journal* **97**: 2848–2492.

Trapp, M., Gutberlet, T., Juranyi, F., Unruh, T., Dem´E, B., Tehei, M. and Peters, J. (2010). Hydration dependent studies of highly aligned multilayer lipid membranes by neutron scattering, *The Journal of Chemical Physics* **133**(16): 164505.

Vilanova, O., Bianco, V. and Franzese, G. (2017). *Multi-Scale Approach for Self-Assembly and Protein Folding*, Cham: Springer International Publishing, 107–128.

Volke, F., Eisenblätter, S., Galle, J. and Klose, G. (1994). Dynamic properties of water at phosphatidylcholine lipid-bilayer surfaces as seen by deuterium and pulsed field gradient proton nmr, *Chemistry and Physics of Lipids* **70**(2): 121–131.

Volkov, V. V., Palmer, D. J. and Righini, R. (2007). Distinct water species confined at the interface of a phospholipid membrane, *Physical Review Letters* **99**: 078302.

Wassall, S. R. (1996). Pulsed field gradient-spin echo nmr studies of water diffusion in a phospholipid model membrane, *Biophysical Journal* **71**: 2724–2732.

Yang, J., Calero, C. and Mart´I, J. (2014). Diffusion and spectroscopy of water and lipids in fully hydrated dimyristoylphosphatidylcholine bilayer membranes, *Journal of Chemical Physics* **140**(10): 104901.

Zhang, Z. and Berkowitz, M. L. (2009). Orientational dynamics of water in phospholipid bilayers with different hydration levels, *The Journal of Physical Chemistry B* **113**(21): 7676–7680.

Zhao, W., Moilanen, D. E., Fenn, E. E. and Fayer, M. D. (2008). Water at the surfaces of aligned phospholipid multibilayer model membranes probed with ultrafast vibrational spectroscopy, *Journal of the American Chemical Society* **130**(42): 13927–13937. PMID: 18823116.

5

Simulation Approaches to Short-Range Interactions between Lipid Membranes

Matej Kanduč
Research Group for Simulations of Energy Materials, Helmholtz-Zentrum Berlin für Materialien und Energie, Hahn-Meitner-Platz 1, 14109 Berlin, Germany

Alexander Schlaich
Laboratoire Interdisciplinaire de Physique, CNRS and Université Grenoble Alpes, UMR CNRS 5588, 38000 Grenoble, France

Bartosz Kowalik, Amanuel Wolde-Kidan, and Roland R. Netz
Freie Universität Berlin, Fachbereich Physik, Arnimallee 14, 14195 Berlin, Germany

Emanuel Schneck
Biomaterials Department, Max Planck Institute of Colloids and Interfaces, Am Mühlenberg 1, 14476 Potsdam, Germany

1 Introduction

As emphasized at several instances in this book, lipid membranes have a wealth of biological functions. But membranes do not act alone. In the congested physiological environment, their functions depend on interactions with other membranes: biological processes involving lipid membranes are characterized by a high degree of spatiotemporal reorganization, including membrane adhesion, vesicle release, or the formation of multilamellar structures.[1,2] In this context, predominantly repulsive membrane interactions prevent close contacts, sustain hydrodynamic pathways for the diffusion of biomolecules between the membrane surfaces, and can induce the release of adhering vesicles. Attractive interactions in conjunction with short-range repulsive interactions, on the other hand, create well-defined average membrane separations that facilitate adhesion by specific binding, membrane fusion, or material exchange. With that, membrane–membrane and membrane–surface interactions do not only govern cell adhesion processes[3] but also the properties of bacterial biofilms[4] and the adsorption of organisms to natural and artificial surfaces.[5]

The characteristics of membrane interactions in terms of strength, scaling, and range are the result of a complex interplay of various force contributions. Long-range interactions, such as the Helfrich repulsion,[6] screened electrostatic forces, and dispersion forces, are nowadays reasonably well understood. The latter two are readily described in the framework of the classical Derjaguin–Landau–Verwey–Overbeek (DLVO) theory.[7] Short-range interactions, on the other hand, elude such descriptions because of the breakdown of standard continuum theories on the length scales of solvent molecules. At these short separations, a universal strong repulsion, commonly termed *hydration repulsion*, dominates the interaction even for uncharged membranes.

The hydration repulsion between biological membranes has an essential biological role in that it creates a barrier against close contact and thereby suppresses uncontrolled membrane adhesion and fusion.[8] Although it generally acts between all kinds of hydrophilic surfaces in water,[9] it was first quantified experimentally for stacks of charge-neutral phospholipid bilayer membranes in terms of pressure–distance curves.[10] Later, it was also reproduced for two individual membrane surfaces with the surface force apparatus (SFA) method.[11] The hydration repulsion exhibits an approximately exponential decay with a decay length of a few angstroms.[12] As a heuristic law, it is nowadays commonly used in modeling the forces between lipid membranes in water.[13,14] As we will see further below, the characteristics of this short-range repulsion depend on the membrane phase and the chemistry of the membrane surfaces.

The physical mechanisms underlying the hydration repulsion have been subject to controversial debates since the 1980s. The repulsion has been attributed to strong lipid–water binding,[15] the unfavorable overlap of interfacial water layers,[16] and to a decrease in the configurational entropy of the lipid molecules upon surface approach,[17] but concluding this debate with only experimental data at hand has turned out to be difficult. Substantial progress has been made only more recently through new mechanistic insights from molecular dynamics simulations, which are the topic of this chapter.

Typical computer simulation studies on lipid membranes are concerned with isolated (i.e., non-interacting) membranes. In the standard simulation setup employing periodic boundary conditions, non-interacting membranes are conveniently separated by a thick-enough water layer ($D_w > 3$ nm), such that the influence of a membrane on its periodic images is weak. In contrast, investigating membrane interactions, such as the hydration repulsion, in simulations requires the determination of interaction pressures as a function of the membrane separation. This is a non-trivial task, because it has to be done in a suitable thermodynamic ensemble that accounts for water exchange with the aqueous environment (see below).

In the remainder of this chapter, we first introduce the basic thermodynamics of membrane interactions and review several approaches to quantify membrane interactions in molecular dynamics simulations. Subsequently, we compare membrane interactions as obtained in MD simulations with experimental data and discuss them with a focus on the hydration repulsion. Finally, we review the most relevant insights from recent simulation studies into the physical mechanisms underlying the short-range repulsion between lipid membranes.

2 Thermodynamics of Membrane Interactions

The interaction between lipid membranes in the aqueous environment is commonly quantified in terms of pressure–distance curves, where the interaction pressure Π is measured as a function of the separation (or water layer thickness) D_w between the membranes. At given temperature T and pressure p, the interaction pressure Π is related to the derivative of the Gibbs free energy, G, with respect to D_w, while the chemical potential of water, $\mu(p, T)$, remains constant,

$$\Pi(D_w) = -\frac{1}{A}\left(\frac{\partial G}{\partial D_w}\right)_{\mu,T}. \tag{1}$$

In this equation, A denotes the surface area, which can either be fixed or can vary with the separation, depending on the boundary conditions imposed.[18]

In experiments, pressure–distance curves $\Pi(D_w)$ are typically obtained by subjecting membrane multilayers to hydrostatic pressures or so-called equivalent pressures of known magnitude.[12,19] The latter are realized under atmospheric pressure by controlled competition for the solvent, i.e., by shifting the chemical potential of the surrounding liquid to lower values. Equivalent pressures can be exerted by bringing the membrane multilayers into contact with aqueous polymer solutions separated by flexible, semi-permeable membranes. The equivalent pressure then coincides with the osmotic pressure afforded by the polymers in solution. Alternatively, the hydration level can be adjusted via vapor exchange with a water reservoir with lowered chemical potential. The water layer thickness D_w is then deduced from the lamellar periodicity D measured by X-ray or neutron diffraction.

3 Simulations of Interacting Membranes

In coarse-grained simulations with implicit solvent, the water chemical potential has no meaning. While this aspect largely facilitates varying the membrane separation, implicit-solvent simulations by construction do not capture interaction mechanisms that are related to the molecular character of the solvent. In the following, we therefore leave solvent-implicit simulations aside and focus on atomistic solvent-explicit simulations, in which for each membrane separation D_w the water chemical potential μ depends on the number N_w of water molecules for given area A between the membrane surfaces.

3.1 Finite-Size Membrane Model Surfaces and Perforated Surfaces

An intuitive way of controlling μ while varying D_w is to let small membrane patches or models thereof interact in a sufficiently large volume filled with explicit water molecules (see Fig. 1A). Since the water molecules between the finite-size membrane objects can exchange with the surrounding bulk water, the chemical potential equilibrates to a constant value throughout the simulation volume. For large-enough volumes, this value is essentially identical to the reference value μ_0 for pure bulk water at ambient conditions. In two studies by Eun et al., graphene plates decorated with atomistically resolved phosphatidylcholine (PC) headgroups were used to mimic membrane surfaces and brought to controlled interaction distances.[20,21] The interaction free energy was globally repulsive and at intermediate distances exhibited an approximately exponential decay with a characteristic length scale $\lambda \approx 0.3$ *nm*, in agreement with experiments. However, in the same distance regime, the decay was modulated with oscillations due to discrete water layering. Such modulations are seen in experiments on interacting rigid surfaces[7,22] but absent in the pressure–distance curves of intrinsically soft lipid bilayers.[12] In other words, the high rigidity of the graphene scaffolds poses limitations to a realistic representation of lipid bilayers. Another limitation for a quantitative comparison with experimental data is that edge effects are inevitable when the lateral extension of the objects is comparable with the distance (see Fig. 1A). More realistic interaction conditions are generated when the simulation model comprises the entire lipid bilayers with periodic boundary conditions. One way of letting such a membrane interact with another surface while allowing for water exchange with a bulk reservoir is the use of surface

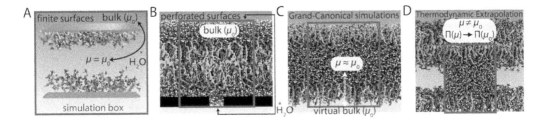

FIGURE 1 Approaches for the treatment of the chemical potential of water in atomistic simulations of interacting lipid membranes. (A) Finite objects interacting in a reservoir of bulk water. (B) Water exchange through perforated surfaces. (C) Grand-Canonical simulations: water exchange with a virtual bulk reservoir. (D) Thermodynamic Extrapolation: extrapolation of the pressure at a given water chemical potential to the pressure at the desired chemical potential. (See color insert for the color version of this figure)

perforations (see Fig. 1B). Using this method, Roark et al. investigated the adsorption of a lipid bilayer to a solid silicon oxide layer with a single pore[23] and obtained the equilibrium surface separation of a few bound water layers, in good agreement with the results of neutron reflectometry experiments. While this method is conceptually suited also to study two interacting bilayers, its practical applicability has still not been thoroughly investigated up to now. To date, it is not clear how fast the system relaxes into an equilibrium state due to slow exchange of water molecules through the pore.[24] Another drawback is that a pore inevitably creates artifacts. Finite-size edge effects and pore artifacts can be excluded with confidence only by using homogeneous, quasi-infinite interacting surfaces. This configuration realistically mimics an infinite stack of bilayers, which corresponds to the most common experimental setup for the study of membrane interactions. The cost of this approach is that the water molecules confined between the membrane surfaces are no longer in contact with a bulk reservoir and, as a consequence, the chemical potential μ in the simulation has to be taken into consideration explicitly. In the following, we discuss two methods of doing this, Grand-Canonical Monte–Carlo (GCMC) simulations and a related method, called Thermodynamic Extrapolation (TE).

3.2 Grand-Canonical Monte–Carlo Simulations

In GCMC simulations,[25] the chemical potential of water in the simulated system is enforced to equal the reference value, μ_0, through frequent deletion and insertion of individual water molecules. This approach is schematically illustrated in Fig. 1C and was employed by Pertsin and co-workers for the study of phospholipid membrane interactions.[26,27] The obtained repulsion was monotonic within the error, indicating that the lipid monolayers used in these studies represent the inherent flexibility of lipid bilayers. An exponential fit to the repulsive pressure yielded a decay constant of $\lambda \approx 0.17\,nm$, which is consistent with experimental observations. GCMC simulations constitute a straightforward approach to quantifying interaction pressures between biological surfaces without edge effects. The bottleneck, however, is the pressure resolution as determined by the precision with which the chemical potential of water can be controlled. When studying hydrophilic surfaces interacting across water, the insertion efficiency for water molecules has so far been too low to reach a pressure resolution below 100 bars, meaning that only conditions of strong surface interactions can be probed.

3.3 Thermodynamic Extrapolation

Over the last years, a novel approach has therefore been established, the thermodynamic extrapolation (TE) method,[28] which is schematically illustrated in Fig. 1D. The basic idea of TE is to deduce the interaction pressure at the reference chemical potential μ_0 through a pressure extrapolation from a simulation at an arbitrary yet precisely known chemical potential μ. For this purpose, simulations are initially performed at constant pressure p, temperature T, and a fixed number of water molecules N_w. In these simulations, the water chemical potential $\mu(N_w, p, T)$ in general deviates from the chemical potential of bulk water, $\mu_0(p, T)$. However, the shift in the chemical potential can be expressed in terms of a pressure difference Δp through the following thermodynamic expansion:

$$\mu(N_w, p + \Delta p, T) = \mu(N_w, p, T) + \left(\frac{\partial \mu}{\partial p}\right)_{N_w, T} \Delta p. \tag{2}$$

Thus, in a first-order approximation, the chemical potential varies linearly with the pressure in the simulation. The proportionality coefficient $(\partial \mu / \partial p)_{N_w, T}$ can be obtained directly in the simulations by determining the chemical potential at several pressures, which is however computationally costly. Alternatively, it can be reformulated with the help of a Maxwell relation,

$$\left(\frac{\partial \mu}{\partial p}\right)_{N_{\mathrm{w}},T} = \left[\frac{\partial}{\partial p}\left(\frac{\partial G}{\partial N_{\mathrm{w}}}\right)_{p,T}\right]_{N_{\mathrm{w}},T} = \left[\frac{\partial}{\partial N_{\mathrm{w}}}\left(\frac{\partial G}{\partial p}\right)_{N_{\mathrm{w}},T}\right]_{p,T} = \left(\frac{\partial V}{\partial N_{\mathrm{w}}}\right)_{p,T} \equiv v_{\mathrm{w}}. \tag{3}$$

The proportionality coefficient equals the partial molecular volume v_{w} of water in the system at constant pressure p and temperature T. Working out v_{w} is much simpler than measuring the chemical potential. It merely requires performing simulations with different numbers of molecules at the same pressure p and calculating the derivative of the volume with respect to N_{w}. Moreover, for interacting phospholipid bilayers, v_{w} is essentially identical to the known partial molecular volume of bulk water, $v_{\mathrm{w}}^0 = 0.030 \text{ nm}^3$, even down to very small separations. With that, the interaction pressure Π at the reference chemical potential μ_0 can be accurately predicted by extrapolating the interaction pressure p in a simulation at an arbitrary chemical potential μ:

$$\Pi = p + \frac{\mu_0 - \mu}{v_{\mathrm{w}}}. \tag{4}$$

Within this procedure, the main task in determining $\Pi(D_w)$ in a computer simulation is to evaluate the separation-dependent chemical potential $\mu(D_w)$ of water between the membrane surfaces. The chemical potential is a measure for the change in free energy upon the addition of a water molecule to the system. It consists of two contributions,

$$\mu = k_{\mathrm{B}} T \ln \rho(\mathbf{r}) + \mu_{\mathrm{ex}}(\mathbf{r}). \tag{5}$$

The first term is the ideal part, which depends on the water density $\rho(\mathbf{r})$ at position $\mathbf{r}$. The second term is the excess chemical potential, which reflects molecular interactions. Both components generally depend on the position, $\mathbf{r}$, but in thermodynamic equilibrium, the total chemical potential μ is uniform and thus position-independent. The excess part of the chemical potential can be determined by an insertion of a water molecule by the Thermodynamic Integration (TI) method,[25] which yields the associated excess free energy. In this procedure, the position of the inserted water molecule has to be spatially constrained to the position r. A desired accuracy of the interaction pressure is $\delta\Pi \simeq \pm 10$ bar, which allows investigating interacting lipid membranes in a wide separation range, covering scenarios of weak and strong interactions corresponding to high and low hydration levels, respectively. In order to achieve such a precision, the chemical potential of water, μ, has to be determined with the precision of $\delta\mu = v_{\mathrm{w}}\delta\Pi \simeq \pm 0.02 \text{ kJ/mol}$ (cf. Eq. (4)), which is readily affordable with the present computer technology. For further technical details related to determining the chemical potential of water or other solvents, we invite the reader to Ref. [29].

---

4 Insights into Short-Range Interactions between Lipid Membranes

4.1 Fluid Lipid Membranes

As pointed out in the introductory chapters of this book, an example for biologically representative and very intensively studied lipid systems are PC lipid bilayers in the fluid L_α phase (see Section 4.2 below for further details). A typical simulation setup for lipid bilayers is shown in Fig. 2A. Here, dilauroyl-phosphatidylcholine (DLPC) lipids (with 12-carbon-atom long chains) in the L_α phase are hydrated with water molecules (not shown) at an isotropic simulation pressure of 1 bar. The membranes are replicated in all three directions via periodic

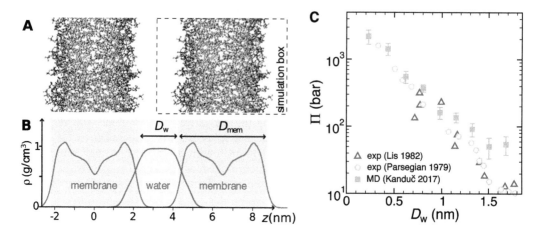

FIGURE 2 (A) Simulation snapshot of interacting phospholipid membranes. (B) Membrane and water mass density profiles along the surface normal. (C) The interaction pressure Π from experiments[12,19] (open symbols) and simulations[30] (full square symbols).

boundary conditions, which mimics a stack of interacting bilayers. The water density profile in Fig. 2B shows a continuous decay away from the membrane surfaces over a range of around 1 nm. This means that the membrane–water interface is not well defined and consequently there is no unique way of deducting the surface–surface separation from the density profiles. In accordance with experimental practice, we define the separation as the width of an equivalent water slab with bulk density, i.e.,

$$D_{w} = N_{w} v_{w}^{0}/A, \tag{6}$$

where N_{w} is the number of water molecules between the membranes, A the membrane surface area, and $v_{w}^{0} = 0.030$ nm^3 is the volume per water molecule in bulk. Note, however, that other definitions, like the distance between the center of mass positions of the headgroups' phosphorous distributions, are equally valid when used consistently.

Experimental pressure–distance curves $\Pi(D_{w})$ obtained in osmotic stress experiments with fluid PC lipid membrane multilayers at atmospheric pressure[12,19] are shown in Fig. 2C by empty symbols. The interaction pressure is positive, i.e., repulsive, meaning that work has to be performed to reduce D_{w}. The pressure reaches magnitudes of several thousand bars for the lowest D_{w} studied and decays approximately exponentially with the separation, with a decay length of $\lambda \approx 0.3$ nm. The figure also shows $\Pi(D_{w})$ as obtained from simulations of DLPC lipids using TE. DLPC lipids were chosen in the simulations, because they remain in the fluid L_{α} phase even at strong dehydration.[12] It is seen that the simulation data are in excellent quantitative agreement with the experiments. This remarkable agreement validates the methodology and demonstrates that the mechanisms through which the hydrated adjacent membranes interact are well captured by the computer model.

As we mentioned in the introduction, pressure–distance curves can be measured with different experimental techniques, in which the bilayers are exposed to different boundary conditions. This raises the question to what extent these boundary conditions influence the bilayer structure and, in turn, the measured pressures. MD simulations have been used to test three sets of boundary conditions, corresponding to the three most important experimental techniques: (i) the osmotic stress method, (ii) the hydrostatic method, and (iii) SFA measurements. In osmotic stress experiments, membranes are subject to an atmospheric pressure. The equivalent simulation ensemble is $N_w p T$ with the fixed number of water molecules N_w, an isotropic pressure of $p = 1$ bar at temperature T. The measured water chemical potential μ in this case depends on D_w. The interacting pressure Π is then evaluated via Eq. (4). We will refer to this set of simulations as 'osmotic stress' simulations in the following. In the second set of MD simulations, the actual pressure is set to the value obtained from the previous set via TE at the value of D_w (strictly speaking, D_w can change slightly because of a minor change in A (see Eq. (6)), but this effect is usually negligible). Consequently, the chemical potential by construction corresponds to the bulk-water chemical potential μ_0 with very good accuracy, as one can verify explicitly. This second set of simulations therefore mimics the experimental technique where a membrane stack is subject to a hydrostatic pressure exerted by a hydraulically driven piston and where water can be exchanged with an external bulk reservoir. We refer to these simulations as use single quotations 'hydrostatic simulations'[19] In both simulation sets, the pressure is isotropic, which is enforced via independent box scaling in lateral and perpendicular directions, respectively. Finally, in order to mimic SFA experiments, the area of the simulation box A is fixed and the pressure p_z is only controlled in the perpendicular direction through one-dimensional volume adjustments, corresponding to the $N_w A p_z T$ ensemble. This ensemble resembles the SFA experiments if one assumes that lipids are strongly grafted to the substrate, such that the area per lipid does not change upon changing the surface separations. We refer to this set of simulations as 'SFA simulations.' In order to assure constant chemical potential μ, simulations need to be performed in two steps, first at ambient pressure of 1 bar and in the second step at the pressure obtained via TE (Eq. (4)).

We first focus on the structural differences of all three simulation ensembles upon dehydration: the variations in the area per lipid, membrane thickness, and membrane volume, which are shown in Fig. 3. As seen in panel A, for the osmotic and hydrostatic cases, the area per lipid A_{lip} shrinks upon approaching two membranes closer together, except for the SFA simulations, where A_{lip} is fixed. Not surprisingly, the area shrinks slightly more in the hydrostatic scenario than in the osmotic one, since the increased pressure compresses the membrane in all directions, but this difference is minor, because of the low volume compressibility. The data from osmotic-stress experiments[12] in Fig. 3A agree quite well with the osmotic simulation predictions. Likewise, the membrane thickness D_{mem} in the simulations can simply be deduced from the bilayer repeat distance D and the water-slab thickness D_w,

$$D_{mem} = D - D_w \tag{7}$$

and the volume per lipid then follows as
$$V_{lip} = \frac{1}{2} A_{lip} D_{mem}. \tag{8}$$

The factor $1/2$ reflects that the membrane thickness D_{mem} accounts for two lipid layers. As seen in Fig. 3B, upon dehydration, the thickness change is most pronounced in the osmotic stress MD, whereas in the hydrostatic simulations the bilayer thickness increase is lower due to high applied pressures. On the contrary, in the SFA simulations, where A_{lip} is fixed, the thickness changes only upon compression at higher applied pressures. The volume per lipid in Fig. 3C, on the other hand, basically only depends on the exerted pressure and is therefore almost unaffected in the osmotic stress scenario, as also observed experimentally (black circles).

Now we can examine to what extent the structural parameters, which significantly depend on the boundary conditions, influence the interaction pressure between the membranes. Figure 3D shows the difference between interaction pressures Π in different ensembles. The difference between the hydrostatic and osmotic simulations is not significant within the error. This indicates that the membrane thickness, which differs significantly between osmotic stress and hydrostatic simulations, has only little influence on the interaction pressure as long as the area per lipid deviates only slightly. We can therefore consider the osmotic stress and hydrostatic methods as equivalent as far as pressure–distance curves are concerned, which constitutes a non-trivial result. However, the situation is

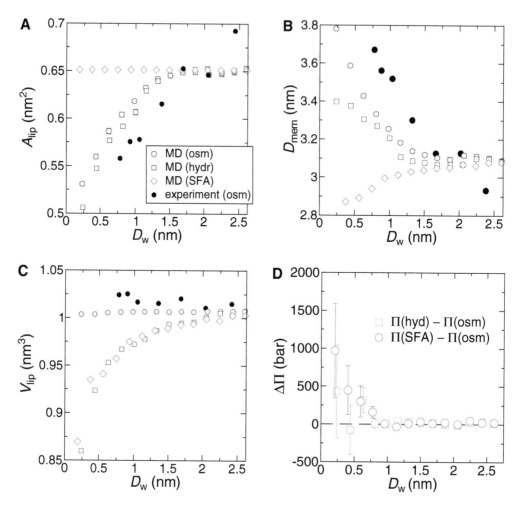

FIGURE 3 Structural properties of DLPC bilayers from three different simulation ensembles (osmotic stress simulations, hydrostatic simulations, and SFA simulations) compared with data from osmotic stress experiments[12] as a function of membrane–membrane separation: (A) area per lipid, (B) bilayer thickness, and (C) volume per lipid. (D) Differences in evaluated interaction pressures between the bilayers in hydrostatic and osmotic stress simulations (orange squares) and between SFA and osmotic stress simulations (blue circles). The data are adopted from Ref. [18]. (See color insert for the color version of this figure)

different for the SFA scenario. Below a membrane separation of about 1 nm, SFA simulations exhibit significantly stronger repulsion (up to 20%) than the osmotic stress simulations. We can attribute this to the substantial deviations in the area per lipid below about $D_w = 1$ nm; see Fig. 3A. That is, a larger area per lipid leads to a slightly larger interaction pressure, which implies that the total force between the membranes cannot be considered as the sum of independent headgroup contributions. Due to the finite size of the simulation box, which in lateral directions typically measures 4–5 nm, atomistic MD simulations do not account for membrane undulations with wavelengths larger than approximately 4 nm, and thus the repulsive undulation force is reduced in the simulation setup. However, the undulation force turns out to be negligible compared with the hydration repulsion for separations below 1.5 nm,[31] which allows the simulations to realistically model the bilayer forces at this low hydration.

We now turn to the analysis of the physical mechanisms responsible for the membrane repulsion at small separation. Fig. 4A shows profiles perpendicular to the membrane plane (i.e., along the z axis) of the binding enthalpy per water molecule and reveals that water is strongly bound to the bilayers. Δz

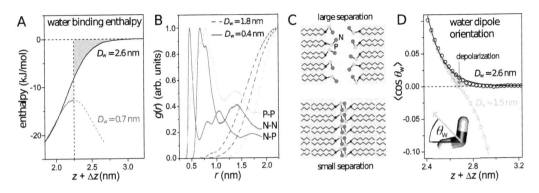

FIGURE 4 Repulsion mechanisms between fluid PC lipid membranes. (A) Binding enthalpy profile per water molecule along the axis perpendicular to the membrane plane. Blue area: enthalpy reduction due to strengthening of water binding upon surface approach. Orange area: enthalpy increase due to removal of strongly bound interfacial water. (B) Radial distribution functions $g(r)$ between phosphorus (P) and nitrogen (N) atoms in the opposing membrane surfaces evidencing a reduction of lipid configurational entropy upon membrane approach. (C) Schematic illustration of the headgroup reorganization upon dehydration. (D) Water orientation profiles $\langle \cos \theta_w \rangle$ at a membrane surface exhibiting depolarization (gray area) with decreasing D_w. (See color insert for the color version of this figure)

is a small shift reflecting variations in the bilayer thickness.[28] At small membrane separations, exemplified by the curve for $D_w = 0.7$ nm, water right in the center of the water layer (denoted by a vertical broken line) is more strongly bound than at large membrane separations ($D_w = 2.6$ nm) at the same distance from the bilayer (this difference is highlighted by the blue area). However, the enthalpy reduction due to this strengthening of water binding is overcompensated by the enthalpy increase due to removal of strongly bound interfacial water (highlighted by the orange area). The net effect contributes to the repulsion mostly at very small membrane separations ($\lesssim 1$ nm).[28]

The headgroup radial distribution functions (RDFs) in Fig. 4B further indicate the contribution of an additional repulsion mechanism related to the lipid configurational entropy.[17] For large membrane separations, the RDFs between the phosphorus (P) and nitrogen (N) atoms of the two opposing membrane surfaces are essentially unstructured and reflect the unperturbed headgroup configurations with N being displaced towards the water with respect to P. For small membrane separations, the N–P distribution is peaked at a distance significantly smaller than the N–N and P–P distributions between like-charged groups. This reorganization, which minimizes the electrostatic energy, is schematically illustrated in Fig. 4C and is inevitably accompanied by an unfavorable loss of configurational entropy, which is evidenced by the pronounced sharp RDF peaks for small membrane separations.[28]

Fig. 4D provides qualitative insights into the role of the structure of interfacial water layers for the appearance of hydration repulsion. Water dipoles are strongly oriented close to the membrane surfaces, whereas the polarization by symmetry vanishes in the center of the water layer. Upon approach of the membrane surfaces, the polarization profiles from the two opposing surfaces interfere destructively, resulting in pronounced depolarization (see gray area in Fig. 4D). The solid lines are predictions of the orientation profiles in the framework of a Landau–Ginzburg theory for an orientational order parameter[16] (see Section 4.4 further below for an extended discussion of this aspect). The good agreement with the data suggests that water polarization effects are indeed operative for interacting lipid membranes.[28]

In summary, the simulations reveal that what is commonly termed *hydration repulsion* is not a single distinct mechanism but rather a combination of several mechanisms whose relative weight depends on the bilayer separation. In the following, in order to disentangle the mechanisms, we compare short-range interactions among several membrane types featuring distinct differences in terms of lipid ordering and chemistry.

4.2 Influence of the Membrane Phase

An important class of phospholipid membranes (composed of lipids with long, saturated alkyl chains) displays a thermotropic phase transition from an ordered gel-like state known as L_β phase at low temperature to a disordered fluid state known as L_α phase at high temperature. This transition named *melting transition* attracted a lot of research attention due to its physiological relevance.[32,33] In this context, it is important to realize that the membrane composition of organisms is typically maintained close to this main transition.[34] In the L_β phase, the alkyl chains assume the so-called all-trans and thus essentially linear conformations (Fig. 5A). In contrast, the key structural characteristic of the L_α phase is a considerable density of gauche bonds, allowing the alkyl chains to assume more random configurations (Fig. 5B). Note that the chemical composition does not change upon the transition, only the conformations of the lipid molecules in the bilayer. Comparing the hydration repulsion in the gel and fluid states should allow one to determine whether direct surface interactions or water ordering, the latter presumably being similar in the gel and fluid states, is the dominating contributor to the repulsion. For quite a long time, however, experiments on the hydration repulsion between gel membranes did not provide coherent results, even among the same types of phospholipids.[35,36,37] It was early on suggested that the discrepancies came from different definitions of the membrane thickness D_{mem} and membrane–membrane separation D_w used in the analysis of the experimental data, which has been settled only recently.[31] Figure 5C compares interaction pressures (as reanalyzed in Ref. [31]) between fluid- and gel-phase PC lipid membranes as a function of the repeat distance $D = D_{mem} + D_w$, which is a model-free experimental observable. As seen immediately, the pressure–distance curves for the two phases are consistent among all experiments and roughly follow exponential decays, albeit with significantly different decay lengths. In the fluid phase, the fitted decay length turns out to be around $\lambda_{fluid} = 0.20$ nm, whereas in the gel phase it is much smaller, $\lambda_{gel} = 0.10$ nm. For pressures above 500 bar, the experimental data deviate from single exponential, as was noted before.[38]

In order to elucidate this behavior, MD simulations were carried out with dipalmitoyl-phosphatidylcholine (DPPC) membranes, which experimentally exhibit the gel-to-fluid melting transition at $T_m = 314$ K.[31] DPPC lipids differ from DLPC only in the chain length. The chains in DPPC are longer and comprise 16 carbon atoms and therefore exhibit a higher transition temperature. Gel phase membranes were created with the help of an 'assisted freezing' method[39] and it was recently confirmed that the melting temperature in the simulations is in good agreement with the experimental value.[40] Snapshots from simulations of fluid and gel phase membranes are presented in Fig. 5A and B. Figure 6 compares pressure–distance curves $\Pi(D_w)$ obtained by TE in the simulations (triangles) with experimental data in (A) gel and (B) fluid states. The difference in the repulsion between gel and fluid membranes remains significant also when the pressures are plotted versus D_w as long as the definition $D_w = N_w v_w^0/A$ is used consistently. Exponential fits to

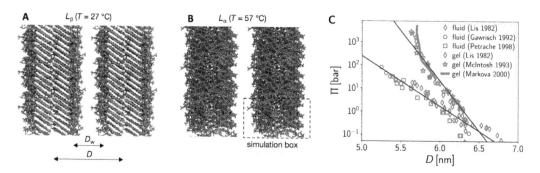

FIGURE 5 (A) Simulation snapshots of interacting DPPC bilayers in (A) the gel L_β phase and in (B) the fluid L_α phase. (C) Experimentally measured pressures as a function of the lamellar repeat distance D for fluid[12,36,37] and gel[12,35,38] membranes. Black lines are exponential fits with decay lengths $\lambda_{fluid} = 0.20$ nm and $\lambda_{gel} = 0.10$ nm.[31]

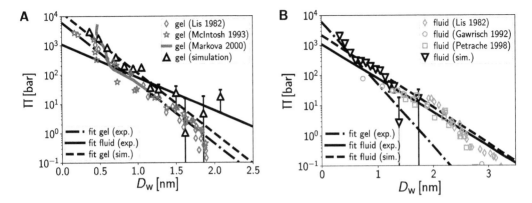

FIGURE 6 Comparison of hydration pressures from simulations (triangles) and experiments as a function of the water slab thickness D_w in the (A) gel and (B) fluid states. Exponential fits to the experimental data give decay lengths $\lambda_{gel} = 0.21$ nm (black dashed line) for the fit range [0, 1.8 nm] and $\lambda_{fluid} = 0.38$ nm (black solid line) for the fit range [0, 2.6 nm]. Fits to the simulation data yield decay lengths $\lambda_{gel} = 0.22$ nm (blue dashed line) and $\lambda_{fluid} = 0.36$ nm (red solid line) for fit ranges [0, 1.3 nm] and [0, 1.4 nm], respectively, which are restricted to the distance range where pressures are strictly positive. (See color insert for the color version of this figure)

the gel-phase pressure–distance data yield decay lengths of 0.21 nm for the experimental data and 0.22 nm for the simulation data. Good agreement is observed also for the membranes in the fluid state with obtained decay lengths of 0.38 nm and 0.36 nm for the experimental and simulation results, respectively.[31] This agreement demonstrates that the simulations capture not only the overall interaction pressure correctly but also the difference in the interaction between membranes in different phases. The observation of two different decay lengths clearly rules out a purely water-mediated mechanism for the hydration repulsion, because in that case not the decay length but the pressure amplitude would differ between the gel and fluid states. The configurational freedom of the lipids is greater in the fluid phase than in the gel phase. At the same time, the repulsion is of longer range in the fluid phase. This suggests that the lipid configurational freedom indeed plays a significant role in the interaction.

4.3 Influence of the Lipid Chemistry

In nature, different types of membranes exhibit largely different lipid compositions.[33] Membranes that undergo frequent reorganization processes like vesicle budding, release, and fusion are rich in PC lipids, while structurally more steady and densely packed multilamellar membrane systems, such as myelin sheaths in vertebrates[41] and the photosynthetic membranes,[42] exhibit high glycolipid contents. The main difference between these two lipid classes is their headgroup chemistry, which for PC lipids is dominated by a single large electric dipole (Fig. 7A), while for glycolipids it consists of multiple small electric dipoles in the form of polar hydroxyl (OH) groups (Fig. 7B). The correlation between high glycolipid contents and the formation of stacked membrane systems suggest an important role of these fundamentally different headgroup architectures. In fact, glycolipids in contrast to PC lipids tend to spontaneously form multilayered aggregates.[43,44]

Fig. 7C shows the pressure–distance curve of glycolipid membranes composed of di-galactosyldiacylglycerol (DGDG) as obtained in experiments[45] together with the corresponding measurements for PC lipids (see also Fig. 2C). While the interaction pressures are repulsive for both DGDG and PC lipids, the striking difference is a much steeper decay in the case of DGDG. An exponential fit to the experimental pressure–distance data yields decay lengths of $\lambda = 0.12$ nm for the DGDG and $\lambda = 0.27$ nm for the PC lipids. The pressure–distance curves of DGDG and PC lipids obtained in the simulations (filled symbols in Fig. 7C) are once again in remarkable quantitative

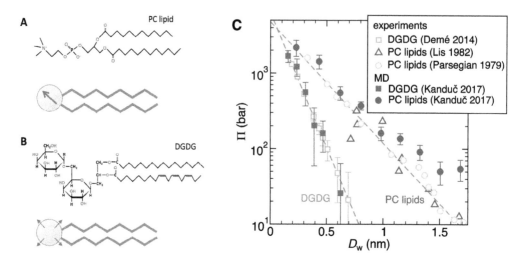

FIGURE 7 Chemical structures of (A) a PC lipid and of (B) the glycolipid DGDG as representatives of two fundamentally different lipid classes found in nature: Lipids with a headgroup chemistry dominated by one large electric dipole and lipids whose headgroups comprise multiple small electric dipoles in the form of OH groups. Both classes are schematically illustrated below the chemical structures. Dipoles are indicated by arrows. (C) Pressure–distance curves of DGDG[45] and PC lipid membranes[19,12] as obtained in experiments (open symbols) and in MD simulations[30] (filled symbols). Straight dashed lines are exponential fits to the experimental data points.

agreement with the experimental data (open symbols) and fully reproduce the difference in the decay lengths [30].

The OH-groups of the saccharide headgroups, like water molecules, can act as hydrogen-bond (HB) donors and acceptors, suggesting that HB formation is an important contributor to the free energy associated with the interaction between DGDG membranes. In classical molecular simulations, HBs are considered as electrostatic interactions that involve two atoms with negative partial charges and a hydrogen atom that is covalently bound to one of them and has a positive partial charge. A widely used criterion was proposed by Luzar and Chandler[46] and is based on a geometric criterion according to which a HB exists if the distance between donor and acceptor atoms is smaller than 0.35 nm and the hydrogen–donor–acceptor angle is smaller than 30°. Using this criterion, the total number of HBs in the system, N^{HB}, can be evaluated. From the number of water molecules N_w between the membranes, we can calculate the equivalent number of HBs in the bulk as $N_w n_{bw}^{HB}$, where $n_{bw}^{HB} = 1.796$ is the number of HBs per water molecule in bulk water. This finally leads to the total number of the excess HBs per headgroup,

$$n_{tot}^{HB} = \frac{N^{HB} - N_w n_{bw}^{HB}}{N_{lip}}, \tag{9}$$

where N_{lip} is the number of lipids in the system. It is insightful to compare this change in the number of HBs with the change of the interaction free energy per lipid G/N_{lip}. The latter can simply be evaluated by integrating the pressure–distance curve, $G = -\int A(D_w)\Pi(D_w)dD_w$. Fig. 8A shows the change in the overall HB number, Δn_{tot}^{HB}, versus the interaction free energy and reveals a linear relation between both quantities. In other words, the repulsion between DGDG membranes can be described entirely in terms of the HB balance.[30] For PC lipids, the situation is very different: as seen from Fig. 8B, there is no simple proportionality between G/N_{lip} and Δn_{tot}^{HB} for PC lipids. Notably, in the limit of large separations (light blue shade in Fig. 8B), G varies, while the HB number has already saturated ($\Delta n_{tot}^{HB} \approx 0$). This behavior reflects the non-HB-related repulsion

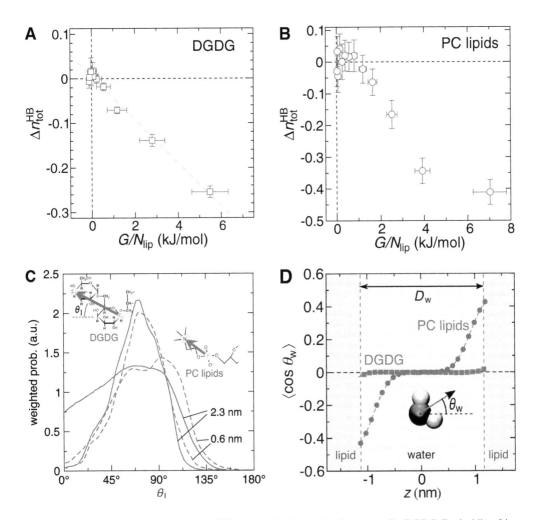

FIGURE 8 (A) Change of the total number of HBs versus the interaction free energy for DGDG. Dashed line: Linear regression through all data points. (B) The same plot for PC lipids. Shaded region: high hydration regime where $\Delta n_{tot}^{HB} \approx 0$. (C) Distributions of DGDG and PC lipid headgroup orientations with respect to the membrane normal for large ($D_w = 2.3$ nm, solid lines) and small ($D_w = 0.6$ nm, dashed lines) separations. Inset: Definition of the headgroup vectors. (D) Water orientation profiles $\langle \cos \theta_w \rangle$ between PC and DGDG lipid membranes at separation $D_w = 2.3$ nm. The blue-shaded area denotes the water layer and the orange area the poorly hydrated lipid region. Inset: Definition of the water dipole angle θ_w. (See color insert for the color version of this figure)

mechanisms between PC membranes discussed in Section 4.1, which also induce longer decay length in the pressure–distance curves.

Regarding the lipid configurational entropy, DGDG and PC lipid membranes exhibit clearly different responses to dehydration, i.e., to mutual approach. Fig. 8C shows distributions of the headgroup angle with respect to the membrane normal, θ_l. The distributions are weighted with a factor $1/\sin(\theta_l)$, so that a constant distribution would correspond to a random orientation. While the angular distribution significantly narrows with decreasing D_w for PC lipid membranes, the corresponding distribution remains essentially unaffected for DGDG membranes. This comparison indicates that entropic repulsion due to a lipid configurational entropy is of minor relevance for the glycolipid membranes. A clear picture also emerges regarding the role of interfacial water orientation. In Fig. 8D, it is seen that DGDG and PC lipid membranes lead to very different profiles of water orientation $\langle \cos \theta_w \rangle$ perpendicular to the membrane surfaces. As discussed already in Section 4.1, for PC lipid membranes, the water dipoles close to the membrane surfaces (indicated by vertical dashed lines) are strongly oriented and significant orientation extends virtually all the way to the center of the water layer. For DGDG bilayers, virtually no water ordering is seen. This dissimilar behavior can be understood from the different electric dipole structure of the headgroups. The large and directionally correlated single electric dipoles of PC lipids induce strong water orientation, while the rather isotropically oriented OH groups of the glycolipids do not. Significant repulsion due to water structuring can thus not be expected for DGDG membranes. These results demonstrate once more that the characteristics of the repulsion are not controlled by the properties of water alone, but to a great extent by the properties of the surfaces. In the present case, it is the surface chemistry that affects the decay length of the repulsive pressure.[30]

We now return to the initial question of why glycolipid membranes tend to form stacks whereas PC lipid membranes typically do not. To this end, we recall that the equilibrium separation between membranes is governed by the balance between repulsive hydration forces and the ubiquitous van der Waals (vdW) attraction between lipid membranes. Since the hydration repulsion is of shorter range for glycolipid membranes, the equilibrium distance is smaller and the free energy minimum deeper than for PC lipids, so that the adhesion of two glycolipid bilayers becomes stronger.

4.4 Repulsion Due to Water Orientation

A first attempt to rationalize the hydration repulsion in terms of an unfavorable overlap of ordering profiles of interfacial water has been proposed by Marčelja and Radić in the 1970s.[16] The basic concept was further refined[47] and frequently discussed in terms of the water molecules' dipole orientation and non-local dielectric response.[48–50] Importantly, the Marčelja–Radić model only considers the indirect (solvent-mediated) contribution to the interaction, which experimentally is not accessible due to the presence of direct membrane–membrane interactions, but which can be obtained in simulations with a suitable definition. Note that the decomposition of the total membrane–membrane interaction pressure into direct and water-mediated indirect contributions is not unique, but nevertheless useful. Figure 9A shows the indirect interaction pressure, Π_{ind} (symbols), between DPPC lipid membranes in gel and fluid phases (see Section 4.2), obtained by switching the direct membrane–membrane interactions off.[31] In both phases, Π_{ind} is repulsive and decays approximately exponentially with separation D_s. The magnitude is slightly larger in the gel phase compared to the fluid phase.

As mentioned before, the definition of the membrane separation is not unique. The Marčelja–Radić model considers the order parameter of water at an interface, and we therefore choose $D_s = D_{\text{app}} - D_0$, with D_{app} the average distance between the phosphorus atoms and $D_0 = 0.47$ nm the equilibrium separation without water. In a Landau–Ginzburg approach, the free energy of the water slab is expanded in terms of a scalar order parameter $m_\perp$,[51] which corresponds to the local dipolar ordering and is oriented at each surface due to surface fields h. Following previous approaches, we use the water polarization as anti-symmetric order parameter for which the profile follows as[47]

$$m_\perp(z) = m_{\perp 0} \frac{\sinh(z/\lambda)}{\sinh(D_s/2\lambda)}, \qquad (10)$$

where λ is a correlation length. Polarization profiles and fits of Eq. (10) to the data points in the fluid phase are shown in Fig. 9B, revealing remarkable agreement for a common value of $\lambda_{\text{ind}} = 0.26$ nm. The polarization surface value is given by $m_{\perp 0} = m_\perp^\infty \tanh(D_s/2\lambda)$. The polarization strength at large separation $m_\perp^\infty$ is proportional to the surface field h. Fig. 9C compares the surface value $m_{\perp 0}$ for gel and fluid phases. While the correlation length λ is found to be

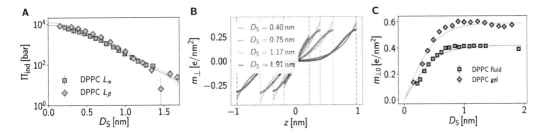

FIGURE 9 Indirect pressures and water orientation profiles. (A) Indirect, water-mediated, interaction pressure between DPPC membranes. Lines denote fits of the Landau–Ginzburg model, Eq. (11). (B) Dipole polarization $m_\perp$. Solid lines denote fits of Eq. (10) to the simulation data, vertical dashed lines denote the interface position. (C) Polarization order parameter at the interface $m_{\perp 0}$ obtained from the fits in (B). Solid lines denote fits of the value at large separation, $m_\perp^\infty$, to the simulation data. (See color insert for the color version of this figure)

the same in both cases, the value of the surface polarization (and thus of the surface fields h) differs significantly. The larger value of h for the gel phase can be attributed to the fact that in the gel phase, the zwitterionic headgroups are packed more densely and aligned more in parallel when compared with the fluid phase, thus inducing a higher degree of water orientation. Taking the derivative of the Landau–Ginzburg free energy with respect to D_s, the indirect pressure follows as[52]

$$\Pi_{\text{ind}} \propto \frac{h^2}{4\lambda \cosh^2(D_s/2\lambda_{\text{ind}})} \approx \Pi_{\text{ind}}^\infty e^{-D_s/\lambda_{\text{ind}}}, \tag{11}$$

where the last result is valid for large distances $D_s \gg \lambda_{\text{ind}} = 0.21$ nm and shows that the pressure is repulsive and decays exponentially. The solid lines in Fig. 9A indicate fits of Eq. (11) to the simulation data points. Indeed, the indirect pressures reveal a common decay length and that the indirect repulsion for the gel phase membranes is higher, which is in line with the stronger surface field. The remarkable agreement between the theoretical expressions and the simulation data in Fig. 9 suggests that the strong indirect repulsion reflects the unfavorable overlap of opposing water orientation profiles. Yet, the shortest observed decay length $\lambda_{\text{ind}} < \lambda$ hints to the presence of additional interactions. Moreover, the interpretation of the total interaction is complex, since the Marčelja–Radić model only takes the water orientation into account. The observed stronger total repulsion between membranes in the gel phase than in the fluid phase is in qualitative agreement with the theoretical prediction. However, the total interaction between membranes consists of many contributions that vary considerably for different systems.

4.5 Influence of Cosolutes

Cosolutes can have a significant influence on the interaction between lipid membranes. The best-known example are ions, which screen electrostatic interactions between charged membranes. But cosolutes also affect interactions between non-charged membranes. The interaction gets altered, for instance, when a cosolute preferentially accumulates at the membrane surface or when it is repelled from the surface. Fig. 10A shows density profiles perpendicular to the membrane plane (i.e., along the z axis) in simulations of dimyristoyl-phosphatidylcholine (DMPC) (with 14-carbon-atom-long chains) membranes interacting across aqueous media containing various types of cosolutes: urea (a denaturant), trimethyl-amine-N-oxide (TMAO, a stabilizing osmolyte), and NaCl (a typical monovalent salt). In the plot, $z = 0$ denotes the interface between the membrane ($z<0$) and the water layer ($z>0$). In agreement with experiments, urea accumulates at the membrane surface, while TMAO is repelled from the surface.[53] The profile of NaCl (in particular

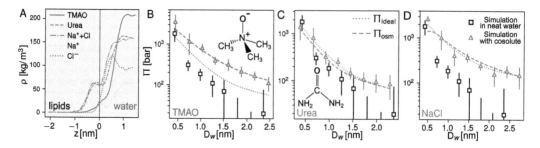

FIGURE 10 Influence of cosolutes on the interaction between DMPC membranes. (A) Density profiles of urea, TMAO, and NaCl perpendicular to the membrane plane in simulations. (B–D) Pressure–distance curves in the absence and presence of 10 wt% cosolute. Dotted lines indicate the pressure Π_{ideal} expected when an ideal-gas contribution is added to the simulation data in the absence of cosolutes. Solid lines indicate the pressure Π_{osm} expected when non-ideal effects (see main text) are taken into account.

the sum of the Na^+ and Cl^- profiles) is similar to that of urea. Irrespective of this behavior, however, the membrane repulsion is enhanced in presence of both urea and TMAO whenever the cosolutes are partitioned between the membranes at a fixed lipid/cosolute stoichiometry.[53] Panels B–D of Fig. 10 show pressure–distance curves obtained in simulations with fixed stoichiometry for urea (B), TMAO (C), and NaCl (D) corresponding to 10 wt% cosolute in the dry mass. It is seen that in all cases, the repulsive pressure is higher than for the membranes interacting across pure water. This additional repulsion has to be attributed to a considerable osmotic pressure afforded by the cosolutes. Strikingly, the increase in the repulsion does not scale inversely proportionally with the cosolute molecular mass. This would be expected for ideal-gas-like cosolute behavior (see Π_{ideal} indicated with dotted lines), because for a lighter molecule, the same mass (10 wt%) represents a higher number of individual molecules contributing to the osmotic pressure. Instead, the non-ideal osmotic coefficient of the osmolyte TMAO together with its decreased available volume between the membrane surfaces, from which it is repelled, gives rise to an over-proportionally high repulsion effect of this molecule. The theoretical pressures Π_{osm} that account for these effects are indicated with solid lines and in good agreement with the simulation data.

5 Summary and Outlook

Interactions between membranes are of great biological relevance and form an important topic in membrane biophysics. Over the last years, molecular dynamics simulations have become the method of choice to elucidate the physical mechanisms of short-range interactions between lipid membranes, which are known to elude standard continuum theoretical descriptions. For a quantitative comparison with experiments, realistic model systems are required, which typically involve quasi-infinite continuous bilayers with periodic boundary conditions. This, in turn, renders the explicit treatment of the chemical potential of water, μ, inevitable. Recent simulation studies utilizing the method of Thermodynamic Extrapolation to account for fixed chemical potentials were able to quantitatively reproduce experimental pressure–distance curves between lipid membranes with various phase states and lipid chemistries. The simulations revealed that what is commonly termed *hydration repulsion* is not a single distinct physical mechanism but the net result of several mechanisms involving water binding enthalpies, lipid configurational freedom, and water structuring. The decay length of the repulsive pressure was found to depend on both the membrane phase and the lipid chemistry, excluding a water-intrinsic mechanism as the sole origin of the repulsion. Instead, the surface properties appear to be an important aspect, if not the dominant one.

Turning back to computer simulations of lipid membranes, the overarching topic of this book, we emphasize that membrane interactions are not only an object of investigation but generally at play in membrane systems. Membrane fusion, stalk formation, and other scenarios commonly investigated in coarse-grained simulations must be considered highly sensitive to this aspect. Only when correctly captured in the simulations can reliable results be obtained. The insights gained from atomistic, solvent-explicit simulations that reproduce experimental data should thus be exploited also to validate and refine coarse-grained simulations.

REFERENCES

1. Mouritsen, O. G., *Life-as a Matter of Fat*, Springer Berlin Heidelberg, 2005.
2. Alberts, B., Bray, D., Lewis, J., Raff, M., Roberts, K., & Watson, J. D., *Molecular Biology of the Cell*, Garland Science, London 4 edition, 2002.
3. Eggens, I., Fenderson, B., Toyokuni, T., Dean, B., Stroud, M., & Hakomori, S., *Specific interaction between Lex and Lex determinants. A possible basis for cell recognition in preimplantation embryos and in embryonal carcinoma cells*, J. Biol. Chem., **264**, 9476–9484, 1989.
4. Lau, P. C. Y., Lindhout, T., Beveridge, T. J., Dutcher, J. R., & Lam, J. S., *Differential lipopolysaccharide core capping leads to quantitative and correlated modifications of mechanical and structural properties in Pseudomonas aeruginosa biofilms*, J. Bacteriol., **191**, no. 21, 6618–6631, 2009.
5. Rosenhahn, A., Schilp, S., Kreuzer, H. J., & Grunze, M., *The role of inert surface chemistry in marine biofouling prevention*, Phys. Chem. Chem. Phys., **12**, no. 17, 4275–4286, 2010.
6. Helfrich, W., *Steric interaction of fluid membranes in multilayer systems*, Z. Naturforsch., **33a**, 305315, 1978.
7. Israelachvili, J. N., *Intermolecular and Surface Forces*, Academic Press Inc., London, 2 edition, 1991.
8. Lipowsky, R., *The conformation of membranes*, Nature, **349**, no. 6309, 475, 1991.
9. Kanduč, M., Schlaich, A., Schneck, E., & Netz, R. R., *Water-mediated interactions between hydrophilic and hydrophobic surfaces*, Langmuir, **32**, no. 35, 8767–8782, 2016.
10. LeNeveu, D. M., Rand, R. P., & Parsegian, V. A., *Measurements of forces between lecithin bilayers*, Nature, **259**, 601603, 1976.
11. Marra, J. & Israelachvili, J. N., *Direct measurements of forces between phosphatidylcholine and phosphatidylethanolamine bilayers in aqueous electrolyte solutions*, Biochemistry, **24**, no. 17, 4608–4618, 1985.
12. Lis, L. J., McAlister, M., Fuller, N., Rand, R. P., & Parsegian, V. A., *Interactions between neutral phospholipid bilayer membranes*, Biophys J., **37**, 657–666, 1982.
13. Leontidis, E., Aroti, A., Belloni, L., Dubois, M., & Zemb, T., *Effects of monovalent anions of the Hofmeister series on DPPC lipid Bilayers part II: Modeling the perpendicular and lateral equation-of-state*, Biophys J., **93**, no. 5, 1591–1607, 2007.
14. Schneck, E., Demé, B., Gege, C., & Tanaka, M., *Membrane adhesion via homophilic saccharide-saccharide interactions investigated by neutron scattering*, Biophys J., **100**, 2151–2159, 2011.
15. Israelachvili, J. N. & Wennerström, H., *Role of hydration and water structure in biological and colloidal interactions*, Nature, **379**, 219, 1996.
16. Marčelja, S. & Radić, N., *Repulsion of interfaces due to boundary water*, Chem. Phys. Lett., **42**, no. 1, 129–130, August 1976.
17. Israelachvili, J. N. & Wennerström, H., *Entropic forces between amphiphilic surfaces in liquids*, J. Phys. Chem., **96**, no. 2, 520–531, 1992.
18. Kanduč, M., Schneck, E., & Netz, R. R., *Hydration interaction between phospholipid membranes: insight into different measurement ensembles from atomistic molecular dynamics simulations*, Langmuir, **29**, no. 29, 9126–9137, 2013.
19. Parsegian, V. A., Fuller, N., & Rand, R. P., *Measured work of deformation and repulsion of lecithin bilayers*, Proc. Natl. Acad. Sci. USA, **76**, no. 6, 2750–2754, 1979.
20. Eun, C. & Berkowitz, M. L., *Origin of the hydration force: water-mediated interaction between two hydrophilic plates*. J. Phys. Chem. B., **113**, 13222, 2009.
21. Eun, C. & Berkowitz, M. L., *Thermodynamic and hydrogen-bonding analyses of the interaction between model lipid bilayers*. J. Phys. Chem. B., **114**, 3013, 2010.

22. Israelachvili, J. N. & Pashley, R. M., *Molecular layering of water at surfaces and origin of repulsive hydration forces*, Nature, **306**, 249, 1983.
23. Roark, M. & Feller, S. E., *Structure and dynamics of a fluid phase bilayer on a solid support as observed by a molecular dynamics computer simulation*, Langmuir, **24**, no. 21, 12469–12473, 2008.
24. Vishnyakov, A., Li, T., & Neimark, A. V., *Adhesion of Phospholipid Bilayers to Hydroxylated Silica: Existence of Nanometer-Thick Water Interlayers*, Langmuir, **33**, no. 45, 13148–13156, 2017.
25. Frenkel, D. & Smit, B., *Understanding Molecular Simulation: From Algorithms to Applications*, Academic Press Inc San Diego., 2002.
26. Pertsin, A., Platonov, D., & Grunze, M., *Direct computer simulation of water-mediated force between supported phospholipid membranes*. J. Chem. Phys., **122**, 244708, 2005.
27. Pertsin, A., Platonov, D., & Grunze, M., *Origin of short-range repulsion between hydrated phospholipid bilayers: a computer simulation study*, Langmuir, **23**, 1388, 2007.
28. Schneck, E., Sedlmeier, F., & Netz, R. R., *Hydration repulsion between biomembranes results from an interplay of dehydration and depolarization*, Proc. Natl. Acad. Sci. USA, **109**, no. 36, 14405–14409, 2012.
29. Schlaich, A., Kowalik, B., Kanduč, M., Schneck, E., & Netz, R. R., *Computational Trends in Solvation and Transport in Liquids*, IAS Series, Simulation Techniques for Solvation-Induced Surface-Interactions at Prescribed Water Chemical Potential, pp. 155–185, Forschungszentrum Juelich, 2015.
30. Kanduč, M., Schlaich, A., De Vries, A. H., Jouhet, J., Maréchal, E., Demé, B., Netz, R. R., & Schneck, E., *Tight cohesion between glycolipid membranes results from balanced water-headgroup interactions*. Nat. Commun., **8**, 14899, 2017.
31. Kowalik, B., Schlaich, A., Kanduč, M., Schneck, E., & Netz, R. R., *Hydration repulsion difference between ordered and disordered membranes due to cancellation of membrane–Membrane and water-mediated interactions*, J. Phys. Chem. Lett., **8**, no. 13, 2869–2874, 2017.
32. Ichimori, H., Hata, T., Yoshioka, T., Matsuki, H., & Kaneshina, S., *Thermotropic and barotropic phase transition on bilayer membranes of phospholipids with varying acyl chain-lengths*, Chem. Phys. Lipids., **89**, no. 2, 97–105, 1997.
33. Van Meer, G., Voelker, D. R., & Feigenson, G. W., *Membrane lipids: where they are and how they behave*, Nat. Rev. Mol. Cell. Bio., **9**, no. 2, 112, 2008.
34. Schneider, M. F., Marsh, D., Jahn, W., Kloesgen, B., & Heimburg, T., *Network formation of lipid membranes: triggering structural transitions by chain melting*, Proc. Nat. Acad. Sci., **96**, no. 25, 14312–14317, 1999.
35. Gawrisch, K., Ruston, D., Zimmerberg, J., Parsegian, V. A., Rand, R. P., & Fuller, N., *Membrane dipole potentials, hydration forces, and the ordering of water at membrane surfaces*, Biophys. J., **61**, no. 5, 1213–1223, 1992.
36. Petrache, H. I., Gouliaev, N., Tristram-Nagle, S., Zhang, R., Suter, R. M., & Nagle, J. F., *Interbilayer interactions from high-resolution x-ray scattering*, Phys. Rev. E., **57**, no. 6, 7014, 1998.
37. McIntosh, T. J. & Simon, S. A., *Contributions of hydration and steric (entropic) pressures to the interactions between phosphatidylcholine bilayers: experiments with the subgel phase*, Biochemistry, **32**, no. 32, 8374–8384, 1993.
38. Markova, N., Sparr, E., Wadsö, L., & Wennerström, H., *A calorimetric study of phospholipid hydration. Simultaneous monitoring of enthalpy and free energy*, J. Phys. Chem. B., **104**, no. 33, 8053–8060, 2000.
39. Schubert, T., Schneck, E., & Tanaka, M., *First order melting transitions of highly ordered dipalmitoyl phosphatidylcholine gel phase membranes in molecular dynamics simulations with atomistic detail*, J. Chem. Phys., **135**, no. 5, 08B607, 2011.
40. Kowalik, B., Schubert, T., Wada, H., Tanaka, M., Netz, R. R., & Schneck, E., *Combination of MD simulations with two-state kinetic rate modeling elucidates the chain melting transition of phospholipid bilayers for different hydration levels*, J. Phys. Chem. B., **119**, no. 44, 14157–14167, 2015.
41. Stoffel, W. & Bosio, A., *Myelin glycolipids and their functions*, Cur. Opin. Neurobio., **7**, no. 5, 654–661, 1997.
42. Boudière, L., Michaud, M., Petroutsos, D., Rébeillé, F., Falconet, D., Bastien, O., Roy, S., Finazzi, G., Rolland, N., Jouhet, J., et al., *Glycerolipids in photosynthesis: Composition, synthesis and trafficking*, Biochimica et Biophysica Acta (BBA): Bio, **1837**, no. 4, 470–480, 2014.

43. Demé, B., Cataye, C., Block, M. A., Maréchal, E., & Jouhet, J., *Contribution of galactoglycerolipids to the 3-dimensional architecture of thylakoids*, FASEB J., **28**, no. 8, 3373–3383, 2014.

44. Ryrif, I. J., Anderson, J. M., & Goodchild, D. J., *The role of the light-harvesting chlorophyll a/b-protein complex in chloroplast membrane stacking*, Eur. J. Biochem., **107**, no. 2, 345–354, 1980.

45. Webb, M. S. & Green, B. R., *Effects of neutral and anionic lipids on digalactosyldiacylglycerol vesicle aggregation*, Biochimica et Biophysica Acta, **1030**, no. 2, 231–237, 1990.

46. Luzar, A. & Chandler, D., *Hydrogen-bond kinetics in liquid water*, Nature, **379**, 55–57, January 1996.

47. Cevc, G., Podgornik, R., & Žekš, B., *The free energy, enthalpy and entropy of hydration of phospholipid bilayer membranes and their difference on the interfacial separation*, Chem. Phy. Lett., **91**, no. 3, 193–196, September 1982.

48. Radić, N. & Marčelja, S., *Solvent contribution to the debye screening length*, Chem. Phy. Lett., **55**, no. 2, 377–379, April 1978.

49. Ruckenstein, E. & Schiby, D., *On the origin of repulsive hydration forces between two mica plates*, Chem. Phy. Lett., **95**, no. 45, 439–443, 1983.

50. Attard, P. & Batchelor, M. T., *A mechanism for the hydration force demonstrated in a model system*, Chem. Phy. Lett., **149**, no. 2, 206–211, August 1988.

51. Landau, L. D., *Zur Theorie der phasenumwandlungen II.* Phys. Z. Sowjetunion, **11**, 26–35, 1937.

52. Schlaich, A., Kowalik, B., Kanduč, M., Schneck, E., & Netz, R. R., *Physical mechanisms of the interaction between lipid membranes in the aqueous environment*, Physica A, **418**, 105–125, January 2015.

53. Pham, Q. D., Wolde-Kidan, A., Gupta, A., Schlaich, A., Schneck, E., Netz, R. R., & Sparr, E., *Effects of urea and TMAO on lipid self-assembly under osmotic stress conditions*, J. Phys. Chem. B., DOI: 10.1021/acs.jpcb.8b02159, 2018.

6

Free-Energy Calculations of Pore Formation in Lipid Membranes

N. Awasthi
Institute for Microbiology and Genetics, Georg-August-Universität Göttingen, Göttingen, Germany

J. S. Hub[1]
Theoretical Physics, Universität des Saarlandes, Saarbrücken, Germany

1 Introduction

For many biophysical processes such as transport of small molecules and ions, cellular signaling, as well as membrane fusion and fission, the formation of polar defects and transmembrane pores is a critical and often rate-limiting step (Bennett and Tieleman, 2014; Fuhrmans, Marelli, Smirnova, and Müller, 2015; Jahn, Lang, and Südhof, 2003; LaRocca, Stivison, Mal-Sarkar, Hooven, Hod, Spitalnik, and Ratner, 2015; Lenertz, Gavala, Hill, and Bertics, 2009; Vorobyov, Olson, Kim, Koeppe, Andersen, and Allen, 2014). Formation of pores also provides a mechanism to control cell death, as employed by T cells and natural killer cells to kill virus-infected cells (Kägi, Ledermann, Bürki, Seiler, Odermatt, Olsen, Podack, Zinkernagel, and Hengartner, 1994; Law, Lukoyanova, Voskoboinik, Caradoc-Davies, Baran, Dunstone, DAngelo, Orlova, Coulibaly, Verschoor, Browne, Ciccone, Kuiper, Bird, Trapani, Saibil, and Whisstock, 2010). Membrane electroporation is an established method for transferring various types of material across membranes, such as RNA and vaccines, with applications in cell biology and medicine (Böckmann, De Groot, Kakorin, Neumann, and Grubmüller, 2008; Neumann, Schaefer-Ridder, Wang, and Hofschneider, 1982). Drugs derived from antimicrobial peptides often act via pore-mediated pathways, and cell-penetrating peptides may deliver cargo across membranes by forming defects in the lipid bilayer (Bechara and Sagan, 2013; Brogden, 2005). The mechanisms underlying antimicrobial and cell-penetrating peptides as well as the formation of fusion pores are far from fully understood. Hence, a quantitative understanding of the process of pore-formation and pore-closure would contribute to a better understanding of transport across cell membranes, and additionally, be beneficial for the design and control of membrane-active peptides.

In this chapter, we discuss the free energies for forming pores over lipid bilayers. Such free energies can be calculated from molecular dynamics (MD) simulations where membrane bilayers can be modeled by all-atom or coarse-grained lipid forcefields. We focus on all-atom lipid models since they offer the possibility to model atomic details such as hydrogen bonds, which are likely to play an important role for the energetics of pore formation. For most lipid membranes, pore formation is beyond the scope of equilibrium MD simulations because pores

---

[1] CONTACT N. Awasthi. Email: neha.awasthi@gmail.com; J.S. Hub. Email: jochen.hub@physik.uni-saarland.de

form spontaneously only on long time scales. Hence, enhanced sampling techniques such as the umbrella sampling method are required to induce pores in lipid bilayers (Torrie and Valleau, 1974). In umbrella sampling, a biasing potential is applied along a preselected reaction coordinate (RC), and this potential ensures sampling along the complete RC from one thermo-dynamic state to another. In our case, this would be a transition from a flat, unperturbed membrane to an open-pore state. Apart from the creation of pores in the simulations, a key motivation of umbrella sampling is that it provides the free-energy profile (or potential of mean force, PMF) along the RC, thereby revealing the free-energy difference between the flat membrane and the open pore, as well as the height of the free-energy barrier (if present) along the opening pathway. Since free-energy differences and barriers determine equilibrium probabil-ities and rates, respectively, obtaining PMFs of pore formation is pivotal for developing a quantitative understanding of pores.

By pulling the membrane along an *ideal* RC, the system would gradually sample the flat-membrane state, the barrier at the transition state (if present), and the open-pore state, and thereby more or less follow the minimum free-energy pathway. However, finding good RCs for complex transitions is far from trivial. Indeed, problems arise when using non-ideal RCs: (i) the PMF may converge poorly and reveal hysteresis between pore-opening and pore-closing pathways, despite the fact that the PMF, being an ensemble property, should not depend on the direction of the pathway; (ii) barriers along the pore-opening transition may be hidden if the barrier is crossed orthogonal to the RC, i.e. the barrier may be integrated out. Below, we discuss the free-energy landscape of membrane pore formation with a special emphasis on the different RCs that have been proposed for calculating PMFs using MD simulations.

2 Membrane Pores from Experiments

2.1 Stable Membrane Pores

Experimentally, transmembrane pores have been investigated in detail using model systems such as vesicles, supported lipid monolayers, or cells. Pores were induced using various stress condi-tions such as surface tension, temperature, and electrochemical gradients (Akinlaja and Sachs, 1998; Melikov, Frolov, Shcherbakov, Samsonov, Chizmadzhev, and Chernomordik, 2001; Tekle, Astumian, Friauf, and Chock, 2001; Zhelev and Needham, 1993). In the experiments, pores were detected using fluorescence-based techniques such as calcein leakage or using electrophysiology. Transient pores have also been observed in giant unilamellar vesicles under tension (Sakuma and Imai, 2015). Such experiments with model systems have provided estimates for the pore size as well as rates of pore formation and closure.

Membranes pores are also observed in experiments in the presence of membrane-active agents such as antimicrobial/cell penetrating peptides or cationic polymers (Brogden, 2005; Zasloff, 2002). Experiments with membrane-active agents have led to discussions about the possible structural models for transmembrane pores, such as the carpet model, the toroidal pore model, or the barrel-stave model. There is extensive literature on the topic of membrane pores in the presence of membrane active peptides review (Brogden, 2005; Wimley, 2010; Zasloff, 2002).

2.2 Metastable Pores Proposed from Experiments

The lifetime of the pores is an important characteristic. It is mainly determined by the presence and height of a free-energy barrier that must be overcome to close the pore. In the absence of such a barrier, pores will close rapidly and hence reveal short lifetimes. In contrast, in the presence of a barrier, the pores are metastable and may reveal long lifetimes.

Nearly four decades ago, Abidor, Arakelyan, Chernomordik, Chizmadzhev, Pastushenko, and Tarasevich (1979) observed the existence of metastable pores in electrophysiology experiments. The experiments revealed rapid transitions between (i) a poorly conducting but long-living state, sometimes referred to as the 'prepore state'; and (ii) a fully formed, highly-conducting, nanometer-sized aqueous pore. Since the pore may rapidly expand between the narrow prepore and the expanded pore, the conductivity of the pore may fluctuate, underlined by the notion of 'flickering pores.' To our knowledge, the term 'prepore' is not yet clearly defined in the literature. Some authors used the term to characterize intermediate, unstable structures during non-equilibrium simulations of pore formation, such as a thin water needle over the hydrophobic membrane core (Böckmann *et al.*, 2008). Other authors used the term to characterize a metastable, narrow aqueous pore, stabilized by a few tilted lipids (Ting, Awasthi, Müller, and Hub, 2018).

Notably, under membrane tension, long-living large pores with a radius of ~ 1 μm were also observed, stabilized by non-equilibrium solvent flow through the pore (Brochard-Wyart, de Gennes, and Sandre, 2000; Karatekin, Sandre, Guitouni, Borghi, Puech, and Brochard-Wyart, 2003; Moroz and Nelson, 1997; Zhelev and Needham, 1993).

---

3 Modeling Pores in Molecular Dynamics Simulations

In parallel with the experimental observation of membrane pores, MD simulations have been used to study transmembrane pores at molecular detail (Gurtovenko, Anwar, and Vattulainen, 2010; Marrink, Jähnig, and Berendsen, 1996). In MD simulations, pores were formed by applying surface tension (Leontiadou, Mark, and Marrink, 2004; Tieleman, Leontiadou, Mark, and Marrink, 2003), electrostatic membrane potentials (Böckmann *et al.*, 2008; Gurtovenko and Vattulainen, 2005; Tarek, 2005; Tieleman *et al.*, 2003), by simulating membrane-active agents such as antimicrobial or cell-penetrating peptides (Herce, Garcia, Litt, Kane, Martin, Enrique, Rebolledo, and Milesi, 2009; Leontiadou, Mark, and Marrink, 2006; Sengupta, Leontiadou, Mark, and Marrink, 2008), or by inserting small charged solutes into the membrane (Neale, Madill, Rauscher, and Pomès, 2013; Tepper and Voth, 2005). Pores were also observed as metastable intermediate structures during spontaneous aggregation of membranes (Marrink, Lindahl, Edholm, and Mark, 2001). From simulations, disordered toroidal pores have been predicted as structures for transmembrane pores in the presence of membrane-active peptides. MD simulations of pore formation have been reviewed repeatedly, hence we refer to reader to the literature for a more complete view on the field (Bennett and Tieleman, 2014; Kirsch and Böckmann, 2016).

Computationally, metastable pores were reported from simulations of stretched membranes (Tolpekina, Den Otter, and Briels, 2004). In tension-free membranes, however, resolving a pore nucleation barrier in the PMF, as required to rationalize a metastable pore, turned out to be challenging (Awasthi and Hub, 2016; Bennett and Tieleman, 2014; Wohlert, den Otter, Edholm, and Briels, 2006); as shown below, these challenges are primarily associated with problems with RCs. Only recently, using the string method in conjunction with self-consistent field theory or using umbrella sampling calculations with a new RC, a pore nucleation barrier could be resolved (Hub and Awasthi, 2017; Ting *et al.*, 2018).

---

4 Reaction Coordinates for Free-Energy Calculations of Pore Formation: A Comparison

PMF calculations are a standard protocol for obtaining the energetics along functionally relevant transitions in biomolecular systems, such as pore formation in membranes. In principle, it should be possible to compute the free-energy landscape for pore formation using MD simulations and enhanced sampling techniques such as umbrella sampling (Torrie and Valleau, 1974). Umbrella sampling requires the definition of one or several RCs (or order parameters), along which the system is steered and PMFs are calculated. In the context of pore formation, the RC should steer the system from the state of an intact membrane to the state with an open transmembrane pore. However, identifying good RCs for complex transitions such as pore formation is often a challenging task (Best and Hummer, 2005; Bolhuis, Chandler, Dellago, and Geissler, 2002; Neale and Pomès, 2016).

Several RCs were implemented for modeling membrane pores, as illustrated in Figure 1. Pulling the system along these RCs is characterized by

- steering the lipids radially (or laterally) from the pore center (Tolpekina *et al.*, 2004; Wohlert *et al.*, 2006)
- pulling a single lipid head group towards the bilayer center, corresponding to a lipid flip-flop transition (Bennett and Tieleman, 2011, 2014; Sapay and Tieleman, 2009; Tieleman and Marrink, 2006);
- pulling water into a membrane-spanning cylinder (Mirjalili and Feig, 2015);
- generating a hydrogen bond chain over the membrane by filling slices of a membrane-spanning cylinder with polar atoms (Hub and Awasthi, 2017).

Upon pulling the membrane along any of these RCs, pores formed first by the penetration of a thin water needle into the membrane, followed by tilting of lipids parallel to the membrane to avoid unfavorable contacts of water with the apolar membrane interior (Bennett and Tieleman, 2014; Böckmann *et al.*, 2008; Tieleman *et al.*, 2003).

To test the performance of the RCs for pore formation, we focus on three aspects: a) are free energies of pore formation obtained from PMFs along the four coordinates in agreement? b) In case that the open pore is metastable, does the PMF along the RC reveal the barrier between the pore-open and pore-closed states, or is the barrier integrated out? b) Do the computed PMFs along an RC suffer from hysteresis owing to slow convergence, i.e. do the PMFs computed from simulation frames taken from pulling simulations conducted in forward and backward directions not agree? For details about simulation parameters, we refer the reader to the reference list of this chapter.

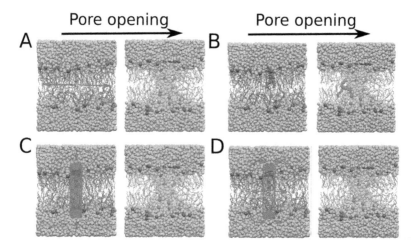

FIGURE 1 Illustration of four reaction coordinates (RCs) suggested for calculating the potential of mean force of transmembrane pore formation. (a) Collective radial coordinate, (b) the flip-flop coordinate, (c) the average water density in a membrane-spanning cylinder, and (d) the chain coordinate. Lipid molecules are visualized as silver sticks, lipid phosphate atoms as gray spheres, and water molecules as cyan spheres. (See color insert for the color version of this figure)

4.1 Steering Lipids Laterally from Pore Center: The Collective Radial Coordinate

The collective radial coordinate ξ_R was proposed by Tolpekina *et al.* (Hu, Sinha, and Patel, 2015; Tolpekina, Den Otter, and Briels, 2004; Wohlert *et al.*, 2006). With this coordinate, a transmembrane pore is created by pushing all lipid molecules radially outwards and parallel to the plane, away from the center of the pore (Figure 1a). Using a switch function, the RC was designed such that lipids close to the pore are pushed more strongly than lipids far away from the pore. The RC is

$$\xi_R = \frac{\Sigma - \Sigma_0}{N - \Sigma_0}, \tag{1}$$

$$\Sigma = \sum_{i=1}^{N} \tanh(r_i/\zeta). \tag{2}$$

Here, N is the total number of lipid molecules, and r_i is the lateral distance between the center of the pore and the center of mass of lipid i. According to Eq. (2), Σ increases as the lipids move laterally from the pore center. Here, the hyperbolic tangent $\tanh(\cdot)$ with parameter ζ serves as the switch function that ensures that moving lipids within a distance of $\sim 2\zeta$ from the pore center influences Σ more strongly than moving lipids at larger distances. Consequently, lipids near the pore center also feel a stronger biasing force when the system is pulled along the RC. The parameter ζ also determines the approximate radius of the fully formed pore; a reasonable value was suggested to be 1 nm. The normalization in Eq. (1) is chosen such that that $\xi_R = 0$ denotes the initial state with no pore but a random distribution of lipids, while $\xi_R \approx 1$ corresponds to a fully established transmembrane pore. Here, Σ_0 is the equilibrium value of Σ, which can be computed prior to a simulation by assuming a random distribution of lipids. This RC was originally implemented for constrained MD simulations, and later also applied for umbrella sampling simulations (Awasthi and Hub, 2016).

4.2 Distance of One Phosphate Group from the Membrane Center: The Flip-Flop Coordinate

Tieleman and coworkers suggested a RC inspired by lipid flip-flop. The RC is defined as the distance d_{ph} between a single lipid phosphate group and the membrane center (Figure 1b) Bennett and Tieleman (2011, 2014); Sapay and Tieleman (2009); Tieleman and Marrink (2006). Using this RC to study pores is mainly motivated from the observation that pulling a phosphate group to the membrane center drags water inside the membrane, thereby triggering the formation of a water pore. Applying this RC for PMF calculations is straightforward since center-of-mass pulling is implemented in many MD suites.

4.3 Average Water Density Inside a Membrane-Spanning Cylinder: Water-Density Coordinate

While the RCs described above steer the lipid molecules to form a pore, Mirjalili and Feig (2015) suggested a coordinate that follows the penetration of water into the membrane (Mirjalili and Feig, 2015). Accordingly, the coordinate is defined as the average water density inside of a membrane-spanning cylinder (Figure 1c), with the cylinder axis aligned with the bilayer normal, and placed symmetrically at the membrane center. The RC can be expressed mathematically using an indicator function $f(\mathbf{r})$ for the cylinder, which takes zero outside and unity inside the cylinder. In order to obtain a RC that is differentiable with respect to the atomic coordinates, the function $f(\mathbf{r})$ must be defined with smooth switch functions at the surface of the cylinder. The RC is given by

$$\rho_{\text{cyl}} = \Gamma_V / V, \tag{3}$$

where $V = \int f(\mathbf{r}) d\mathbf{r}$ is the volume of the cylinder, and Γ_V denotes the number of water molecules inside the cylinder. The latter is given by

$$\Gamma_V = \sum_{i=1}^{N_w} f(\mathbf{R}_i) \tag{4}$$

where N_w denotes the total number of water molecules, and $\mathbf{R}_i$ are the Cartesian coordinates of the water oxygen atom i. It is important to note that ρ_{cyl} does not correspond to a three-dimensional density field, but it is instead a scalar quantity that is given by the *number* of water molecules inside the cylinder. Below, we normalize ρ_{cyl} with the bulk water density, such that $\rho_{\text{cyl}}/\rho_{\text{bulk}} \approx 1$ corresponds to a fully filled cylinder.

4.4 Generating a Continuous Polar Defect: The Chain Coordinate

This RC was designed to differentiate between (i) polar defects that partly penetrate in the membrane and (ii) a continuous defect spanning the entire membrane (Hub and Awasthi, 2017). To this end, the RC was defined using a membrane-spanning cylinder that is decomposed into N_s slices along the membrane normal (Figure 1d, slice thickness ~1 Å). Then, the RC is given by the *number* of slices that are occupied by polar heavy atoms. This definition ensures that the RC is modulated purely by adding polar atoms to empty slices, but hardly by adding polar atoms to previously filled slices. Consequently, by pulling the system along the RC towards the open-pore state, the slices are filled one-by-one by polar atoms, thereby forming a continuous hydrogen bond chain over the entire membrane. Hence, we refer to the RC as 'chain coordinate' in this work. In turn, by pulling the system back to the closed-pore state, slices are fully depleted from polar atoms, thereby breaking the continuous hydrogen bond chain over the membrane.

The RC was defined as follows:

$$\xi_{\text{chain}} = N_s^{-1} \sum_{s=0}^{N_s-1} \delta_s(N_s^{(p)}) \tag{5}$$

Here, N_s denotes the number of slices, and $N_s^{(p)}$ is the number of polar heavy atoms inside slice s of the membrane-spanning cylinder. The function δ_s is an indicator function that takes zero if no polar atoms are in slice s, and takes a value close to unity if one more polar atoms are in slice s:

$$\delta_s(N_s^{(p)}) \approx \begin{cases} 0 & \text{if } N_s^{(p)} = 0 \\ 1 & \text{if } N_s^{(p)} \geq 1 \end{cases} \tag{6}$$

This property is critical to distinguish between the cases of (i) few slices occupied by many polar atoms, as found in partial defects spanning part of the membrane, and (ii) structures in which every slice is occupied by at least one polar atom, as found in a continuous membrane-spanning polar defect. To ensure that ξ_{chain} is differentiable with respect to the atomic coordinates, δ_s and $N_s^{(p)}$ were formulated using differentiable switch functions (Hub and Awasthi, 2017).

Critically, the x–y position (in the membrane plane) of the membrane-spanning cylinder is not fixed but instead dynamically defined depending on the position of the aqueous defect. In other words, the cylinder follows the defect as the defect explores the membrane plane. This property avoids that the system may move along the RC by shifting the defect laterally out of the cylinder, which was identified as a common source for hysteresis between pore-opening and pore-closing pathways.

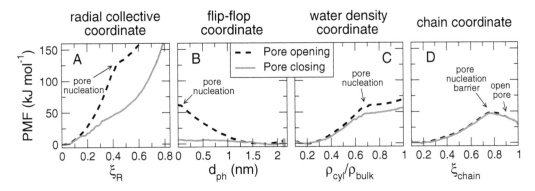

FIGURE 2 Potentials of mean force (PMF) of transmembrane pore formation in a DMPC membrane, computed along four different RCs. (a) Collective radial coordinate, (b) flip-flop coordinate, (c) water density, and (d) the chain coordinate. PMFs along pore-opening and pore-closing pathways are shown in black and gray, respectively. For a discussion, see text. Lipids were modeled with the force field by Berger et al. (1997), and simulation parameters were chosen as described previously (Awasthi and Hub, 2016).

4.5 The Reaction Coordinate Greatly Influences the PMFs of Pore Formation

In order to evaluate the performance of the different RCs, we compare the PMFs calculated for a DMPC bilayer with 128 lipid molecules at 300 K as shown in Figure 2a–d. All PMFs were derived using umbrella sampling windows of 150 ns each, using only the last 50 ns for analysis and omitting the first 100 ns for equilibration. PMFs along pore-opening (black broken lines) and along pore-closing pathways (gray solid lines) are plotted in Figure 2a–d as a function of each RC, i.e., ξ_R, d_{ph}, ρ_{cyl}/ρ_{bulk}, and ξ_{chain}. Evidently, the choice of the RC has a great impact (i) on the estimated free-energy difference between the open pore and the flat membrane, (ii) on undesirable hysteresis between PMFs along pore-opening and pore-closing pathways, and (iii) on the appearance of a nucleation barrier, i.e., whether the open pore is identified as being metastable.

In the following, we discuss three important differences between the PMFs shown in Figure 2. First, the free energies of the open pore (relative to the flat membrane) as given by the PMFs differ greatly between the four RCs (Table 1, second column). In particular, the PMF along ξ_R suggests strongly increased free energies as compared to the PMFs along the other three RCs. This indicates that the collective radial coordinate perturbs the membrane more strongly than strictly required to form a pore.

Second, comparing the PMFs computed along pore-opening and pore-closing pathways reveals that the PMFs converge on different time scales. Evidently, the simulations with the radial

TABLE 1

Free energies of a transmembrane pore and the pore nucleation barrier (if resolved) in a system of 128 DMPC lipids at 300 K modeled with the Berger force field, derived from the PMFs along four reaction coordinates for pore formation.

Reaction coordinate	Pore free energy[a] (kJ/mol^{-1})	Nucleation barrier (kJ/mol^{-1})	Hysteresis
Radial collective, ξ_R	>125	Not resolved	Major
Flip-flop, d_{ph} (nm)	60	Not resolved	Major
Water-density, ρ_{cyl}/ρ_{bulk}	57	Not resolved	Minor
Chain coordinate, ξ_{chain}	35	48	None

[a] Estimated from PMFs shown in Figure 2a–d. In case of hysteresis, the values were taken from the PMFs along the pore-opening pathway.

collective and the flip-flop coordinate strongly suffer from hysteresis. Visual inspection of the trajectories reveals the structural reason for hysteresis. Namely, upon pulling the system from the pore-open state along ξ_R or along d_{ph} back towards the flat-membrane state, the aqueous defects simply do not close. In other words, pulling along the ξ_R or d_{ph} RCs is insufficient to break the hydrogen bond network over the membrane, as would be required to close the pore, within 150 ns of simulation. For instance, when pulling the restrained phosphate group from the membrane center ($d_{ph} = 0$ nm) back to the head group region ($d_{ph} \approx 2$ nm), the restrained lipid is simply replaced with other lipids, thereby stabilizing the aqueous defect. In contrast, PMFs along the water density coordinate exhibit only minor hysteresis, and the PMFs along the chain coordinate exhibit virtually no hysteresis, suggesting that the PMFs are fully converged within 150 ns. This suggests that it is critical to steer the water molecules, and not purely the lipids, in order to break the continuous hydrogen bond network and, hence, to close the pore.

Third, the PMF along the chain coordinate exhibits a barrier for pore nucleation, demonstrating that the open pore forms a metastable state (Figure 2d). The barrier is compatible with equilibrium simulations on the same system that, starting from an open pore, did not show a pore-closing event within $\sim$10 μs of simulation (Awasthi and Hub, 2016). In contrast, PMFs along the other three RCs lack this barrier, suggesting that the barrier was integrated out because the system crosses the barrier in a direction orthogonal to these three RCs (Table 1, third column).

What is the underlying reason for problems with hysteresis and with the loss of the nucleation barrier? This question is addressed in Figure 3, which analyzes umbrella sampling windows restrained along the ξ_R, d_{ph}, and ρ_{cyl} coordinates, by projecting each the simulation frames of umbrella window onto the chain coordinate ξ_{chain}. The key findings is that, upon pulling the membrane along the radial coordinate, the flip-flop coordinate, or the water density coordinate, the system may sample structures of a flat membrane, of a partial defect, or of the open pore (Figure 3, horizontal bars); however, the system hardly samples the transition state of pore formation. In other words, restraining the system along ξ_R, d_{ph}, or ρ_{cyl} does not restrain the system close to the transition state of pore formation. This undesirable property eventually manifests in sampling problems and hysteresis and in the integrated out nucleation barrier.

To conclude, all the RCs described here have been used successfully to study pore formation in membranes. Simulations employing these coordinates gave atomic-level insight into the structure of pores and the mechanism of pore formation, and they provided the energetics of pore formation at least on a semiquantitative level. However, only the chain coordinate is capable of

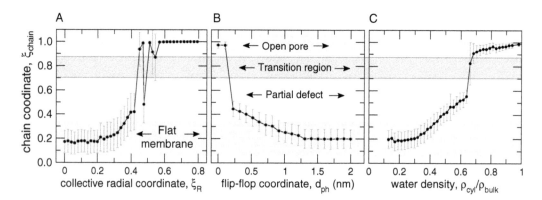

FIGURE 3 Rationalizing problems with hysteresis and with the lack of a nucleation barrier in a membrane of DMPC: Umbrella sampling simulations restrained along (a) the collective radial collective coordinate ξ_R, (b) the flip-flop coordinate d_{ph}, and (c) the water density coordinate ρ_{cyl}, projected onto the chain coordinate ξ_{chain}. Dots and bars indicate the average and standard deviation of ξ_{chain}. The analysis demonstrates that restraining the DMPC membrane along ξ_R, d_{ph}, or ρ_{cyl} does not restrain the system close to the transition state of pore formation. Adapted with permission from Hub and Awasthi (2017). Copyright (2017) American Chemical Society.

resolving the nucleation barrier, which is critical for estimating the life time of the pore. In addition, only PMF calculations along the chain coordinate converge rapidly and do not suffer from hysteresis in this system.

5 Longer Lipid Tails Increase the Free Energy of Pore Formation

Sapay et al. systematically analyzed the energetics of pore formation in PC membranes as function of lipid tail length (Bennett et al., 2014; Sapay and Tieleman, 2009). Using PMF calculations along the lipid flip-flop coordinate, the authors found that the free energy of pore formation generally increases with the thickness of the membrane (Figure 4). Later simulations with the chain coordinate confirmed these trends (see also Figure 5a) (Ting *et al.*, 2018). These findings are rationalized by the fact that pore formation in thicker membranes requires the formation of larger aqueous defects before a complete transmembrane pore may form. Further, by decomposing the PMFs into enthalpic and entropic contributions, Bennett *et al.* found that pore formation in PC membranes is opposed by a large loss of entropy, but favored by a gain of enthalpy.

6 Relative Volume of Lipid Head to Lipid Tails Determines the Metastability of the Pore

Since the pore rim represents a region of high local curvature, it is not surprising that the intrinsic shape of the lipid molecules may influence the energetics of pore formation. Indeed, Ting *et al.* (2018) showed that lipids with a large head group-to-tail volume ratio form metastable pores, whereas lipids with a small head group-to-tail volume ratio form unstable pores. These findings were obtained by PMF calculations along the chain coordinate with atomistic MD simulations, as well as with a minimal coarse-grained lipid model in conjunction with the string method and self-consistent field theory.

The PMFs along the chain coordinate ξ_{chain} in Figure 5 demonstrate the effect of the head-to-tail volume ratio on metastability. PMFs are shown for five PC (Figure 5a) and five PG lipids (Figure 5b) of increasing tail length and tail unsaturation. Evidently, pores in membranes with short saturated tails such as DLPC, DMPC, DLPG, and DMPG are metastable, demonstrated by the free-energy minimum at $\xi_{chain} \approx 1$ and the nucleation barrier (or transition state) at $\xi_{chain} \approx 0.85$

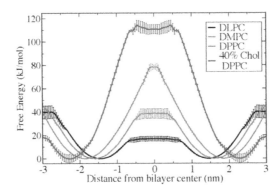

FIGURE 4 PMFs for pore formation along the flip-flop coordinate for phosphatidylcholine (PC) membranes with increasing tail length and as function of cholesterol content. Longer tails and the addition of cholesterol increase the free energy of pore formation. Adapted with permission from Bennett and Tieleman (2014). Copyright (2014) American Chemical Society. (See color insert for the color version of this figure)

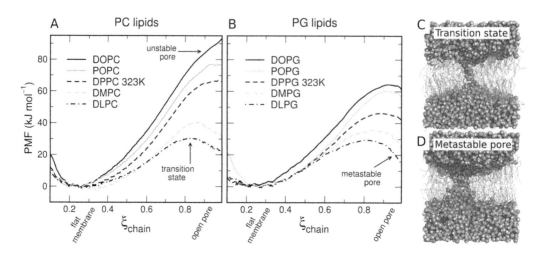

FIGURE 5 (a and b) PMFs for pore formation calculated using the chain coordinate, ξ_{chain}, for membrane bilayers of different PC and phosphatidylglycerol (PG) lipid molecules with increasing tail length and tail unsaturation. The PMFs were taken from Ting *et al.* (2018), using extended simulation time for DPPC and POPC compared to our previous work. For all PMFs, $\xi_{\text{chain}} \sim 0.2$ denotes flat membranes and $\xi_{\text{chain}} \sim 1$ denotes open pores. Pronounced nucleation barriers are observed for DLPC, DMPC, DLPG, and DMPG, indicating metastable pores, and shallow barriers for long-tailed PG lipids. For long-tailed PC lipids, no barriers are observed, indicating unstable pores. Lipids were modeled with the Charmm36 force field (Pastor and MacKerell, 2011). PMFs for DPPC and DPPG were computed at 323 K, and all other PMFs at 300 K. (c) Typical simulation snapshots of a DMPG membrane at the transition state, revealing a thin water needle, and (d) of a fully formed metastable pore. Lipids are shown as sticks, solvent as spheres.

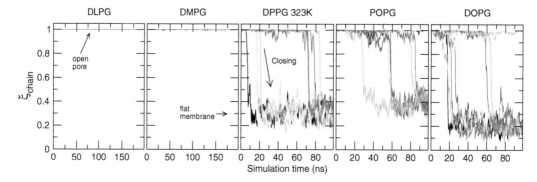

FIGURE 6 Trajectories of free MD simulations starting from frames with an open pore in PG membranes with increasing tail length and increasing tail unsaturation. Eight independent simulations shown in different colors were conducted for each lipid type (from left to right): DLPG, DMPG, DPPG, POPG, and DOPG. No bias was applied to the simulations. Lipids were modeled with the Charmm36 force field, and simulation parameters were chosen as described previously (Pastor and MacKerell, 2011; Ting *et al.*, 2018). To visualize the spontaneous closing of the pores during the free simulations, the trajectories are projected onto the RC ξ_{chain}, where $\xi_{\text{chain}} \approx 1$ and $\xi_{\text{chain}} \approx 0.3$ correspond to open pores and flat unperturbed membranes, respectively (arrows). No pore closed in the bilayers of DLPG and DMPG within 200 ns, confirming the metastability of the pores, and compatible with the pronounced nucleation barriers revealed by the PMF (compare with Figure 5b). Pores in bilayers of DPPG, POPG, and DOPG closed on the tens of nanosecond time scale, compatible with the shallow nucleation barriers in these membranes. (See color insert for the color version of this figure)

(Figure 5a and b dot-dashed black, dashed gray curves). This transition state is characterized by a thin water column spanning the complete bilayer, a structure that has been observed in previous studies (Figure 5c) (Awasthi and Hub, 2016; Bennett and Tieleman, 2014). In PC membranes, lipids with longer tails (DPPC) and longer, unsaturated tails (POPC and DOPC) form unstable pores, as evident from the absence of a nucleation barrier (Figure 5a dashed black, solid gray, and solid black curves). Moreover, owing to the increased volume of PG over PC head groups, shallow nucleation barriers for lipids with longer and unsaturated tails are observed for DPPG, POPG, and DOPG lipids (Figure 5 dashed black, solid gray, and solid black curves respectively). Hence, the increased head-to-tail volume ratio of PG as compared to PC lipids correlates with an increased tendency to form metastable pores.

As a simple test of metastability, one can monitor the pore closing times in free simulations starting from an open pore. Figure 6 presents eight unbiased MD simulations of the each of the five PG membranes starting from an open pore. We observe that no pore closed within 200 ns for DLPG and DMPG, confirming the metastability of the pores in DLPG and DMPG, and in line with the pronounced nucleation barriers. In contrast, for DPPG, POPG, and DOPG, the pores close on a time scale of tens of nanoseconds, compatible with the more shallow nucleation barrier. These free simulations, corroborated by previous studies of pore closure simulations in PC membranes (Ting *et al.*, 2018), demonstrate that the PMFs along the chain coordinate indeed reflect the minimum free-energy path of pore formation.

7 Practical Consideration for PMF Calculations with the Chain Coordinate

The chain RC requires a number of parameters, which should be chosen correctly to ensure that the PMFs converge rapidly and do not reveal any hysteresis (Hub and Awasthi, 2017). Specifically,

- the thickness of the slices should be chosen such that polar heavy atoms in neighboring slices may form stable hydrogen bonds, even in the presence of some thermal fluctuations. A suitable value is 1 Å;
- the fraction to which a slice is considered as 'filled' upon the addition of the first heavy atom. Here, 0.75 was found to be suitable;
- the radius of the cylinder R_{cyl} is an important parameter. If R_{cyl} is too large, two laterally displaced partial defects in opposite leaflets (one partial defect connected with the upper, one connected with the lower water compartment) may form, thereby filling all cylinder slices but not forming a continuous defect. In consequence, slight hysteresis between opening and closing pathways may appear. In previous work, we suggested $R_{cyl} = 1.2$ nm as suitable. However, we found that for very soft membranes, such as membranes of DLPG or DLPC, a smaller R_{cyl} is required to strictly avoid hysteresis. As such, we recommend a value of $R_{cyl} = 0.8$ nm or 1.0 nm for future work.

The chain coordinate has been implemented as an extension of GROMACS 2016 and 2018 (Abraham et al., 2015). The source code of the implementation is available upon request from the authors of this chapter.

8 Summary and Outlook

Pioneering simulations of pore formation used the collective radial and the flip-flop coordinate to induce pores. These simulations have provided unprecedented insight into mechanisms, atomic structures, and energetics involved in pore formation over lipid membranes. The simulations further revealed that a quantitative understanding of pores requires the calculation of free energies of pore formation.

In order to study the influence of factors such as lipid composition, electric fields, tension, or membrane-active agents on pore formation, it is critical to make sure that the PMFs are fully converged and that they do not suffer from hysteresis effects. Otherwise, given that the computed PMFs are modulated by such factors, it would remain unclear whether such factors influence (i) the magnitudes of hysteresis, or (ii) the true underlying free-energy surface. For example, the Lennard–Jones cutoff leads to different magnitudes of hysteresis with the flip-flop coordinate, but the underlying free-energy landscape is hardly influenced by the cutoff (Huang and García, 2014; Hub and Awasthi, 2017).

In addition, to understand how the lifetime of pores depends on external factors, it is necessary to detect the presence and the height of a free-energy barrier that may separate the open pore from the flat membrane, i.e. to detect whether the pore is metastable.

Therefore, we have presented an overview of four different RCs that have been proposed for PMF calculations of pore formation using MD simulations: the radial collective coordinate, the flip-flop coordinate, the water density coordinate, and the chain coordinate. We found that the PMFs computed with umbrella sampling greatly depend on the choice of the RC. Specifically, we found that the radial collective RC may perturb the membrane more strongly than required for pore formation. In addition, the radial collective and the flip-flop coordinate may suffer from pronounced hysteresis between opening and closing pathways, suggesting that PMFs computed along these RCs converge slowly. The water density coordinate revealed greatly reduced hysteresis. However, none of the three coordinates reveal the nucleation barrier in a DMPC membrane, inconsistent with the metastability of the open pore in free simulations, suggesting that the barrier was integrated out. These problems recently prompted the development of the chain coordinate, which was designed to probe the formation and rupture of the continuous hydrogen bond network over the membrane. We found that PMF calculations with the chain coordinate converge rapidly, they do not suffer from hysteresis, and they do not integrate out the pore nucleation barrier (if a barrier is present).

These recent developments may be readily used to probe how electric fields, membrane-active peptides, lipid composition, membrane curvature, or other factors shape the free-energy landscape of pore formation, with implications on membrane fusion and fission, virus–host interactions, and biotechnological applications.

FUNDING

Support by the Deutsche Forschungsgemeinschaft is gratefully acknowledged (grant numbers SFB 803/A12 and HU 1971-3/1).

REFERENCES

Abidor, I., Arakelyan, V., Chernomordik, L., Chizmadzhev, Y., Pastushenko, V., and Tarasevich, M., 1979. 246 - electric breakdown of bilayer lipid membranes i. The main experimental facts and their qualitative discussion, *Bioelectrochem. Bioenerg.*, 6 (1), 37–52.

Abraham, M.J., Murtola, T., Schulz, R., Páll, S., Smith, J.C., Hess, B., and Lindahl, E., 2015. GROMACS: high performance molecular simulations through multi-level parallelism from laptops to supercomputers, *SoftwareX*, 1, 19–25.

Akinlaja, J. and Sachs, F., 1998. The breakdown of cell membranes by electrical and mechanical stress, *Biophys. J.*, 75 (1), 247–254.

Awasthi, N. and Hub, J.S., 2016. Simulations of pore formation in lipid membranes: reaction coordinates, convergence, hysteresis, and finite-size effects, *J. Chem. Theory Comput.*, 12 (7), 3261–3269, pMID: 27254744.

Bechara, C. and Sagan, S., 2013. Cell-penetrating peptides: 20 years later, where do we stand? *FEBS Lett.*, 587 (12), 1693–1702.

Bennett, W.D., Sapay, N., and Tieleman, D.P., 2014. Atomistic simulations of pore formation and closure in lipid bilayers, *Biophys. J.*, 106 (1), 210–219.

Bennett, W.D. and Tieleman, D.P., 2011. Water defect and pore formation in atomistic and coarse-grained lipid membranes: pushing the limits of coarse graining, *J. Chem. Theory Comput.*, 7 (9), 2981–2988.

Bennett, W.D. and Tieleman, D.P., 2014. The importance of membrane defects lessons from simulations, *Acc. Chem. Res.*, 47 (8), 2244–2251.

Berger, O., Edholm, O., and Jähnig, F., 1997. Molecular dynamics simulations of a fluid bilayer of dipalmitoylphosphatidylcholine at full hydration, constant pressure, and constant temperature, *Biophys. J.*, 72, 2002–2013.

Best, R.B. and Hummer, G., 2005. Reaction coordinates and rates from transition paths, *Proc. Natl. Acad. of Sci. USA*, 102 (19), 6732–6737.

Böckmann, R.A., De Groot, B.L., Kakorin, S., Neumann, E., and Grubmüller, H., 2008. Kinetics, statistics, and energetics of lipid membrane electroporation studied by molecular dynamics simulations, *Biophys. J.*, 95 (4), 1837–1850.

Bolhuis, P.G., Chandler, D., Dellago, C., and Geissler, P.L., 2002. Transition path sampling: throwing ropes over rough mountain passes, in the dark, *Annu. Rev. Phys. Chem.*, 53 (1), 291–318.

Brochard-Wyart, F., de Gennes, P., and Sandre, O., 2000. Transient pores in stretched vesicles: role of leak-out, *Physica A*, 278 (1), 32–51.

Brogden, K.A., 2005. Antimicrobial peptides: pore formers or metabolic inhibitors in bacteria? *Nat. Rev. Microbiol.*, 3 (3), 238–250.

Fuhrmans, M., Marelli, G., Smirnova, Y.G., and Müller, M., 2015. Mechanics of membrane fusion/pore formation, *Chem. Phys. Lipids*, 185, 109–128.

Gurtovenko, A.A., Anwar, J., and Vattulainen, I., 2010. Defect-mediated trafficking across cell membranes: insights from in silico modeling, *Chem. Rev.*, 110 (10), 6077–6103.

Gurtovenko, A.A. and Vattulainen, I., 2005. Pore formation coupled to ion transport through lipid membranes as induced by transmembrane ionic charge imbalance: atomistic molecular dynamics study, *J. Am. Chem. Soc.*, 127 (50), 17570–17571.

Herce, H., Garcia, A., Litt, J., Kane, R., Martin, P., Enrique, N., Rebolledo, A., and Milesi, V., 2009. Arginine-rich peptides destabilize the plasma membrane, consistent with a pore formation translocation mechanism of cell-penetrating peptides, *Biophys. J.*, 97 (7), 1917–1925.

Hu, Y., Sinha, S.K., and Patel, S., 2015. Investigating hydrophilic pores in model lipid bilayers using molecular simulations: correlating bilayer properties with pore-formation thermodynamics, *Langmuir*, 31 (24), 6615–6631.

Huang, K. and García, A.E., 2014. Effects of truncating van der waals interactions in lipid bilayer simulations, *J. Chem. Phys.*, 141 (10), 09B605_1.

Hub, J.S. and Awasthi, N., 2017. Probing a continuous polar defect: a reaction coordinate for pore formation in lipid membranes, *J. Chem. Theory Comput.*, 13 (5), 2352–2366, pMID: 28376619.

Jahn, R., Lang, T., and Südhof, T.C., 2003. Membrane fusion, *Cell*, 112 (4), 519–533.

Kägi, D., Ledermann, B., Bürki, K., Seiler, P., Odermatt, B., Olsen, K.J., Podack, E.R., Zinkernagel, R. M., and Hengartner, H., 1994. Cytotoxicity mediated by T cells and natural killer cells is greatly impaired in perforin-deficient mice, *Nature*, 369, 31–37.

Karatekin, E., Sandre, O., Guitouni, H., Borghi, N., Puech, P.H., and Brochard-Wyart, F., 2003. Cascades of transient pores in giant vesicles: line tension and transport, *Biophys. J.*, 84 (3), 1734–1749.

Kirsch, S.A. and Böckmann, R.A., 2016. Membrane pore formation in atomistic and coarse-grained simulations, *Biochim. Biophys. Acta — Biomembranes*, 1858 (10), 2266–2277, biosimulations of lipid membranes coupled to experiments.

LaRocca, T., Stivison, E., Mal-Sarkar, T., Hooven, T., Hod, E., Spitalnik, S., and Ratner, A., 2015. CD59 signaling and membrane pores drive Syk-dependent erythrocyte necroptosis, *Cell Death Dis.*, 6 (5), e1773.

Law, R.H., Lukoyanova, N., Voskoboinik, I., Caradoc-Davies, T.T., Baran, K., Dunstone, M.A., DAngelo, M.E., Orlova, E.V., Coulibaly, F., Verschoor, S., Browne, K.A., Ciccone, A., Kuiper, M.

J., Bird, P.I., Trapani, J.A., Saibil, H.R., and Whisstock, J.C., 2010. The structural basis for membrane binding and pore formation by lymphocyte perforin, *Nature*, 468 (7322), 447–451.

Lenertz, L.Y., Gavala, M.L., Hill, L.M., and Bertics, P.J., 2009. Cell signaling via the P2X7 nucleotide receptor: linkage to ros production, gene transcription, and receptor trafficking, *Purinerg. Signal.*, 5 (2), 175–187.

Leontiadou, H., Mark, A.E., and Marrink, S.J., 2004. Molecular dynamics simulations of hydrophilic pores in lipid bilayers, *Biophys. J.*, 86 (4), 2156–2164.

Leontiadou, H., Mark, A.E., and Marrink, S.J., 2006. Antimicrobial peptides in action, *J. Am. Chem. Soc.*, 128 (37), 12156–12161.

Marrink, S., Jähnig, F., and Berendsen, H., 1996. Proton transport across transient single-file water pores in a lipid membrane studied by molecular dynamics simulations, *Biophys. J.*, 71 (2), 632.

Marrink, S.J., Lindahl, E., Edholm, O., and Mark, A.E., 2001. Simulation of the spontaneous aggregation of phospholipids into bilayers, *J. Am. Chem. Soc.*, 123, 8638–8639.

Melikov, K.C., Frolov, V.A., Shcherbakov, A., Samsonov, A.V., Chizmadzhev, Y.A., and Chernomordik, L.V., 2001. Voltage-induced nonconductive pre-pores and metastable single pores in unmodified planar lipid bilayer, *Biophys. J.*, 80 (4), 1829–1836.

Mirjalili, V. and Feig, M., 2015. Density-biased sampling: a robust computational method for studying pore formation in membranes, *J. Chem. Theory Comput.*, 11 (1), 343–350.

Moroz, J. and Nelson, P., 1997. Dynamically stabilized pores in bilayer membranes, *Biophys. J.*, 72 (5), 2211–2216.

Neale, C., Madill, C., Rauscher, S., and Pomès, R., 2013. Accelerating convergence in molecular dynamics simulations of solutes in lipid membranes by conducting a random walk along the bilayer normal, *J. Chem. Theory Comput.*, 9 (8), 3686–3703.

Neale, C. and Pomès, R., 2016. Sampling errors in free energy simulations of small molecules in lipid bilayers, *Biochim. Biophys. Acta — Biomembranes*, doi: 10.1016/j.bbamem.2016.03.006.

Neumann, E., Schaefer-Ridder, M., Wang, Y., and Hofschneider, P., 1982. Gene transfer into mouse lyoma cells by electroporation in high electric fields, *EMBO J.*, 1 (7), 841–845.

Pastor, R. and MacKerell Jr, A., 2011. Development of the CHARMM force field for lipids, *J. Phys. Chem. Lett.*, 2 (13), 1526–1532.

Sakuma, Y. and Imai, M., 2015. From vesicles to protocells: the roles of amphiphilic molecules, *Life*, 5 (1), 651–675.

Sapay, N.W.F.D.B. and Tieleman, D.P., 2009. Thermodynamicsof flip-flop and desorption for a systematic series of phosphatidylcholine lipids, *Soft Matter*, 5, 3295–3302.

Sengupta, D., Leontiadou, H., Mark, A.E., and Marrink, S.J., 2008. Toroidal pores formed by antimicrobial peptides show significant disorder, *Biochim. Biophys. Acta – Biomembranes*, 1778 (10), 2308–2317.

Tarek, M., 2005. Membrane electroporation: a molecular dynamics simulation, *Biophys. J.*, 88 (6), 4045–4053.

Tekle, E., Astumian, R., Friauf, W., and Chock, P., 2001. Asymmetric pore distribution and loss of membrane lipid in electroporated DOPC vesicles, *Biophys. J.*, 81 (2), 960–968.

Tepper, H.L. and Voth, G.A., 2005. Protons may leak through pure lipid bilayers via a concerted mechanism, *Biophys. J.*, 88 (5), 3095–3108.

Tieleman, D.P., Leontiadou, H., Mark, A.E., and Marrink, S.J., 2003. Simulation of pore formation in lipid bilayers by mechanical stress and electric fields, *J. Am. Chem. Soc.*, 125 (21), 6382–6383.

Tieleman, D.P. and Marrink, S.J., 2006. Lipids out of equilibrium: energetics of desorption and pore mediated flip-flop, *J. Amer. Chem. Soc.*, 128 (38), 12462–12467.

Ting, C.L., Awasthi, N., Müller, M., and Hub, J.S., 2018. Metastable prepores in tension-free lipid bilayers, *Phys. Rev. Lett.*, 120 (12), 18103.

Tolpekina, T., Den Otter, W., and Briels, W., 2004. Nucleation free energy of pore formation in an amphiphilic bilayer studied by molecular dynamics simulations, *J. Chem. Phys.*, 121, 12060–12066.

Torrie, G.M. and Valleau, J.P., 1974. Monte Carlo free energy estimates using non-Boltzmann sampling: application to the sub-critical Lennard-Jones fluid, *Chem. Phys. Lett.*, 28, 578–581.

Vorobyov, I., Olson, T.E., Kim, J.H., Koeppe, R.E., Andersen, O.S., and Allen, T.W., 2014. Ion-induced defect permeation of lipid membranes, *Biophys. J.*, 106 (3), 586–597.

Wimley, W.C., 2010. Describing the mechanism of antimicrobial peptide action with the interfacial activity model, *ACS Chem. Biol.*, 5 (10), 905–917.

Wohlert, J., den Otter, W.K., Edholm, O., and Briels, W.J., 2006. Free energy of a trans-membrane pore calculated from atomistic molecular dynamics simulations, *J. Chem. Phys.*, 124 (15), 154905.

Zasloff, M., 2002. Antimicrobial peptides of multicellular organisms, *Nature*, 415 (6870), 389–395.

Zhelev, D.V. and Needham, D., 1993. Tension-stabilized pores in giant vesicles: determination of pore size and pore line tension, *Biochim. Biophys. Acta – Biomembranes*, 1147 (1), 89–104.

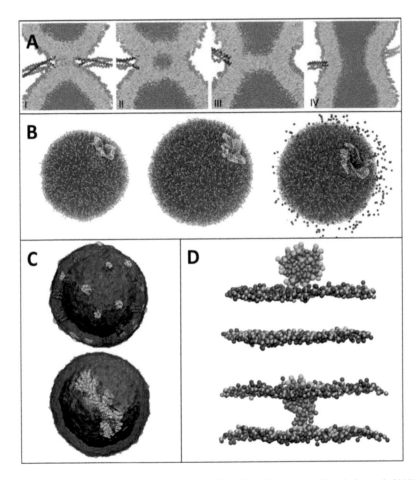

FIGURE 2.2 (A) Two vesicles undergoing fusion mediated by SNARE proteins (Risselada et al. 2011). Lipids are shown in orange (heads) and gray (tails), and the SNARE proteins are shown in red, green, and blue. The solvent inside the vesicles is shown in blue. (B) Snapshots of a liposome undergoing hypo-osmotic shock due to the interaction with a mechano-sensitive channel (Louhivuori et al. 2010). (C) Aggregation of transmembrane proteins induced by hydrophobic mismatch (Parton et al. 2011). The lipid bilayer is shown in blue and the proteins are shown in yellow. (D) A 2-nm gold nanoparticle coated with hydrophobic octanethiol ligands (red beads) and 11-mercaptoundecane-sulfonate ligands (terminating with green anionic beads) that interacts with a lipid bilayer (Simonelli et al. 2015).

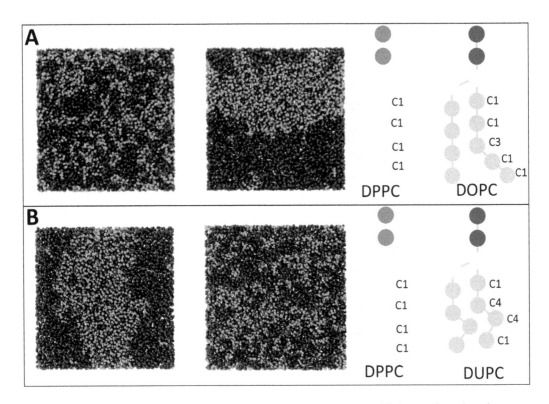

FIGURE 2.3 Phase changes in lipid–cholesterol bilayers. (A) Left snapshot: Equilibrium configuration of a ternary bilayer composed of DPPC (green), DOPC (blue) and cholesterol (red), with current Martini parameterization. Right snapshot: Equilibrium configuration of the same ternary bilayer, but with a more repulsive C1-C3 interaction. (B) Left snapshot: Equilibrium configuration of a ternary bilayer composed of DPPC (green), DUPC (blue) and cholesterol, with current Martini parameterization. Right snapshot: Equilibrium configuration of the same ternary bilayer, but with a less repulsive C1–C4 interaction. From Davis et al. (2013).

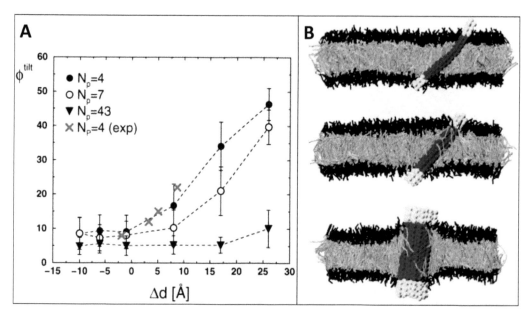

FIGURE 2.5 (A) Protein tilt angle vs. hydrophobic mismatch $\Delta d = d_p - d_L^{(0)}$, where d_p and $d_L^{(0)}$ are the hydrophobic thicknesses of the protein and the pure DMPC bilayer, respectively, for three values of protein size, N_p. Proteins are simulated as a bundle of N_P amphiphilic chains. The red crosses are from fluorescence spectroscopy experiments for the M13 natural peptide (Koehorst et al. 2004). (B) Configuration snapshots that correspond, from top to bottom, to $Np = 4$, 7 and 43, respectively. The hydrophobic and hydrophilic beads of the protein are in blue and yellow, respectively. The hydrophobic and hydrophilic beads of the lipids are in green and black, respectively. From Venturoli et al. (2005).

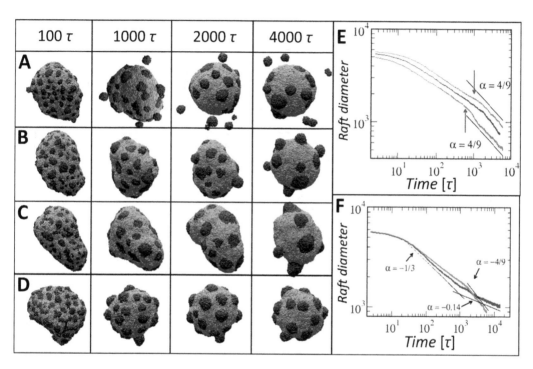

FIGURE 2.6 Time sequence of snapshots of a multi-component lipid vesicles undergoing phase separation for: (A) a vesicle with high line tension, (B) a vesicle with intermediate line tension, (C) a vesicle with low line tension, and (D) a vesicle with low line tension and asymmetric transbilayer lipid distribution. (E) Average domain size vs. time for the case of A (red), B (blue) and C (green). (F) Average domain size vs. time for the case of snapshots series in C (red) and D (blue). From Laradji and Kumar (2004, 2005, 2006).

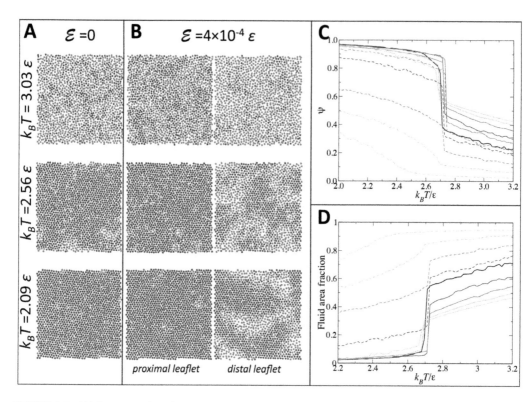

FIGURE 2.8 (A) Sequence of equilibrium snapshots at different temperatures of a non-supported lipid bilayer ($E = 0$). (B) Sequence of equilibrium snapshots at different temperatures of an SLB at a finite bilayer–substrate interaction, $E = 4 \times 10^{-4}\varepsilon$. In (B), both proximal and distal leaflets are shown. Blue and red dots correspond to chains in the gel and fluid state, respectively. (C) Lipid density of the proximal leaflet (solid lines) and distal leaflet (dotted lines) for different bilayer–substrate interactions. Red corresponds to $E = 1 \times 10^{-4}\varepsilon$, blue corresponds to $E = 2 \times 10^{-4}\varepsilon$, green corresponds to $E = 4 \times 10^{-4}\varepsilon$, orange corresponds to $E = 6 \times 10^{-4}\varepsilon$, and cyan corresponds to $E = 8 \times 10^{-4}\varepsilon$. The black curve corresponds to a non-supported lipid bilayer. (D) Translational order parameter of both leaflets. Same color coding as in (C) is used. From Poursoroush et al. (2017).

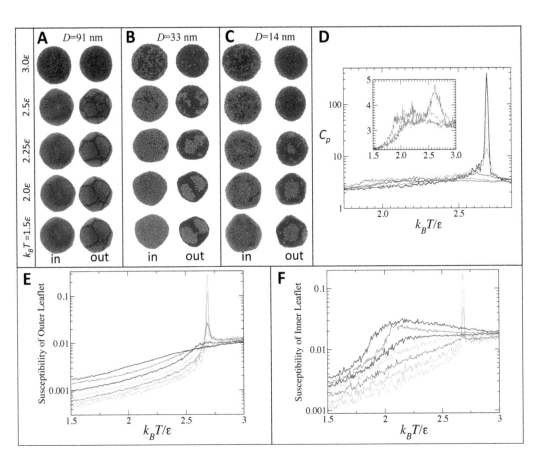

FIGURE 2.9 Snapshots of liposomes with three different diameters, $D = 91$ nm (A), $D = 33$ nm (B), and $D = 14$ nm (C), at different temperatures. For each diameter, both the inner (In) and outer (Out) leaflets are shown. (D): Specific heat vs. temperature, where the color of the shown curves corresponds to liposome diameters $D = 128$ nm (black), 91 nm (green), 64 nm (magenta), 45 nm (orange), 33 nm (blue), 23 nm (red), and 14 nm (violet). Inset shows zoomed-in specific heats for liposomes with diameters 33 nm (blue), 23 nm (red), and 14 nm (violet). (E) and (F): Susceptibility vs. temperature for the outer and inner leaflet, respectively. Color code as described in (D). From Spangler et al. (2012).

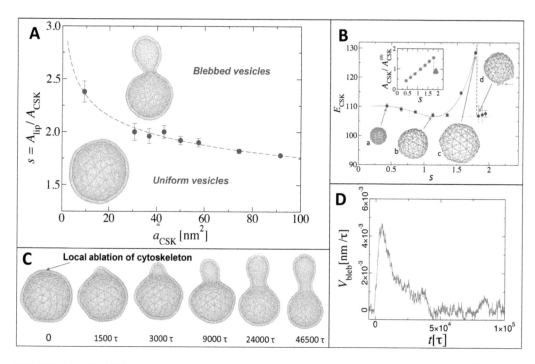

FIGURE 2.10 (A) Blebbing-formation phase diagram in terms of relaxed corral area (horizontal axis), $a_{CSK}^{(0)}$ and the mismatch parameter, $s = A_{lip}/A_{CSK}^{(0)}$ (vertical axis), where A_{lip} is the area of the lipid vesicle and $A_{CSK}^{(0)} = N_{cor}a_{CSK}^{(0)}$ is the relaxed area of the cytoskeleton (N_{cor} is the number of cytoskeleton corrals). $s > 1$ therefore corresponds to a stretched cytoskeleton. Red and yellow in the configurations represent the lipid membrane and the cytoskeleton, respectively. (B) Cytoskeleton elastic energy versus the mismatch parameter, s, along with configurations of the cytoskeleton. The orange arrow on snapshot (d) indicates where the bleb is located. The inset of (B) shows that, in the uniform phase, the area of the cytoskeleton increases linearly with increasing s, since in this phase the whole vesicle is apposed to the cytoskeleton. However, in the blebbing phase, the cytoskeleton is relaxed, hence $A_{CSK} \approx A_{CSK}^{(0)}$. (C) Time sequence of configurations of a vesicle undergoing blebbing following a localized ablation (blue arrow) at $t = 0$. (D) Front velocity of the bleb growth shown in (C). From Spangler et al. (2011).

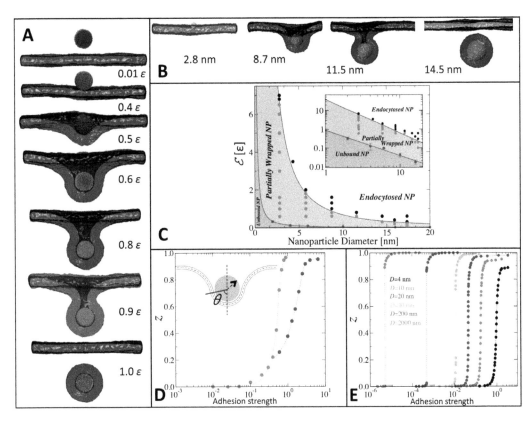

FIGURE 2.11 (A) Sequence of equilibrium snapshots of an 8.7-nm nanoparticle (green) and a tensionless membrane (brown) for different adhesion strengths. (B) Snapshots of nanoparticles with diameter ranging between 2.8 nm (left) and 14.5 nm (right), for the same adhesion strength corresponding to 0.6ε. (C) Wrapping-endocytosis phase diagram in terms of nanoparticle diameter and adhesion strength. Black symbols (in white region) represent the state of endocytosed nanoparticles. Red symbols (in cyan region) correspond to nanoparticles that are partially wrapped by the bilayer. Nanoparticles are unbound from the bilayer in the green region. The inset shows the same data in log–log plot. The slope of the solid line is −1.3, and that of the dashed line is −1.8. (D) Degree of wrapping, defined by $z = (1 - \cos\theta)/2$, where θ is the wrapping angle, vs. adhesion strength for nanoparticles of diameter 2.8 nm (blue) and 7.8 nm (red). (E) Degree of wrapping vs. adhesion strength as obtained from a continuum theory that takes into account the thickness of the membrane and the finite range of the nanoparticle. From Spangler et al. (2016).

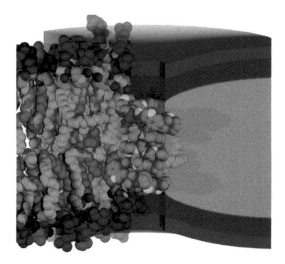

FIGURE 3.2 An illustration of a material continuum deformation around a gramicidin channel. The hydrophobic length of the channel 'pinches' the surrounding leaflets.

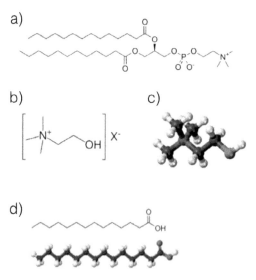

FIGURE 4.1 Dimyristoyl phosphatidylcholine (DMPC) phospholipid macromolecule. a) DMPC chemical structure. b) Headgroup: choline (quaternary ammonium salt) containing the N,N,N-trimethylethanolammonium cation and with an undefined counteranion X $^-$. c) Choline detailed structure as ball-and-stick, with carbon in black, hydrogen in white, oxygen in red, nitrogen in blue. d) Tailgroup: two myristoyl chains, in skeletal and ball-and-stick representations.

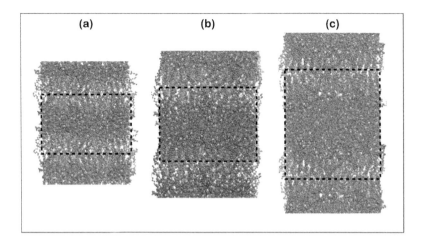

FIGURE 4.2 Snapshots of a DMPC bilayer with fixed hydration levels (a) $\omega = 7$, (b) 20, (c) 34. Other hydration levels considered in Calero et al. (2016) were $\omega = 4$, 10, and 15. Gray and red beads represent phospholipid tails and headgroups, respectively. Blue and white beads represent oxygen and hydrogen atoms of water. The dashed line indicates the size of the simulation box.

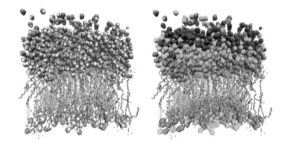

FIGURE 4.10 Snapshot of the hydrated DMPC phospholipid membrane, represented by the gray carbon backbone of the lipids. The system has periodic boundary conditions, reproducing water between stacked membranes. (Left panel) An initial configuration with water oxygens represented in red and water hydrogens in white. (Right panel) The same configuration where now oxygen and hydrogen atoms are colored based on their distance from the closest lipid atom: (1) red for those at distance up to 3 Å, (2) green for those at larger distances up to 9 Å, (3) blue for the rest in 9–15 Å range. Careful inspection reveals that there are water molecules with atoms with different colors. These molecules across regions are excluded from our analysis.

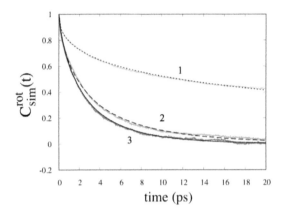

FIGURE 4.11 Rotational dipolar correlation function $C_{sim}^{rot}(t)$ as function of time from MD simulations of water between stacked membranes at T=288.6 K and 1 atm. From top to bottom: Correlation of (1) water molecules within 3 Å average distance from the membrane, (2) at larger distances up to 9 Å, (3) at distances above 9 and up to 15 Å. Data from simulations are represented as colored points (blue for 1, green for 2, red for 3), fit with stretched exponential decay equation (6) with continuous lines (dotted for 1, dashed for 2, solid for 3). Fitting parameters are given in Table I.

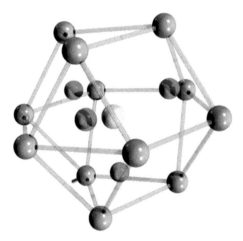

FIGURE 4.12 Pictorial representation of the second neighbor shell in cubic ice (or cuboctahedron). Gray spheres indicate the position of oxygen atoms, while the green lines emphasize the geometrical structure. The central oxygen is depicted by the yellow sphere, while the red spheres represent the first shell of neighboring oxygens.

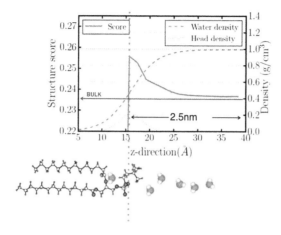

FIGURE 4.13 Upper panel: Structure of water and lipids along the z-axis perpendicular to the average surface plane of the membrane. The right vertical axes of the plot represent the density profile of lipid heads (in orange) and water (in gray) along z. The z of maximum heads density coincides approximately with the largest variation of water density and is marked by a vertical dotted line in red. The left vertical axes represent the average score S of water (in red) along z. S is zero inside the membrane and is higher than its bulk value (blue arrow) within at least 2.5 nm distance from the surface. Lower panel: Schematic representation of a configuration with a lipid head near the surface location (vertical dotted line in red) and water molecules mimicking their density profile as indicated in the upper panel.

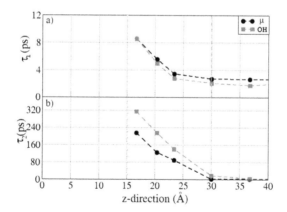

FIGURE 4.15 Relaxation times (a) τ_1 and (b) τ_2 for the dipole (black circles) and OH (red squares) vector correlation functions as a function of the distance from the membrane surface. All the relaxation times, with $\tau_2 \gg \tau_1$, rapidly increase where the variations in S with respect to bulk is appreciable (region between the two dotted vertical lines). The distance is calculated with respect to the center of the bilayer ($z = 0$). Adapted from Martelli and Frontiers (2018).

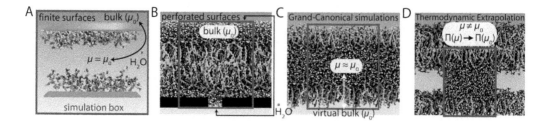

FIGURE 5.1 Approaches for the treatment of the chemical potential of water in atomistic simulations of interacting lipid membranes. (A) Finite objects interacting in a reservoir of bulk water. (B) Water exchange through perforated surfaces. (C) Grand-Canonical simulations: water exchange with a virtual bulk reservoir. (D) Thermodynamic Extrapolation: extrapolation of the pressure at a given water chemical potential to the pressure at the desired chemical potential.

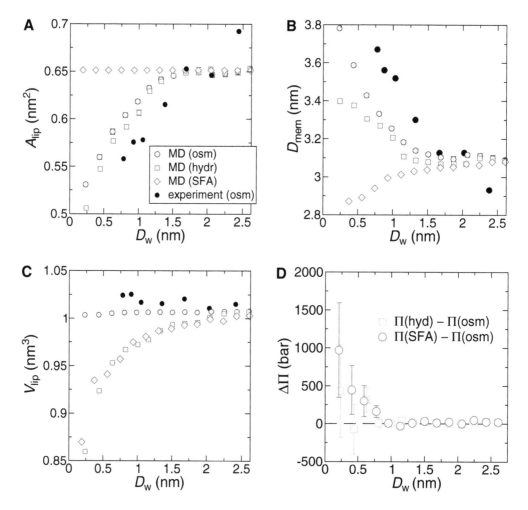

FIGURE 5.3 Structural properties of DLPC bilayers from three different simulation ensembles (osmotic stress simulations, hydrostatic simulations, and SFA simulations) compared with data from osmotic stress experiments[12] as a function of membrane–membrane separation: (A) area per lipid, (B) bilayer thickness, and (C) volume per lipid. (D) Differences in evaluated interaction pressures between the bilayers in hydrostatic and osmotic stress simulations (orange squares) and between SFA and osmotic stress simulations (blue circles). The data are adopted from Ref. [18].

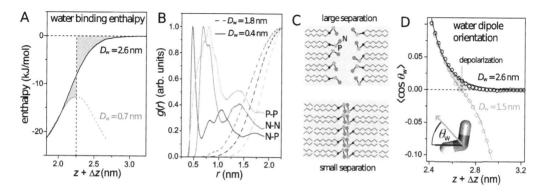

FIGURE 5.4 Repulsion mechanisms between fluid PC lipid membranes. (A) Binding enthalpy profile per water molecule along the axis perpendicular to the membrane plane. Blue area: enthalpy reduction due to strengthening of water binding upon surface approach. Orange area: enthalpy increase due to removal of strongly bound interfacial water. (B) Radial distribution functions $g(r)$ between phosphorus (P) and nitrogen (N) atoms in the opposing membrane surfaces evidencing a reduction of lipid configurational entropy upon membrane approach. (C) Schematic illustration of the headgroup reorganization upon dehydration. (D) Water orientation profiles $\langle \cos \theta_w \rangle$ at a membrane surface exhibiting depolarization (gray area) with decreasing D_w.

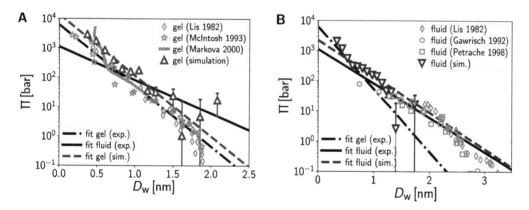

FIGURE 5.6 Comparison of hydration pressure from simulations (triangles) and experiments as a function of the water slab thickness D_w in the (A) gel and (B) fluid states. Exponential fits to the experimental data give decay lengths $\lambda_{gel} = 0.21$ nm (black dashed line) for the fit range [0, 1.8 nm] and $\lambda_{fluid} = 0.38$ nm (black solid line) for the fit range [0, 2.6 nm]. Fits to the simulation data yield decay lengths $\lambda_{gel} = 0.22$ nm (blue dashed line) and $\lambda_{fluid} = 0.36$ nm (red solid line) for fit ranges [0, 1.3 nm] and [0, 1.4 nm], respectively, which are restricted to the distance range where pressures are strictly positive.

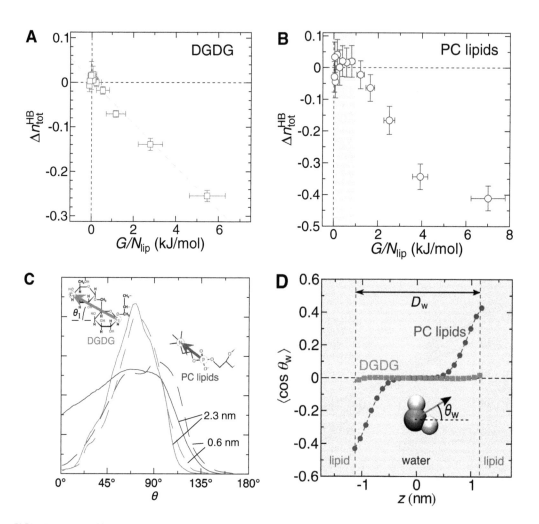

FIGURE 5.8 (A) Change of the total number of HBs versus the interaction free energy for DGDG. Dashed line: Linear regression through all data points. (B) The same plot for PC lipids. Shaded region: high hydration regime where $\Delta n_{tot}^{HB} \approx 0$. (C) Distributions of DGDG and PC lipid headgroup orientations with respect to the membrane normal for large ($D_w = 2.3$ nm, solid lines) and small ($D_w = 0.6$ nm, dashed lines) separations. Inset: Definition of the headgroup vectors. (D) Water orientation profiles $\langle \cos\theta_w \rangle$ between PC and DGDG lipid membranes at separation $D_w = 2.3$ nm. The blue-shaded area denotes the water layer and the orange area the poorly hydrated lipid region. Inset: Definition of the water dipole angle θ_w.

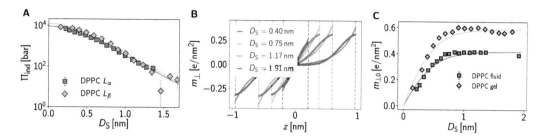

FIGURE 5.9 Indirect pressures and water orientation profiles. (A) Indirect, water-mediated, interaction pressure between DPPC membranes. Lines denote fits of the Landau–Ginzburg model, Eq. (11). (B) Dipole polarization $m_\perp$. Solid lines denote fits of Eq. (10) to the simulation data, vertical dashed lines denote the interface position. (C) Polarization order parameter at the interface $m_{\perp 0}$ obtained from the fits in (B). Solid lines denote fits of the value at large separation, $m_\perp^\infty$, to the simulation data.

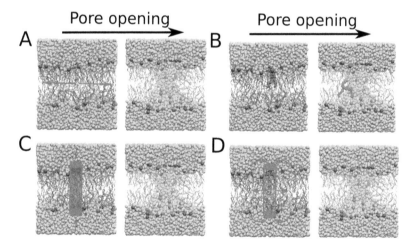

FIGURE 6.1 Illustration of four reaction coordinates (RCs) suggested for calculating the potential of mean force of transmembrane pore formation. (a) Collective radial coordinate, (b) the flip-flop coordinate, (c) the average water density in a membrane-spanning cylinder, and (d) the chain coordinate. Lipid molecules are visualized as silver sticks, lipid phosphate atoms as gray spheres, and water molecules as cyan spheres.

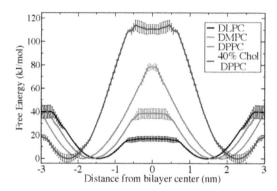

FIGURE 6.4 PMFs for pore formation along the flip-flop coordinate for phosphatidylcholine (PC) membranes with increasing tail length and as function of cholesterol content. Longer tails and the addition of cholesterol increase the free energy of pore formation. Adapted with permission from Bennett and Tieleman (2014). Copyright (2014) American Chemical Society.

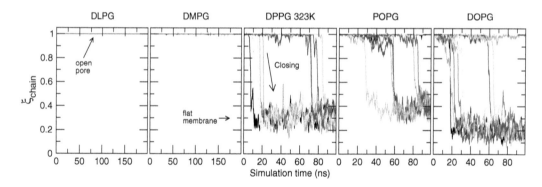

FIGURE 6.6 Trajectories of free MD simulations starting from frames with an open pore in PG membranes with increasing tail length and increasing tail unsaturation. Eight independent simulations shown in different colors were conducted for each lipid type (from left to right): DLPG, DMPG, DPPG, POPG, and DOPG. No bias was applied to the simulations. Lipids were modeled with the Charmm36 force field, and simulation parameters were chosen as described previously (Pastor and MacKerell, 2011; Ting *et al.*, 2018). To visualize the spontaneous closing of the pores during the free simulations, the trajectories are projected onto the RC ξ_{chain}, where $\xi_{chain} \approx 1$ and $\xi_{chain} \approx 0.3$ correspond to open pores and flat unperturbed membranes, respectively (arrows). No pore closed in the bilayers of DLPG and DMPG within 200 ns, confirming the metastability of the pores, and compatible with the pronounced nucleation barriers revealed by the PMF (compare with Figure 5b). Pores in bilayers of DPPG, POPG, and DOPG closed on the tens of nanosecond time scale, compatible with the shallow nucleation barriers in these membranes.

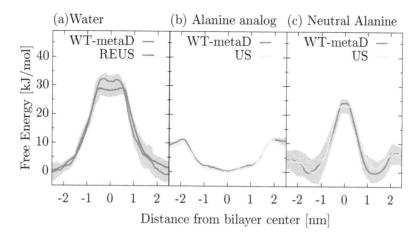

FIGURE 7.1 Free energy profiles for the translocation of (a) a single water molecule, (b) the side chain analog of alanine, and (c) the unionized form of alanine through a DOPC bilayer. Uncertainties were estimated by bootstrapping for US and REUS simulations and by the method proposed in Ref.[18] for WT-metaD calculations. For these three solutes, all biasing methods yield similar results. Adapted with permission from J. Chem. Theory Comput. 14, 1762. Copyright 2018 American Chemical Society

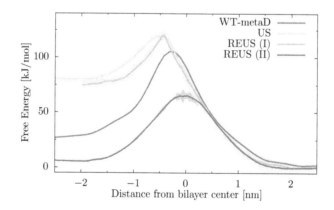

FIGURE 7.4 Free energy profiles for the translocation of zwitterionic alanine. WT-metaD, US, and initial REUS simulations yield PMFs that are not symmetric, which illustrates their failure to converge. A second set of REUS simulations, with initial conditions chosen as described in the main text, does result in a converged, symmetric profile with a smooth barrier at the center of the bilayer. Adapted with permission from J. Chem. Theory Comput. 14, 1762. Copyright 2018 American Chemical Society

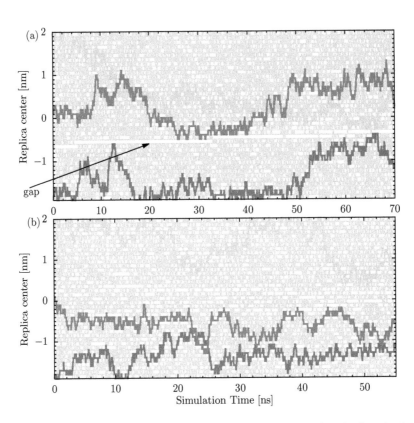

FIGURE 7.5 Replica exchange patterns for zwitterionic alanine, illustrating the motion of trajectories through replica space. In the initial REUS (I) simulation, there is a persistent gap (indicated by the arrow) that separates two sets of replicas with very limited exchange (a). The REUS (II) simulation, which used a different set of initial conditions, does not show such an obstruction (b). In both panels, we highlight 3 out of 40 trajectories (the ones starting in the $z = 2$, 0.1, and −1.9 nm replicas) for clarity. Adapted with permission from J. Chem. Theory Comput. 14, 1762. Copyright 2018 American Chemical Society

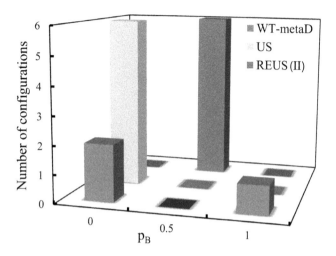

FIGURE 7.8 Committor distribution functions for zwitterionic alanine. For every tested $z = 0$ configuration obtained from WT-metaD and US simulations, all unbiased trajectories that start from this configuration evolve towards the same side of the membrane. Configurations with $z = 0$ generated in the REUS (II) simulation, on other hand, have no preference for either side of the bilayer. Adapted with permission from J. Chem. Theory Comput. 14, 1762. Copyright 2018 American Chemical Society

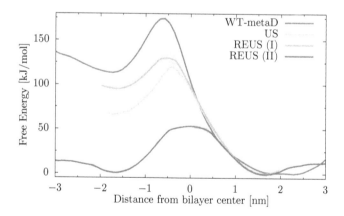

FIGURE 7.9 Free energy profiles for the translocation of a cationic arginine residue. As illustrated previously in Figure 4 for the case of zwitterionic alanine, WT-metaD, US, and initial REUS simulations did not converge, as is evident from their asymmetric PMFs. The REUS (II) profile, obtained after extensive equilibration, is symmetric and has its maximum at $z = 0$. Adapted with permission from J. Chem. Theory Comput. 14, 1762. Copyright 2018 American Chemical Society

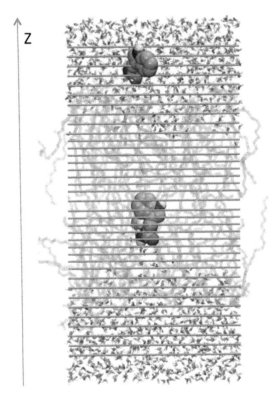

FIGURE 8.1 An atomically detailed model of a snapshot in time of a permeating trajectory. The Z-axis is normal to the plane and is the coarse coordinate used in the diffusion solubility model. The red and white stick models are water molecules. The green sticks are phospholipid (DOPC) molecules. The blocked trypotophan molecules (space filling models) and the meaning of the orange lines (milestones) are discussed in Section 4.

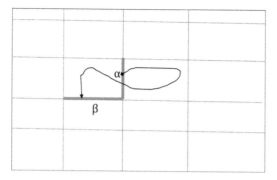

FIGURE 8.2 A schematic illustration of a Milestoning trajectory (black arrowed curve) that starts at milestone α and end at milestone β (red line segments). Note that the trajectory is terminated only after 'touching' a milestone different from the milestone it was initiated on.

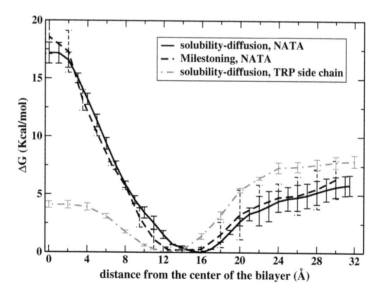

FIGURE 8.3 The potential of mean force computed both with Milestoning and Umbrella sampling are similar with a large permeation barrier for translocation of NATA in the middle of the membrane and favorable binding in the glycerol region of the phospholipid molecules. (Reprinted with permission from A. E. Cardenas, G. J. Jas, K. Y. DeLeon, W. A. Hegefeld, K. Kuczera, and R. Elber, "Unassisted transport of N-acetly-L-tryptophanamide through DOPC membrane: Experiment and simulation", J. Phys. Chem. B, 116,2739-2750. Copyright 2012 American Chemical Society, Ref. [23]).

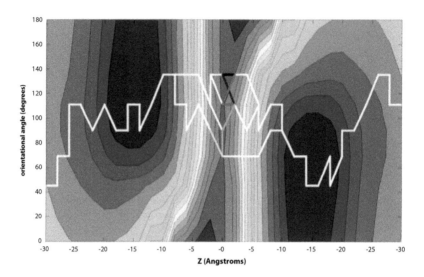

FIGURE 8.4 The free energy for permeation of NATA as a function of the location of the center of mass of the permeant along the Z-axis and its orientation. Changes in the free energy are shown with the color-filled contours: lower free energy in blue and higher free energy in red. An orientational preference is clearly observed but the larger changes of free energy occur along the Z-axis. The figure also shows paths of maximum flux for permeation. The transition state for the paths displayed are colored with different gray shadings (Reproduced with permission from A. E. Cardenas and R. Elber, "Computational Study of Peptide Permeation Through Membrane: Searching for Hidden Slow Variables", Mol. Phys., 111,3565-3578(2013) Ref. [7] with the permission of Taylor & Francis)

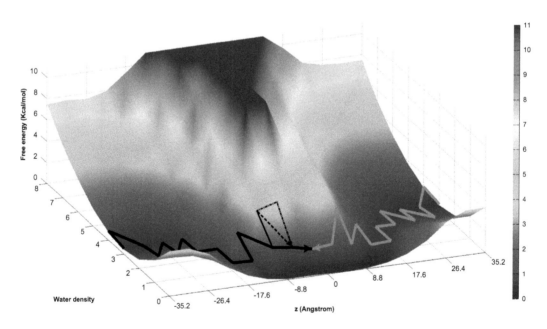

FIGURE 8.6 Free-energy surface for water permeation as a function of water occupancy in each cell and the position along the membrane axis. The figure also shows maximum flux path for water permeation to the membrane center from the left-side water (in black) and right-side water phases (in mauve). The three maximum flux paths are shown at the left (they only differ from each other in the segments shown with dashed lines). These maximum flux paths clearly involve more than one water moving inside the membrane core. (Reproduced from A. E. Cardenas and R. Elber, "Modeling Kinetics and Equilibrium of Membranes with Fields: Milestoning Analysis and Implication to Permeation", J. Chem. Phys. 141,054101(2014) (Ref. 25), with the permission of AIP Publishing).

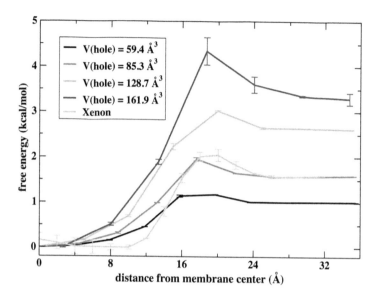

FIGURE 8.7 Free-energy profile for hole formation along the distance from the center of the membrane. The black, red, green, and blue lines show its dependence on the size of the cubic grid used in the density discretization (see Figure 5). The orange line shows the PMF for permeation of a xenon atom. The diameter of a xenon atom is 4.3 Å, while the size of the cubic grid for the red line (the closest to the PMF for xenon) is 4.4 Å (enough to enclose a xenon atom). (Reproduced from Ref. 25, with the permission of AIP Publishing).

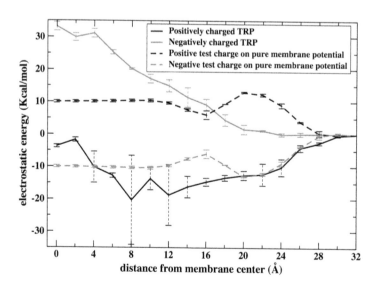

FIGURE 8.9 Electrostatic potential energy of a permeant charge moving inside the membrane. The solid lines are the energy felt by the positive- (solid black line) and negative-charged permeant (solid red line). The dashed line is the results obtained by placing a charge probe in a pure lipid membrane. The pure membrane prefers to transport a negatively charged molecule while the membrane perturbed by the presence of the permeant prefers to move a positively charged molecule. (Reprinted with permission from A. E. Cardenas, R. Shrestha, L. J. Webb, and R. Elber, "Membrane Permeation of a Peptide: It is Better to be Positive", J. Phys. Chem. B, 119, 6412-6420. Copyright 2015 American Chemical Society. Ref. [3].)

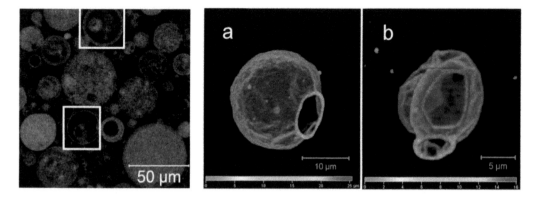

FIGURE 9.1 Confocal microscopy images of NP-membrane complexes. On the left: bodipy-labeled Au NPs (2 nm diameter), functionalized by a mixture of hydrophobic and anionic ligands, co-localize with the bilayer of multi-lamellar DOPC vesicles. The NPs are shown to be present in outer as well as in inner vesicles, suggesting spontaneous passive translocation (adapted from Van Lehn *et al.*[13]). On the right: SiO$_2$ NPs (18 nm diameter, negative zeta potential) stabilize unusual curvatures and holes in DOPC giant unilamellar vesicles (adapted from Zhang *et al.*[11])

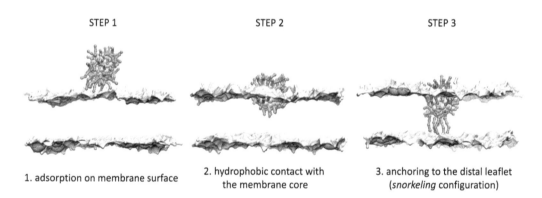

FIGURE 9.2 The 3 steps of NP–membrane interaction, as simulated by MD at CG level. Step 1 corresponds to surface adsorption of the monolayer-protected NP. In Step 2, the NP establishes a hydrophobic contact with the membrane core. In Step 3, the NP is anchored to both membrane leaflets. Water and lipid tails are not shown, lipid headgroups are shown as gray surfaces, the Au NP core is yellow, hydrophobic ligands are green, and anionic ligands are pink.

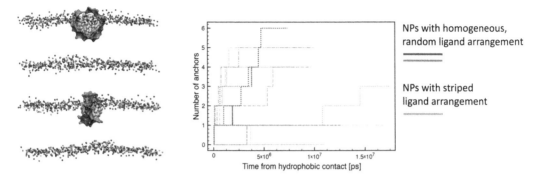

FIGURE 9.3   Left: Au NPs with a striped arrangement of hydrophobic and anionic ligands interact with a POPC lipid bilayer. Only the hydrophobic ligands are shown in red, while lipid headgroups are represented as blue/gray beads. Lipid tails and water are not shown. The NP inserts in the membrane with a fixed orientation of its hydrophobic stripe. Right: the kinetics of anchoring for NPs with different ligand patterns. The NPs with a random ligand arrangement, corresponding to a homogeneous distribution of charges and hydrophobic groups on their surface, anchor to the membrane with a faster kinetics.

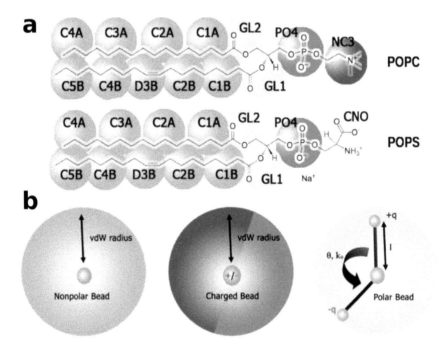

FIGURE 11.1   Coarse-grained (CG) mapping scheme and bead types in WEPMEM (Water-Explicit Polarizable MEMbrane) model. (a) CG mapping scheme of POPC (upper) and POPS (lower) lipid. POPC lipid consists of choline (NC3), phosphate (PO4), and glycerol (GL1/GL2) groups as headgroups. POPS lipid consists of serine (CNO), phosphate (PO4), and glycerol (GL1/GL2) groups as headgroups. Both lipids have a palmitoyl tail (C1A, C2A, C3A, C4A) and an oleoyl tail (C1B, C2B, D3B, C4B, C5B). (b) The color of beads represents different bead types. Yellow beads are polarizable, green beads are hydrophobic, red beads are negatively charged, and blue beads are positively charged. Reproduced from Ganesan, S.J., Xu, H. and Matysiak, S., Effect of lipid headgroup interactions on membrane properties and membrane-induced cationic β-hairpin folding, Phys. Chem. Chem. Phys., 18, 17836–17850, 2017. With permission from the PCCP Owner Societies.

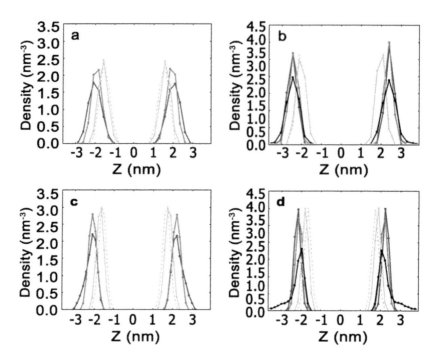

FIGURE 11.2 Density distribution of POPC and POPS lipids with atomistic GROMOS force field and WEPMEM model: (a) atomistic POPC lipid; (b) WEPMEM POPC lipid; (c) atomistic POPS lipid; (d) WEPMEM POPS lipid. Red: choline/serine group, blue: phosphate group, solid yellow: glycerol group in oleoyl tail, dashed yellow: glycerol group in palmitoyl tail, black: counterion. The center of the bilayer is located at $z = 0$. Reproduced from Ganesan, S.J., Xu, H. and Matysiak, S., Effect of lipid headgroup interactions on membrane properties and membrane-induced cationic β-hairpin folding, Phys. Chem. Chem. Phys., 18, 17836–17850, 2017. With permission from the PCCP Owner Societies.

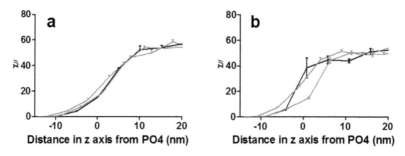

FIGURE 11.3 Susceptibility profile of (a) POPC and (b) POPS lipid. Black curves are calculated from WEPMEM model, gray curves are from MARTINI model, and red curves are from GROMOS force field. Reproduced from Ganesan, S.J., Xu, H. and Matysiak, S., Effect of lipid headgroup interactions on membrane properties and membrane-induced cationic β-hairpin folding, Phys. Chem. Chem. Phys., 18, 17836–17850, 2017. With permission from the PCCP Owner Societies.

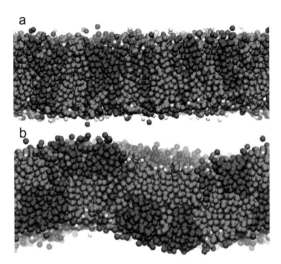

FIGURE 11.5 Side view snapshots of 1:1 mixed 960 POPC/POPS lipids bilayer demonstrating the difference in bilayer curvature. (a) The bilayer with ion size of 1.0 σ; (b) The bilayer with ion size of 0.8σ. Blue regions represent POPC lipids, red regions represent POPS lipids, and small yellow beads are monovalent ions. Reprinted with permission from Ganesan, S.J., Xu, H., and Matysiak, S., Influence of monovalent cation size on nanodomain formation in anionic—zwitterionic mixed bilayers, *J. Phys. Chem. B*, 121, 787–799. Copyright 2017 American Chemical Society.

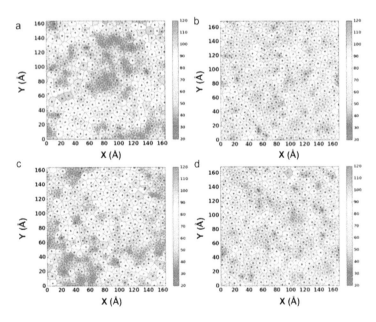

FIGURE 11.6 Voronoi diagram of 1:1 mixed 960 POPC/POPS lipids bilayer showing the area of each lipid. (a) The lipid bilayer upper leaflet with ion size of 0.8 σ; (b) The lipid bilayer lower leaflet with ion size of 0.8σ; (c) The lipid bilayer upper leaflet with ion size of 1.0σ; (d) The lipid bilayer lower leaflet with ion size of 1.0σ. Each Voronoi cell corresponds to a single lipid shown in red (POPS lipid) or blue (POPC lipid) dots. Color bar represents the area of each lipid in Å^2. Reprinted with permission from Ganesan, S.J., Xu, H., and Matysiak, S., Influence of monovalent cation size on nanodomain formation in anionic–zwitterionic mixed bilayers, *J. Phys. Chem. B*, 121, 787–799. Copyright 2017 American Chemical Society.

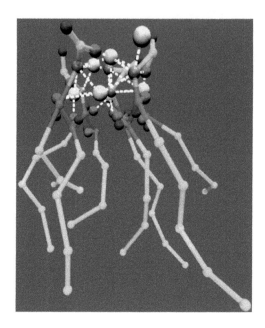

FIGURE 11.7 A typical snapshot of cation–PS complex demonstrating interactions in POPS lipid headgroup. Yellow beads are monovalent cations, tan beads are phosphate groups, and magenta beads are glycerol groups. Drude-like dummy particles with positive charges are shown in red, and with negative charges shown in blue. The white dashed lines indicate the binding of monovalent cations to lipid headgroups. Reprinted with permission from Ganesan, S.J., Xu, H., and Matysiak, S., Influence of monovalent cation size on nanodomain formation in anionic—zwitterionic mixed bilayers, *J. Phys. Chem. B*, 121, 787–799. Copyright 2017 American Chemical Society.

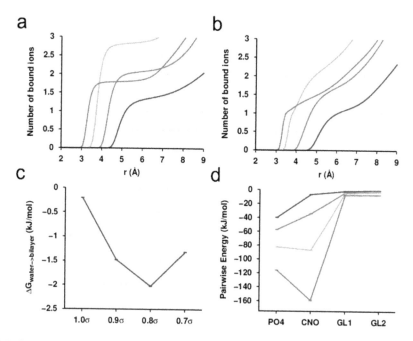

FIGURE 11.9 Ion coordination number near lipid headgroups, ion transfer free energy between water and bilayer interface, and ion-to-lipid headgroup beads pairwise interaction energy. (a) Ion coordination number near lipid phosphate group; (b) ion coordination number near lipid serine group; (c) ion transfer free energy between water and bilayer interface; (d) ion-to-lipid headgroup beads (PO4: phosphate, CNO: serine, GL1/GL2: glycerol groups) pairwise interaction energy. Color scheme in a, b, and d: blue curves are for ion size of 1.0 σ, red curves for ion size of 0.9σ, green curves for ion size of 0.8σ, and magenta curves for ion size of 0.7σ. Reprinted with permission from Ganesan, S.J., Xu, H., and Matysiak, S., Influence of monovalent cation size on nanodomain formation in anionic—zwitterionic mixed bilayers, *J. Phys. Chem. B*, 121, 787–799. Copyright 2017 American Chemical Society.

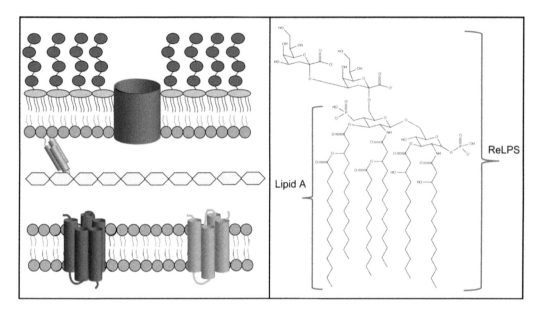

FIGURE 12.1 A schematic diagram of the cell envelope of Gram-negative bacteria (left). The LPS molecules are purple, blue, and dark green, phospholipids are paler green, peptidoglycan is yellow, outer membrane protein is magenta, inner membrane proteins are orange and dark yellow, and Braun's lipoprotein is blue. The chemical structure of Re-LPS from E. coli (right).

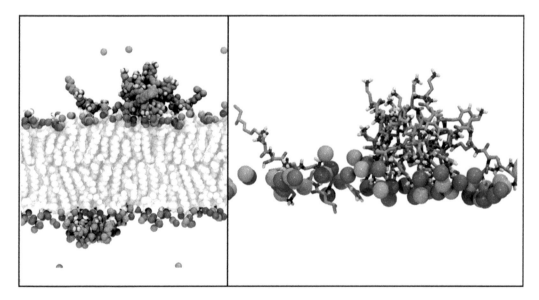

FIGURE 12.2 Atomistic simulations of polymyxin B1 in a bilayer of LPS did not show any penetration of the lipopeptide into the bilayer.[12] Phosphorus atoms of the LPS molecules are shown as gold spheres, Ca^{2+} ions are pink spheres, LPS tails are white spheres, and polymyxin is shown in stick representation.

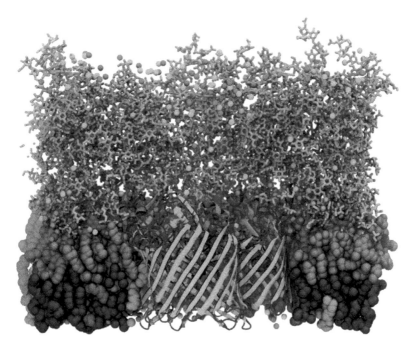

FIGURE 12.3 Representative snapshot of an OmpF trimer (barrel, green; helix, red; loop and turn, blue) embedded in Outer Membrane (OM) of E. coli LPS with R1 core and five repeating units of O6-antigen. The outer leaflet contains Lipid A, represented as pink spheres, core sugars as orange stick model, and O-antigen polysaccharides as gray stick model. The inner leaflet contains 1-palmitoyl(16:0)-2-palmitoleoyl(16:1 cis-9)-phosphatidylethanolamine (blue spheres), 1-palmitoyl(16:0)-2-vacenoyl(18:1 cis-11)-phosphatidylglycerol (orange spheres), and 1,10-palmitoyl-2,20-vacenoyl cardiolipin with a net charge of −2e (magenta spheres). Ca^{2+} ions are represented as cyan small spheres, K^+ ions as green small spheres, and Cl^- ions as magenta small spheres. For clarity, some portions of the system have been removed.[21]

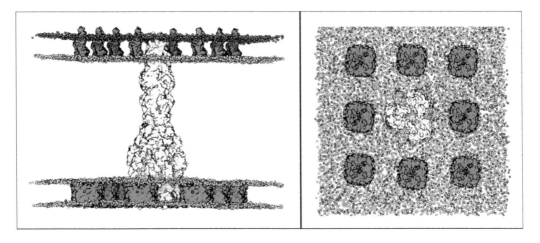

FIGURE 12.4 Two views of one of the simulation systems incorporating both the inner and outer membranes of E. coli reported by Hsu et al.[33] LPS headgroups are purple, phospholipid headgroups are cyan, AqpZ is red, OmpA is blue, and the TolC-AcrABZ efflux pump is yellow.

7

Free Energy Calculation of Membrane Translocation: What Works When, and Why?

Nihit Pokhrel and Lutz Maibaum

1 Introduction

One of the principal roles of biological membranes is to prevent the uncontrolled flux of material across the lipid bilayer. This allows, for example, the interior of a cell to form a chemical environment that is distinct from that in the extracellular space. However, not all molecules are hindered from crossing the membrane to the same extent; especially small and uncharged solutes exhibit appreciable permeation rates. Being able to predict whether a molecule can enter a cell by passive diffusion across the membrane is critical, for example, in the development and assessment of drug candidates.[1]

Using the language of chemical reactions, one can think of membrane translocation as a process that starts with the solute on one side of the membrane (the reactant, or initial, state) and ends with the solute on the other side (the product, or final, state). The path in-between those states is parametrized by a reaction coordinate, the precise nature of which is typically unknown. The rate at which a reaction occurs is often dominated by high free energies of intermediates, which act as a barrier separating the reactant and product states.

Quantifying the magnitude of such free energy barriers is the first step to predict the permeation rates of small solutes. Computer simulations offer a promising route to calculate the free energy profile, also called the potential of mean force (PMF), that governs the translocation process. However, they also face a significant challenge: spontaneous translocation occurs only rarely, especially if the energetic barriers involved are large. Even though advances in computing hardware and new algorithms allow conventional molecular dynamics (MD) simulations at atomistic resolution to extend to multiple microseconds, this is by far insufficient to directly observe a sufficient number (if any!) of translocation events. This is a common problem when performing computer simulations, and a large number of methods have been developed over time to address this issue. In one way or another, these methods bias the system of interest to sample configurations unlikely to be seen in an unbiased simulation, and do so in a way that one can obtain the underlying free energy profile of the unbiased system. Examples of such methods are umbrella sampling,[2] metadynamics (and its variations),[3] replica exchange (and its variations),[4] adaptive biasing force,[5] and thermodynamic integration.[6] These methods improve the statistical sampling and reduce the amount of computational resources required to calculate free energy profiles.

This wealth of available methods is marvelous, but it also raises the question which of them one should choose for a given application, such as calculating the free energy profile of membrane translocation. Only a small number of studies have compared the relative performance of these methods by applying them to the same system. For example, Bochicchio and coworkers found that metadynamics and umbrella sampling performed similarly well when computing the PMF for polypropylene and polyethylene oligomers moving across a phospholipid bilayer.[7] In another study, Lee and colleagues computed the free energy profiles of urea, benzoic acid, and codeine in their neutral forms to compare umbrella sampling, replica exchange umbrella sampling, adaptive biasing force, and multiple-walker adaptive biasing force, and found no clear advantage for any of these methods.[8] For the charged solute n-propylguanidinium, the side chain analog of arginine, Neale and coworkers found that the virtual replica exchange method converges faster than umbrella sampling.[9] We have recently used the well-tempered metadynamics (WT-metaD),[10] umbrella sampling (US),[2] and replica exchange umbrella sampling (REUS)[11] methods to study the translocation of a variety of small molecules through a dioleoylphosphatidylcholine (DOPC) bilayer. We find that while all three methods perform similarly well for solutes that are charge-neutral or that have only a small dipole moment, there are significant convergence problems in the case of charged or highly polar solutes, and that those are most easily detected and overcome by REUS simulations.[12]

These results show that the strengths of electrostatic interactions are likely to affect the relative efficiency among various methods. This by itself gives valuable guidance as to which method one should choose for a given system. It also creates an opportunity to learn something about the basic physical properties from the difference in performance of these methods.

In this chapter, we will explore these topics in detail. We begin with a brief overview of the WT-metaD, US, and REUS methods of free energy calculations. In Section 3, we will show-case how these methods can be used to compute the PMF for various solutes. We demonstrate that observables unrelated to the free energy, in particular the so-called committor of a configuration, can be used to assess the quality of the sampling obtained in simulations. We conclude in Section 4 with some general advice for performing free energy calculations of membrane translocation.

2 Methods for Free Energy Calculations

Our goal is to calculate the equilibrium free energy F as a function of a variable that describes the progress along the translocation pathway. The variable that is most commonly used is the normal component z of the distance between the center of mass (COM) of the solute and that of the membrane. For typical phospholipid bilayers, z ranges from approximately –2 to 2 nm, with $z = 0$ corresponding to a solute position in the middle of the bilayer. The free energy profile $F(z)$ is also known as the PMF.

The variable z is itself a function of the atomic coordinates whose dynamics is calculated in an MD simulation. Such quantities are frequently called collective variables (CVs). To calculate $F(z)$, a simulation must properly sample all representative configurations over the entire range of interest of the CV. This is often unfeasible, especially when the free energy profile contains significant barriers. For example, a hydrophilic solute is unlikely to spontaneously cross the hydrophobic core of a lipid bilayer within typical simulation time scales. However, using accelerating schemes that bias the system to explore a larger range of z than it would usually do, one can calculate these barriers. Below we briefly review three such methods: WT-metaD, US, and REUS.

Even with these methods, the calculation of a free energy profile is computationally expensive. In some cases, one might have *a priori* knowledge about the form of $F(z)$. In particular, if the lipid bilayer is symmetric (i.e., has the same composition in each leaflet) and the environments above and below the bilayer are the same, then $F(z)$ must be an even function. In principle, it is then sufficient to calculate the free energy only for one half-space of z, and fill in the other using the symmetry argument. As we will see, this is not recommended in practice, because it eliminates a simple but powerful opportunity to assess the convergence of the calculation.

2.1 Well-Tempered Metadynamics

Metadynamics is a biasing technique that overcomes sampling problems by adding a history-dependent bias potential to the system potential energy.[13] In regular intervals, a Gaussian-shaped potential is added to the system energy, centered at the current value of the collective variable z. This potential drives z away from its current value, which facilitates the exploration of the entire range of the CV. Over time, these Gaussians add up to the biasing potential

$$V_G(z, t) = \sum_{t' < t} \omega_0 \exp\left(-\frac{(z - z(t'))^2}{2\sigma^2}\right), \tag{1}$$

where ω_0 is the height of the Gaussian potential, σ is its width, and $z(t')$ is the value of the collective variable at time t'.

In the well-tempered variant of metadynamics (WT-metaD), the height of the added Gaussian decreases exponentially with the already deposited bias potential at the current value of the collective variable. In other words, the constant ω_0 in (1) is replaced with[10]

$$\omega = \omega_0 \exp\left(-\frac{V_G(z(t'), t')}{k_B \Delta T}\right), \tag{2}$$

where ΔT is an input parameter that effectively increases the temperature at which the CV is sampled. This simulation scheme has been shown to yield the sought-after free energy profile in the limit of long simulation times through[14]

$$F(z) = -\frac{T + \Delta T}{\Delta T} \lim_{t \to \infty} V_G(z, t) \tag{3}$$

up to an irrelevant additive constant.

2.2 Umbrella Sampling

US adds a biasing potential to the system's Hamiltonian to enhance the sampling of configurations that are high in energy.[2] In this case, the biasing potential is static. We choose a sequence of N 'windows' that span the range of interest of the collective variable z. In the ith window, the system is biased to remain close to a predetermined value z_i by using a harmonic umbrella potential

$$V_i(z) = \frac{1}{2}k(z - z_i)^2, \tag{4}$$

where k is the stiffness of the potential.

For each of those windows, we perform a standard MD simulation, from which we obtain the probability distribution $P_i^b(z)$ of the collective variable in the biased system. We can recover the distribution $P_i(z)$ of the unbiased system over the range of values observed in the ith window via

$$P_i(z) = P_i^b(z)e^{\beta V_i(z)}e^{-\beta C_i}, \tag{5}$$

where C_i is an unknown constant to be determined shortly. Having obtained such estimates for the probability distribution for each window, we still need to combine this information to obtain the free energy profile $F(z)$ over the entire range of z. One efficient way to do so is the Weighted

Histogram Analysis Method (WHAM),[15,16] which expresses the probability distribution $P(z)$ as a linear superposition of the estimates from each window,

$$P(z) = \sum_{i=1}^{N} w_i(z) P_i(z). \tag{6}$$

The weights $w_i(z)$ are chosen to minimize the statistical error of the reconstructed distribution. It can be shown that these weights then satisfy the coupled equations

$$w_i(z) = \frac{S_i e^{-\beta V_i(z) + \beta C_i}}{\sum_{j=1}^{N} S_j e^{-\beta V_j(z) + \beta C_j}} \tag{7}$$

$$e^{-\beta C_i} = \int dz\, e^{-\beta V_i(z)} P(z) \tag{8}$$

which must be solved self-consistently for the unknown C_i. Here, S_i is the number of independent samples obtained from the ith simulation. Once the unbiased distribution (6) is obtained, we can compute the free energy profile

$$F(z) = -k_B T \ln P(z). \tag{9}$$

2.3 Replica Exchange Umbrella Sampling

REUS is very similar to US: again we perform N different simulations, also called replicas, each with a different bias potential such as the one shown in (4). In REUS, however, these simulations are not independent: at pre-determined time intervals two replicas exchange their current configuration with a probability that is determined by the detailed balance condition. The latter guarantees that each replica still samples configurations according to the equilibrium distribution of the biased system. The reconstruction of the free energy profile proceeds in the same way as in US, for example by using the WHAM algorithm.

REUS is different from regular replica exchange simulations, in which the replicas are unbiased in z but are instead held at different temperatures. In that case, simulations at the temperature of interest benefit from the enhanced sampling at higher temperatures. In REUS, trajectories in the ith replica are bound to only explore states close to z_i, and the exchange with other replicas allows the system to overcome free energy barriers in orthogonal directions if those barriers are smaller in other regions of the collective variable. For example, a solute at the center of a bilayer might have very limited rotational freedom. In a regular US simulation, this solute might be trapped there in a specific orientation for a long time. In REUS, on the other hand, the solute is essentially allowed to diffuse to the bilayer surface, rotate, and return to the bilayer center.

As we will see, REUS has another advantage over US in addition to the enhanced sampling: by monitoring the pattern of accepted exchanges between replicas, one can identify sampling problems that occur when the space of replicas partitions into subsets that rarely exchange configurations with each other. This provides a valuable diagnostic tool that is not available in US simulations.

2.4 Error Estimation

Every free energy profile obtained from computer simulations should be accompanied by an estimate of the statistical uncertainty. Because all of the above methods are guaranteed to converge to the correct result in the limit of infinite simulation times, a simple and practical way to establish convergence is to show that the estimate of $F(z)$ becomes independent of simulation time. In the case of metadynamics, it is furthermore assumed that the observed dynamics of the collective variable will become diffusive when the added bias potential completely compensates any variation in the system's free energy profile, which can serve as an additional test for convergence.

Quantifying the uncertainty is more difficult. A straightforward but computationally expensive way to do so is to repeat the entire calculation multiple times, and then compute the uncertainty in $F(z)$ using the standard tools of error analysis. A more convenient and potentially more efficient way to obtain error bars is to divide the data of a single simulation into smaller segments in such a way that they can be considered uncorrelated, and to compute uncertainties based on these segments. Bootstrapping is a method that samples many such partitionings, and can be readily applied to US and REUS simulations. One freely available implementation of this algorithm is the tool g_wham, which is part of the Gromacs software package, and which provides an easy-to-use way to calculate free energy profiles and their uncertainties by combining the WHAM method with bootstrapping analysis.[17]

The ever-increasing nature of the bias potential $V_G(z)$ complicates this analysis in the case of metadynamics simulations. However, one can replace (3) with a time-independent estimator of the free energy profile $F(z)$, and compute its uncertainties from the variance of such estimates at different simulation times.[18] As long as there are no systematic sampling errors, this method provides accurate error bars for the PMF.

2.5 The Committor and Its Distribution

The free energy profile $F(z)$ is a thermodynamic description of a process that proceeds along the collective variable z. It is the starting point for classical theories of chemical kinetics, which typically consider reactions in which $F(z)$ has two minima that correspond to the reactant state A and the product state B, and that are separated by a peak. This peak forms a barrier whose crossing determines the reaction rate. The location of the peak is sometimes thought of as the transition state, which separates the reactant and the product states.

This picture is accurate only if z is indeed the reaction coordinate, i.e., it describes accurately and completely the progress of the process under consideration. Whether that is the case is usually not clear *a priori* in complex systems such as biological membranes. Kinetic theories of chemical dynamics have developed quantitative metrics to describe the progress of a reaction.[19,20] Rather than relying on a projection of the free energy onto an arbitrary CV, they consider the ensemble of paths that go from A to B. Where a single state lies along a reaction is described by the committor p_B, defined as the probability that a trajectory initiated from that state will visit the product state B before the reactant state A. By definition, a state that has $p_B = 1/2$ is a transition state. The set of all transition states is called the transition state ensemble.

For the translocation process of hydrophilic solutes across a phospholipid bilayer, the stable states A and B correspond to solute positions above and below the membrane, respectively. If the bilayer is symmetric, then one might expect that a solute that lies at the center of the membrane has equal probability of moving to the upper or the lower membrane/water interface when let go. In other words, at least some configurations with $z = 0$ should have a committor value of $1/2$.

All three biasing methods discussed above generate configurations with $z = 0$, and each of those has an associated committor value p_B. We will see in the next section that the probability distribution $P(p_B)$ of these committor values contains valuable information. If the set of configurations with $z = 0$ were the transition state ensemble, then the distribution $P(p_B)$ would be non-zero only at $p_B = 1/2$. In general, this set will also contain configurations that are not transition states, but at least for a symmetric bilayer the distribution $P(p_B)$ must be symmetric to reflect the equivalence of the upper and the lower sides of the membrane. In either case, one expects a significant peak at $p_B = 1/2$. As we will see, the different biasing methods yield different committor distribution functions due to less-than-perfect sampling, and we can use the shape of the estimated $P(p_B)$ to assess whether simulations have accurately sampled the transition state ensemble.

In practice, we estimate the committor distribution for the ensembles of configurations with $z = 0$ generated by WT-metaD, US, and REUS simulations. From each method, we select a small number of such configurations, and use them together with randomized initial velocities as starting points for four unbiased MD simulations. We then record in how many of these four trajectories the solute reaches the lower ($z < 0$) membrane/water interface before the upper interface. The resulting fraction serves as an estimate for the p_B-value of a configuration. We then calculate a histogram of these committors to obtain an estimate for the distribution function $P(p_B)$ for each sampling method.

It should be noted that both the number of configurations tested and the number of trajectories used to compute a configuration's committor value should be much larger if one wants to calculate an accurate estimate of the committor distribution function.[20] The significant computational cost of bilayer simulations and the long time scale of solute motion across the membrane limit us to relatively small numbers. However, we are interested only in large, qualitative differences between the distribution functions obtained from the three different methods, for which this approach is sufficient.

2.6 Simulation Details

Many popular MD codes such as Gromacs,[21] NAMD,[22] and LAMMPS[23] have some built-in support for free energy calculations. Their capabilities can be significantly enlarged with the Plumed plugin,[24] which provides a flexible framework to define a diverse set of collective variables and enables multiple types of biasing methods, including those discussed in this section. The initial construction of the system to be simulated has become much more user friendly with the arrival of tools such as the Charmm-GUI web server, which can generate input files for multiple MD engines.[25,26]

In the next section, we illustrate the use and potential pitfalls of the three biasing methods for the calculation of translocation free energy profiles. All our simulations were performed using Gromacs 4.6.7 with the Plumed 2.1 plugin. Temperature and pressure were maintained at 320 K and 1 atm using the Nose-Hoover thermostat and Berendsen barostat, respectively. Long range electrostatic interactions were computed using the fourth order PME method[27] with a Fourier spacing of 0.12 nm. Both real-space coulombic and van der Waals interactions were calculated up to a cutoff distance of 1 nm. Bond lengths were constrained using the LINCS algorithm.[28] A symmetric dioleoylphosphatidylcholine (DOPC) bilayer was constructed using the united-atom Berger forcefield,[29] water molecules were treated using the rigid simple point charge (SPC) model,[30] and all the permeants were modeled using the all-atom OPLS-AA forcefield.[31] Additional information about these simulations can be found in Ref. 12.

3 Translocation of Specific Solutes

In this section, we demonstrate the usage of the three biasing methods for the calculation of the translocation free energy profiles of different solutes through a DOPC bilayer. We begin with the transport of a single water molecule, which is crucial for a cell's ability to control its osmotic pressure. We then consider three different forms of alanine, which differ in their electric dipole moments and therefore in their hydrophobicities. Finally, we study arginine as an example of a cationic amino acid. These examples illustrate different aspects of membrane free energy calculations. We will see that for some solutes all three sampling methods perform similarly well, while for others there are significant differences.

3.1 Water

The movement of water molecules across the cell membrane is of paramount biological importance, because it provides a mechanism that allows the cell to control the concentrations of electrolytes and to maintain an osmotic balance. While some of this movement is facilitated by membrane proteins, we here focus only on the passive diffusion of individual water molecules through the lipid bilayer.

Following the simulation methodology outlined in the previous section, we calculated the translocation free energy profile of a water molecule through a DOPC bilayer using both the WT-metaD and the REUS methods. Here the water molecule plays the role of a solute. Results are shown in Figure 1 (a), which shows that both methods give the same result: as the water molecule enters the bilayer, the free energy increases rapidly at first, and then plateaus once the molecule reaches the central core of the membrane. The height of this plateau is approximately 27 kJ/mol relative to bilayer/water interface, which is consistent with previous results.[32,33,34] The flattening of the PMF in the central part of the bilayer coincides with a region of lower lipid tail density and increased disorder.[34,35]

Note that these free energy profiles were calculated over the entire range of the solute position z; the symmetry of the PMF emerged naturally and was not put in manually, for example by averaging over positive and negative z. The fact that the calculated $F(z)$ has such high symmetry is by itself a good indicator for the convergence of the simulations.

The length of the WT-metaD trajectory that gave rise to the data in Figure 1(a) was 420 ns, while the REUS result is based on 40 replicas of 8 ns each, which gives a similar total simulation time of 320 ns. These results show that both methods yield the same result with similar uncertainties in a comparable amount of computing times. As such, neither method has a clear advantage over the other.

3.2 Alanine

The insertion and translocation free energy profiles of peptides are of particular interest for multiple reasons. First, they determine the affinity of proteins to the bilayer, for example through the energetic cost (or gain) of inserting a transmembrane helix. Second, there has been an increased interest in peptide therapeutics, i.e., drugs in the form of peptides that need to cross membranes to become active.[36] Third, the so-called cell-penetrating peptides have been found to significantly increase the uptake of attached drug molecules.[37] In all these applications, it is desirable to have

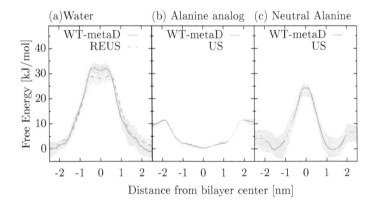

FIGURE 1 Free energy profiles for the translocation of (a) a single water molecule, (b) the side chain analog of alanine, and (c) the unionized form of alanine through a DOPC bilayer. Uncertainties were estimated by bootstrapping for US and REUS simulations and by the method proposed in Ref. [18] for WT-metaD calculations. For these three solutes, all biasing methods yield similar results. Adapted with permission from J. Chem. Theory Comput. 14, 1762. Copyright 2018 American Chemical Society. (See color insert for the color version of this figure)

quantitative information about the PMF of translocation on a per-residue basis. Here we illustrate the use of biased sampling methods to obtain such PMFs, using alanine as an example.

There are multiple reasonable choices for the specific chemical form of an amino acid that one can use in such calculations. Here we use three such choices that are frequently found in the literature: the side-chain analog, the fully neutral form, and the zwitterionic form. We will see that these three forms of alanine behave very differently in free energy calculations, and that the cause for these differences lies in the difference of the molecular dipole moments.

3.2.1 Side-Chain Analog of Alanine

One way to model a single residue is by considering only its side chain, and replacing the α-carbon with a single hydrogen atom.[38] In the case of alanine, this so-called side chain analog is simply a methane molecule. Unlike water, methane is hydrophobic, and one would therefore expect a qualitatively different free energy profile. This is indeed the case, as illustrated in Figure 1(b): the free energy is minimal at the center of the bilayer, where it is approximately 10 kJ/mol lower than in the surrounding water.

The PMFs shown in this figure were obtained using the WT-metaD and the US methods. Both methods yield symmetric free energy profiles, which are the same within the statistical uncertainty. Again, it seems that both methods are equally capable of accelerating the PMF calculation.

3.2.2 Neutral Alanine

Considering only the amino acid side chains is insufficient to predict the insertion free energies of membrane proteins, which also depends on the protein backbone.[39] A single amino acid in its unionized form contains both a carboxyl ($-COOH$) and an amino ($-NH_2$) group. In this form, alanine is charge neutral and has a weak dipole moment.

Figure 1(c) shows the free energy profiles of alanine obtained from WT-metaD and US simulations. Both methods yield symmetric profiles that exhibit a substantial maximum in the bilayer core, reflecting the hydrophobic character of this solute. There are shallow minima in the free energy at the lipid/water interface, which indicate that alanine is weakly attracted to the bilayer surface. As before, both methods perform similarly well when computing such profiles.

In practice, one would typically be satisfied when a PMF calculation converged to a point of acceptably small uncertainties and, in case of a symmetric bilayer, yields a symmetric free energy profile. However, to highlight the differences with other, more polar or charged solutes, we showcase how committor analysis can be used to gain further confidence in the correctness of a PMF calculation.

As described in Section 2.5, the committor p_B of a configuration is the probability that the solute will reach the lower membrane/water interface before the corresponding upper interface in a trajectory that is started from this configuration with random initial velocities. If the bilayer is symmetric, then one would expect that a large number of configurations in which the solute is at the center of the bilayer (i.e., $z = 0$) have a committor value of 1/2.

Both metadynamics and umbrella sampling calculations generate sample configurations with $z = 0$; the former by continually driving the solute across the bilayer, and the latter in a window in which the solute is constrained to remain close to the bilayer center. We can test whether those configurations have the expected committor value by using them as starting points for multiple short trajectories, and by monitoring to which side of the bilayer the solute moves in these trajectories. This calculation is illustrated in Figure 2: from the underlying simulation, which is either performed using WT-metaD (a) or US (b), we choose two configurations with $z = 0$, and initiate four unbiased trajectories from each. In all cases, we find that the solute moves towards the upper interface in two out of the four trajectories, and towards the lower interface in the remaining trajectories. Our best estimate for the committor of all four tested configurations is therefore indeed $p_B = 1/2$. It should be noted that this perfect agreement is fortuitous: while the committor of a configuration is a well-defined property, its estimate as obtained from unbiased trajectories follows the binomial distribution. One should not expect an exactly equal number of trajectories to reach the two membrane/water interfaces, and neither should a slight

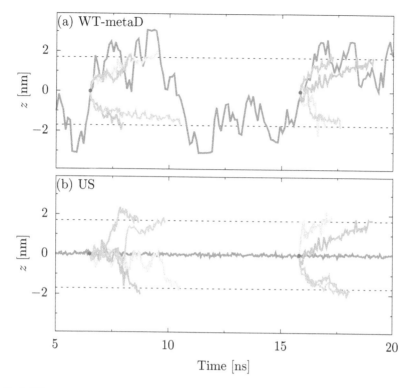

FIGURE 2 Both WT-metaD (a) and US (b) methods sample the transition state of translocation in simulations of neutral alanine. Starting with $z = 0$ configurations generated from a metadynamics trajectory (a) or from the window anchored at the membrane center (b), we initiate four short, unbiased MD simulations. For each of the tested configurations, half of those trajectories evolve towards one side of the bilayer, and the other half to the opposite side. We conclude that these configurations have a committor value $p_B = 1/2$.

imbalance be considered proof that a configuration is not a transition state. If in doubt, one can increase the number of trajectories to obtain a more accurate estimate of the committor value.

By extending this process to more than the two configurations shown in Figure 2, we can calculate the distribution $P(p_B)$ of committor values of the $z = 0$ configurations generated by the two sampling methods. The result for alanine is shown in Figure 3, where we used six configurations as the starting points for four trial trajectories each. We see that both WT-metaD and US generate ensembles of states that have a committor distribution sharply peaked at $p_B = 1/2$. This is consistent with the physical intuition that a solute at the center of a symmetric bilayer has no preference for either side of the membrane. However, as we will see below, this is not always automatically the case.

3.2.3 Zwitterionic Alanine

Under biologically relevant conditions, the carboxyl group of alanine is deprotonated while the amino group is protonated, which yields the zwitterionic form of alanine. While still charge neutral, this form has a significant dipole moment.

As shown in Figure 4, this seemingly minor change in solute electrostatics has a dramatic effect on the calculated free energy profiles: the PMFs calculated using WT-metaD, US, and in an initial REUS simulation (denoted REUS (I)), all appear to be different, which is a clear warning sign that these simulations have not converged. Equally disturbing, none of the obtained PMFs are symmetric with respect to the center of the membrane, which we know they should be for a symmetric bilayer. What happened?

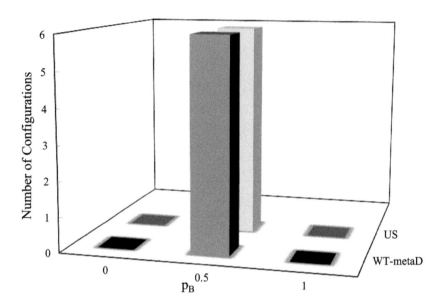

FIGURE 3 Committor distribution for neutral alanine. The committor calculation was performed for 6 WT-metaD and 6 US configurations at $z = 0$. The single peak at $p_B = 1/2$ shows that alanine has no bias towards either side of the membrane in all tested configurations.

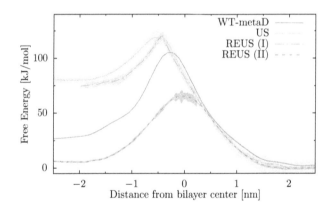

FIGURE 4 Free energy profiles for the translocation of zwitterionic alanine. WT-metaD, US, and initial REUS simulations yield PMFs that are not symmetric, which illustrates their failure to converge. A second set of REUS simulations, with initial conditions chosen as described in the main text, does result in a converged, symmetric profile with a smooth barrier at the center of the bilayer. Adapted with permission from J. Chem. Theory Comput. 14, 1762. Copyright 2018 American Chemical Society. (See color insert for the color version of this figure)

Before we address this question, it is worth noting that the mere realization that something is not right is an important insight. The error bars in the PMFs, calculated using standard error analysis methods, are apparently far too small, so we cannot trust them to give an accurate measure of the uncertainties. How then are we to judge the results of a free energy calculation? In practice it is often unfeasible to perform the same calculation using multiple different methods, as we have done here. A cheaper and simpler test is to use the prior knowledge that for a symmetric bilayer, the PMF must also be symmetric. This provides a valuable test for convergence, and is an advantage that should not be given up lightly. For example, one might argue that it should be sufficient to compute the free energy profile only for $z \geq 0$, and fill in the remainder by simply taking the mirror image with respect to the $z = 0$ axis. This would indeed work if the calculation was perfectly converged. However, testing this convergence is most easily done by calculating the free energy over the entire range of z. The same warning applies when one calculates the entire PMF, which is then manually symmetrized – the convergence should be checked before further processing the free energy profile.

A first glimpse as to why all three simulation methods might fail to converge over typical simulation time scales can be obtained by visually inspecting the generated trajectories: unlike in the case of unionized alanine, the zwitterionic form drags with it both lipids and water molecules as it is driven across the membrane by the biasing algorithm. This is caused by strong electrostatic interactions between the solute and the zwitterionic headgroups of the phospholipids as well as the polar water molecules. As we will see, this collective motion creates long-lasting defects in the membrane, and their relaxation exceeds the simulation time scale.

Compared to WT-metaD and US, REUS provides additional useful information to diagnose this issue: the pattern of configuration exchanges between neighboring replicas. In principle, each trajectory should diffuse through the complete space of replicas due to the ongoing exchange. Failure to do so is an important indicator of convergence problems. This is demonstrated in Figure 5(a), which shows the exchange pattern of 40 trajectories diffusing through the space of replicas, which span the bilayer from −1.9 nm to 2 nm. While exchange occurs frequently between most replicas, there is a suspicious gap, visible as a white line, which indicates that the replicas on either side only rarely

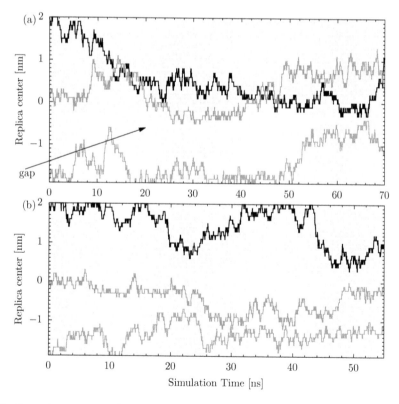

FIGURE 5 Replica exchange patterns for zwitterionic alanine, illustrating the motion of trajectories through replica space. In the initial REUS (I) simulation, there is a persistent gap (indicated by the arrow) that separates two sets of replicas with very limited exchange (a). The REUS (II) simulation, which used a different set of initial conditions, does not show such an obstruction (b). In both panels, we highlight 3 out of 40 trajectories (the ones starting in the $z = 2$, 0.1, and −1.9 nm replicas) for clarity. Adapted with permission from J. Chem. Theory Comput. 14, 1762. Copyright 2018 American Chemical Society. (See color insert for the color version of this figure)

switch configurations. Evidently there is a large difference between the configurations in those replicas that significantly decreases the acceptance probability of exchange attempts.

The nature of this difference becomes apparent when visualizing the trajectories of replicas on either side of the gap: replicas above the gap contain configurations in which the alanine is in close contact with headgroups from the upper leaflet's phospholipids, as illustrated in the upper and left panel of Figure 6. Below the gap, on the other hand, are configurations in which the alanine is either surrounded by headgroups from either the lower leaflet's lipids (bottom and right panel) or forms a defect in the membrane that affects the upper and the lower leaflets similarly (center panel). This difference is imposed by the initial conditions used in the various replicas.

The effect of this gap on the replica exchange dynamics can be seen in the three highlighted trajectories shown in Figure 5(a): tracking a simulation across exchanges, we see a movement that resembles that of a random walk, but that cannot typically cross the gap – it acts almost like a hard wall. The gap thereby separates the replica space into two distinct regions that only rarely exchange configurations with each other. The gap is not, however, entirely static: every so often it seems to move upward, which indicates that a replica from above the gap with a one-sided bilayer deformation has converted to a more symmetric conformation below the gap. This process is extremely slow: over the 70 ns simulation time shown in Figure 5(a), the gap has shifted by only three replicas towards the center of the bilayer. Nevertheless, this motion reflects the unswerving progress towards equilibration of the system.

Having realized that it is the slow conversion from one-sided to symmetric defect configurations that impedes the convergence of the REUS calculation, it seems possible to side-step this process by using a configuration in which the solute is in close contact with lipids from both leaflets as an initial condition. We therefore performed a new calculation, referred to as REUS (II), that uses the configuration shown in the central panel of Figure 6 as the starting point for

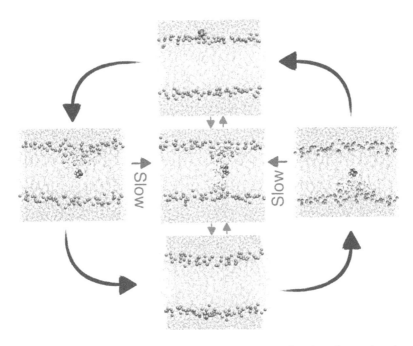

FIGURE 6 Typical configurations encountered in simulations of the translocation of strongly polar and charged solutes. Such solutes tend to localize at one of the membrane/water interfaces (top and bottom panels). When driven inside the bilayer, they create a one-sided deformation of the bilayer (left and right). The transition state, however, is symmetric with respect to the two bilayer leaflets (center). The transition to this symmetric defect is very slow, because it requires crossing of a free energy barrier in a direction other than z, and that is therefore not accelerated by the biasing method.

replicas close to the bilayer center. This simulation's exchange pattern, shown in Figure 5(b), no longer shows a long-lasting gap; all replicas appear to exchange equally with their neighbors in a random fashion. Not only does the exchange pattern now behaves as one would expect for an equilibrated simulation, the resulting free energy profile satisfies the required condition of symmetry as shown in Figure 4; the PMF obtained from the REUS (II) simulation is an even function of z, and has a smooth maximum at the bilayer center that corresponds to a free energy barrier of 60 kJ/mol.

So far we have demonstrated how one can use the exchange pattern of REUS simulations to detect possible convergence problems. This method cannot be used for WT-metaD or US calculations, which is a considerable advantage for the replica exchange method. A generally applicable method to test proper equilibration of solute configurations near the bilayer center can be developed using the committor p_B. In Figure 2, we showed that for unionized alanine, unbiased simulations initiated from $z = 0$ configurations obtained in WT-metaD and US simulations had equal chance of progressing toward the upper and the lower membrane/water interface. Figure 7 shows that this is not the case for zwitterionic alanine: for the two shown starting configurations obtained from a metadynamics simulation, alanine returns to the side of the membrane from which it just came in four out of four tested trajectories. Similarly, in a US simulation with a harmonic bias potential at the center of the bilayer, and that was started with an initial configuration in which the solute was in close contact with

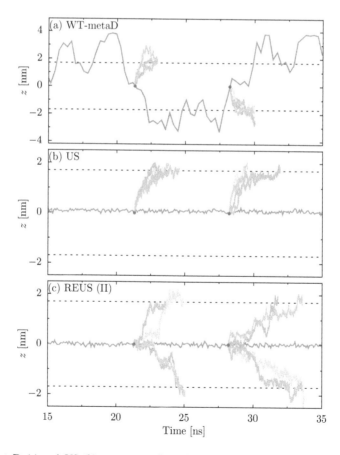

FIGURE 7 WT-metaD (a) and US (b) generate configurations with $z = 0$ that are not transition states. Such configurations have a strong tendency to return to the side of the bilayer from which they just came (a) or to the side from which the initial condition in that window originated (b). The second set of REUS simulations, on the other hand, generates configurations that show no bias toward either side of the membrane, and that are therefore transition states.

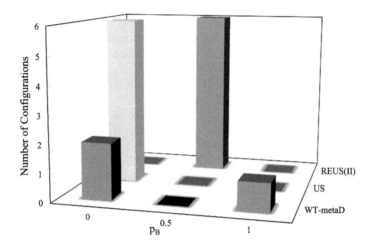

FIGURE 8 Committor distribution functions for zwitterionic alanine. For every tested $z = 0$ configuration obtained from WT-metaD and US simulations, all unbiased trajectories that start from this configuration evolve towards the same side of the membrane. Configurations with $z = 0$ generated in the REUS (II) simulation, on other hand, have no preference for either side of the bilayer. Adapted with permission from J. Chem. Theory Comput. 14, 1762. Copyright 2018 American Chemical Society. (See color insert for the color version of this figure)

lipids from the upper leaflet, all tested trajectories return to the upper membrane/water interface. Clearly, in both cases, the $z = 0$ configurations retain some memory of how the alanine got to the center of the bilayer. This is also the case for the REUS (I) simulation. The better equilibrated REUS (II) simulation, on the other hand, yields configurations in which the solute is near the bilayer center, and that have equal probability of evolving towards the upper and the lower sides of the membrane in unbiased simulations, as shown in Figure 7(c).

Extending this analysis to multiple, randomly selected configurations, we can estimate the committor distribution function, $P(p_B)$, for the ensemble of $z = 0$ configurations generated by the three biasing methods. The results are shown in Figure 8; the configurations from WT-metaD and US simulations have a strong bias to either one or both sides of the bilayer, while those obtained in the REUS (II) simulations have an unbiased committor value of $p_B = 1/2$ and are therefore transition states. These important states do not seem to be sampled in either metadynamics or umbrella sampling calculations, which is likely the reason for their failure to converge. We reiterate that neither the number of tested configurations nor the number of unbiased trial simulations is sufficient to obtain a quantitative estimate of the committor distribution function, but they are adequate to test the ability of a simulation method to sample transition states at all.

3.3 Arginine

Finally, let us look at arginine, a cationic amino acid. In part due to their prevalence in cell penetrating peptides, the translocation mechanisms of such positively charged amino acids has received much attention.[9,40–42] It has been found that motion of arginine through a lipid bilayer is accompanied by the formation of water defect in the hydrophobic core that keeps the amino acid hydrated even when it is near the center of the bilayer. Lipid headgroups line this membrane-spanning pore, which causes a major structural change in the bilayer. The calculation of translocation free energy profiles is hindered by the presence of barriers in degrees of freedom other than the position z, in particular due to the coupling of solute orientation and bilayer deformation.[43]

In the previous section, we argued that the difficulties encountered in PMF calculations of zwitterionic alanine are due to strong electrostatic interactions between the solute and the lipids

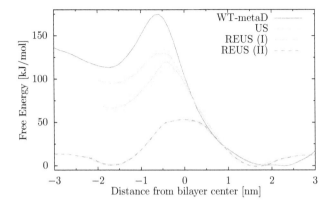

FIGURE 9 Free energy profiles for the translocation of a cationic arginine residue. As illustrated previously in Figure 4 for the case of zwitterionic alanine, WT-metaD, US, and initial REUS simulations did not converge, as is evident from their asymmetric PMFs. The REUS (II) profile, obtained after extensive equilibration, is symmetric and has its maximum at $z = 0$. Adapted with permission from J. Chem. Theory Comput. 14, 1762. Copyright 2018 American Chemical Society. (See color insert for the color version of this figure)

or water. If that was the case, then one would expect the same problems, perhaps even in an exacerbated way, to arise for arginine, which contains a positively charged guanidino group in addition to the zwitterionic carboxyl and amino group pair. This is indeed the case, as illustrated in Figure 9: WT-metaD, US, and initial REUS simulations yield free energy profiles that are far from symmetric and therefore evidently not converged, even though the calculated uncertainties seem to be rather small. It is worth noting that these PMFs were obtained from long simulations extending over a microsecond or longer, and still they show clear signs of non-convergence.

A look at the replica exchange pattern reveals the same underlying problem as we saw for zwitterionic alanine: there is a persistent gap that separates replica space into two regions with very few exchanges in-between.[12] This gap moves upward even more slowly in the arginine calculations, indicating that the relaxation time scales are even longer for the charged residue. As before, the principal difference between configurations above and below the gap is the symmetry of the defect that the solute creates in the bilayer. It is not surprising that the solution to the problem is therefore also the same: by starting a new set of REUS simulations with initial conditions that reflect a symmetric defect for replicas near the bilayer center, one obtains the symmetric free energy profile denoted REUS (II) in Figure 9. It exhibits shallow local minima for arginine positions close to the lipid headgroups ($z \approx \pm 1.7$nm), which implies that arginine is attracted to the surface of the membrane. At the hydrophobic center of the bilayer, on the other hand, there is a significant barrier of approximate 50 kJ/mol that hinders translocation.

A committor analysis for the configurations generated by the various biasing methods also paints a familiar picture: the configurations generated in WT-metaD, US, and a first set of REUS simulations that contain arginine at the bilayer center have a strong bias towards one side of the membrane.[12] In other words, these configurations are not transition states. Repeating this calculation for the second REUS simulation, we find that all tested configurations from the replica centered at $z = 0$ have a committor value close to 1/2. It is this simulation, which includes sampling of the actual transition states, that yields the correct free energy profile.

4 Implications and Recommendations

We have seen that all three tested biasing methods perform well for uncharged or weakly polar solutes, but struggle when one wants to compute translocation free energy profiles of strongly polar or charged molecules. In REUS simulations, the underlying cause for this difficulty can be readily seen in the exchange pattern: the slow conversion of configurations in which a solute close to the bilayer center is initially surrounded by lipids from only one leaflet towards a more symmetric state. This process is illustrated in Figure 6 by the horizontal arrows, and its progress is reflected by the slow dynamics of the gap in Figure 5(a). Once this relaxation has occurred, a second REUS calculation yields reliable PMFs.

While there is no equivalent to the exchange pattern in US calculations, it is likely that the same mechanism is at work here. In both US and REUS simulations, one computes trajectories in which the solute is restrained in the bilayer center by a harmonic bias potential. The simulations will converge as long as the length of these trajectories is longer than the time scale of the required lipid reorganization. Metadynamics simulations, on the other hand, behave differently: because the solute is continually driven across the bilayer, it is unlikely to remain within the hydrophobic core for a sufficiently long time to allow the formation of a symmetric defect or pore. Every time it diffuses towards the bilayer/water interface and then re-enters the lipid tail region, the solute is likely to drag with it lipid headgroups and water molecules from the proximal side, thereby creating yet another one-sided defect. An exaggerated illustration of this process is the counterclockwise cycle in Figure 6. However, this side-stepping of the actual transition state can likely be avoided by a judicious choice of the metadynamics parameters.

The behavior depicted in this figure is in fact a common concern in free energy calculations of many complex systems: because we are in interested in $F(z)$, we bias the system along the coordinate z. However, there are additional barriers in other coordinates; for example, the lipid reorganization that is required for the horizontal transitions in Figure 6. Because we do not bias these 'hidden' variables, the calculation of free energy profiles remains challenging.

An obvious solution to this problem is to also bias these collective variables in addition to the position z that is of primary interest. This can be accomplished by introducing multi-dimensional biasing potentials. An intriguing alternative for metadynamics calculations is to simulate multiple trajectories in parallel, each using a different CV. Every so often these trajectories are allowed to exchange configurations similar to REUS calculations. This bias exchange metadynamics (BE-metaD[44]) scheme facilitates the calculation of free-energy profiles in one CV by driving the exploration of additional variables.

It is usually not known in advance which variables should be biased to accelerate the convergence of free energy calculations. Our results from the previous section show that breaking and forming solute–lipid and solute–water contacts is a bottleneck that impedes equilibration; therefore, biasing the number of such contacts should be beneficial. That this is indeed the case was shown by Ghaemi and coworkers, who studied the translocation of ethanol through a phospholipid bilayer using BE-metaD in multiple CVs, including the coordination numbers that measure the solvation of the solute by lipid heads, tails, and water.[45]

For the small solutes that we have considered here, we have used their center of mass (COM) for the definition of the distance variable z. For more complex solutes, this is not necessarily the ideal choice. Hinner and coworkers have shown that the convergence rate of PMF calculations for a voltage-sensitive dye molecule depends on which solute atom is used in the definition of z.[46] PMF calculations for even bigger molecules can benefit from using multiple CVs that measure the distance between different groups of the solute and the membrane.[47] Alternatively, one can bias the orientation of the solute molecule with respect to the bilayer. If the solute itself has significant conformational flexibility, then one should consider including such internal CVs as well in the biasing scheme.

There is a large number of biasing methods and CVs to choose from, which might seem overwhelming at first. One can take comfort in the fact that any combination will at least in principle converge to the correct result if only the simulations are performed over long enough times. As we have seen, however, determining that the calculation is converged can be rather difficult. The standard algorithms to estimate statistical uncertainties fail if there are systematic sampling deficiencies, which seems to occur frequently. How then is one to assess convergence? For the practically important case of symmetric bilayers, testing the symmetry of the obtained free energy profile is likely the easiest and cheapest way. It requires the calculation of the PMF on both side of the membrane, and therefore taking the shortcut of computing $F(z)$ in only one half-space is not recommended. A harder to obtain, but also more informative, measure is the committor distribution function. If one has any *a priori* knowledge of the transitions states, then these functions can be used to test whether those states are indeed sampled in a biased simulation. Computing $P(p_B)$ is not a free energy calculation, but instead requires a large number of short, unbiased simulations. It provides additional, valuable information about the sampling process. As we have seen, the committor distribution function can be used to find barriers in hidden collective variables that hamper free energy calculations.

The calculation of membrane binding and translocation free energies is a very active field of computational chemistry, and one that has seen dramatic progress over the past decade. New biasing methods, better collective variables, faster computers, more accurate molecular mechanics force fields, and simulation software that has become more user friendly all contribute to this exciting development. It is a sign of success that these calculations are now regularly performed during the development phase of new compounds.

REFERENCES

1. R. V. Swift and R. E. Amaro. Back to the future: Can physical models of passive membrane permeability help reduce drug candidate attrition and move us beyond QSPR? *Chem. Biol. Drug Dse.*, 81(1):61, 2013.
2. G. M. Torrie and J. P. Valleau. Nonphysical sampling distributions in Monte Carlo free-energy estimation: Umbrella sampling. *J. Comput. Phys.*, 23(2):187, 1977.
3. A. Laio and M. Parrinello. Escaping free-energy minima. *Proc. Natl. Acad. Sci. USA*, 99(20):12562–12566, 2002.
4. Y. Sugita and Y. Okamoto. Replica exchange molecular dynamics method for protein folding simulation. *Chem. Phys. Lett.*, 314(1):141, 1999.
5. E. Darve and A. Pohorille. Calculating free energies using average force. *J. Chem. Phys.*, 115 (20):9169, 2001.
6. J. G. Kirkwood. Statistical mechanics of fluid mixtures. *J. Chem. Phys.*, 3(1935):300, 2004.
7. D. Bochicchio, E. Panizon, R. Ferrando, L. Monticelli, and G. Rossi. Calculating the free energy of transfer of small solutes into a model lipid membrane: Comparison between metadynamics and umbrella sampling. *J. Chem. Phys.*, 143(14):144108, 2015.
8. C. T. Lee, J. Comer, C. Herndon, N. Leung, A. Pavlova, R. V Swift, C. Tung, C. N. Rowley, R. E. Amaro, C. Chipot, Y. Wang, and J. C. Gumbart. Simulation-based approaches for determining membrane permeability of small compounds. *J. Chem. Inf. Mod.*, 314(4):721, 2016.
9. C. Neale, C. Madill, S. Rauscher, and R. Pomés. Accelerating convergence in molecular dynamics simulations of solutes in lipid membranes by conducting a random walk along the bilayer normal. *J. Chem. Theory Comput.*, 9(8):3686, 2013.
10. A. Barducci, G. Bussi, and M. Parrinello. Well-tempered metadynamics: A smoothly converging and tunable free-energy method. *Phys. Rev. Lett.*, 100(2):20603, 2008.
11. Y. Sugita, A. Kitao, and Y. Okamoto. Multidimensional replica-exchange method for free-energy calculations. *J. Chem. Phys.*, 113(15):6042, 2000.
12. N. Pokhrel and L. Maibaum. Free energy calculations of membrane permeation: Challenges due to strong headgroup-solute interactions. *J. Chem. Theory Comput.*, 14(3):1762, 2018.
13. A. Laio and F. L. Gervasio. Metadynamics: A method to simulate rare events and reconstruct the free energy in biophysics, chemistry and material science. *Rep. Prog. Phys.*, 71(12):126601, 2008.
14. J. F. Dama, M. Parrinello, and G. A. Voth. Well-tempered metadynamics converges asymptotically. *Phys. Rev. Lett.*, 112:240602, 2014.
15. S. Kumar, J. M. Rosenberg, D. Bouzida, R. H. Swendsen, and P. A. Kollman. The weighted histogram analysis method for free-energy calculations on biomolecules. *J. Comput. Chem.*, 13(8):1011, 1992.
16. B. Roux. The calculation of the potential of mean force using computer simulations. *Comp. Phys. Comm.*, 91(1–3):275–282, 1995.

17. J. S. Hub, B. L. D. Groot, and D. V. D. Spoel. g_wham–A free weighted histogram analysis implementation including robust error and autocorrelation estimates. *J. Chem. Theory Comput.*, 6:3713, 2010.

18. P. Tiwary and M. Parrinello. A time-independent free energy estimator for metadynamics. *J. Phys. Chem. B*, 119(3):736, 2015.

19. P. G. Bolhuis, D. Chandler, C. Dellago, and P. L. Geissler. TRANSITION PATH SAMPLING: Throwing ropes over rough mountain passes, in the dark. *Annu. Rev. Phys. Chem.*, 53(1):291, 2002.

20. C. Dellago, P. G. Bolhuis, and P. L. Geissler. *Transition Path Sampling*, Vol. 123, p. 1. John Wiley & Sons, Inc., Hoboken, NJ, 2003.

21. D. V. D. Spoel, E. Lindahl, B. Hess, G. Groenhof, A. E. Mark, and J. C. Herman. GROMACS: Fast, flexible, and free. *J. Comput. Chem.*, 26(16):1701, 2005.

22. J. C. Phillips, R. Braun, W. Wang, J. Gumbart, E. Tajkhorshid, E. Villa, C. Chipot, R. D. Skeel, L. Kalé, and K. Schulten. Scalable molecular dynamics with NAMD. *J. Comput. Chem.*, 26(16):1781–1802, 2005.

23. S. Plimpton. Fast parallel algorithms for short-range molecular dynamics. *J. Comput. Phys.*, 117(1):1, 1995.

24. M. Bonomi, D. Branduardi, G. Bussi, C. Camilloni, D. Provasi, P. Raiteri, D. Donadio, F. Marinelli, F. Pietrussi, R. A. Broglia, and M. Parrinello. PLUMED: A portable plugin for free-energy calculations with molecular dynamics. *Comp. Phys. Commun.*, 180(10):1961, 2009.

25. E. L. Wu, X. Cheng, S. Jo, H. Rui, K. C. Song, E. M. Dávila-Contreras, Y. Qi, J. Lee, V. Monje-Galvan, R. M. Venable, J. B. Klauda, and W. Im. CHARMM-GUI membrane builder toward realistic biological membrane simulations. *J. Comp. Chem.*, 35(27):1997–2004, 2014.

26. J. Lee, X. Cheng, J. M. Swails, M. S. Yeom, P. K. Eastman, J. A. Lemkul, S. Wei, J. Buckner, J. C. Jeong, Y. Qi, S. Jo, V. S. Pande, D. A. Case, C. L. Brooks, A. D. MacKerell, J. B. Klauda, and W. Im. CHARMM-GUI input generator for NAMD, GROMACS, AMBER, OpenMM, and CHARMM/OpenMM simulations using the CHARMM36 additive force field. *J. Chem. Theo. Comp.*, 12(1):405–413, 2016.

27. T. Darden, D. York, and L. Pedersen. Particle mesh Ewald: an N.log(N) method for Ewald sums in large systems. *J. Chem. Phys.*, 98(12):10089, 1993.

28. B. Hess, H. Bekker, H. J. C. Berendsen, and J. G. E. M. Fraaije. LINCS: A linear constraint solver for molecular simulations. *J. Comput. Chem.*, 18(12):1463, 1997.

29. O. Berger, O. Edholm, and F. Jähnig. Molecular dynamics simulations of a fluid bilayer of dipalmitoylphosphatidylcholine at full hydration, constant pressure, and constant temperature. *Biophys. J.*, 72(5):2002, 1997.

30. H. J. C. Berendsen, J. P. M. Postma, W. F. van Gunsteren, and J. Hermans. *Interaction Models for Water in Relation to Protein Hydration*, Vol. 14, p. 331. Springer, Dordrecht, Netherlands, 1981.

31. G. A. Kaminski, R. A. Friesner, J. Tirado-Rives, and W. L. Jorgensen. Evaluation and reparametrization of the OPLS-AA force field for proteins via comparison with accurate quantum chemical calculations on peptides. *J. Phys. Chem. B*, 105(28):6474, 2001.

32. N. Sapay, W. F. D. Bennett, and D. P. Tieleman. Thermodynamics of flip-flop and desorption for a systematic series of phosphatidylcholine lipids. *Soft Matter*, 5(5):3295, 2009.

33. M. Orsi and J.W. Essex. Passive permeation across lipid bilayers: A literature review. In Mark S. P. Sansom and Philip C. Biggin, editors, *Molecular Simulations and Biomembranes from Biophysics to Function*, p. 76. Royal Society of Chemistry, Cambridge, UK, 2010.

34. S. J. Marrink and Herman J. C. Berendsen. Simulation of water transport through a lipid membrane. *J. Phys. Chem.*, 98(15):415, 1994.

35. J. A. Marqusee and K. A. Dill. Solute partitioning into chain molecule interphases: Monolayers, bilayer membranes, and micelles. *The J. Chem. Phys.*, 85(1):434, 1986.

36. K. Fosgerau and T. Hoffmann. Peptide therapeutics: Current status and future directions. *Drug Discovery Today*, 20(1):122–128, 2015.

37. S. Deshayes, M. C. Morris, G. Divita, and F. Heitz. Cell-penetrating peptides: Tools for intracellular delivery of therapeutics. *Cell. Mol. Life Sci.*, 62(16):1839–1849, 2005.

38. A. C. V. Johansson and E. Lindahl. The role of lipid composition for insertion and stabilization of amino acids in membranes. *J. Chem. Phys.*, 130(18):185101, 2009.

39. W. C. Wimley and S. H. White. Experimentally determined hydrophobicity scale for proteins at membrane interfaces. *Nat. Struct. Mol. Biol.*, 3:842–848, 1996.
40. J. L. MacCallum, W. F. D. Bennett, and D. P. Tieleman. Transfer of arginine into lipid bilayers is nonadditive. *Biophys. J.*, 101(1):110, 2011.
41. J. L. MacCallum, W. F. D. Bennett, and D. P. Tieleman. Distribution of amino acids in a lipid bilayer from computer simulations. *Biophys. J.*, 94(9):3393, 2008.
42. S. Dorairaj and T. W. Allen. On the thermodynamic stability of a charged arginine side chain in a transmembrane helix. *Proc. Natl. Acad. Sci. USA*, 104(12):4943, 2007.
43. C. Neale and R. Pomès. Sampling errors in free energy simulations of small molecules in lipid bilayers. *Biochim. Biophys. Acta, Biomembr.*, 1858(10):2539, 2016.
44. S. Piana and A. Laio. A bias-exchange approach to protein folding. *J. Phys. Chem. B*, 111(17):4553–4559, 2007.
45. Z. Ghaemi, M. Minozzi, P. Carloni, and A. Laio. A novel approach to the investigation of passive molecular permeation through lipid bilayers from atomistic simulations. *J. Phys. Chem. B*, 116(29):8714–8721, 2012.
46. M. J. Hinner, S. J. Marrink, and A. H. D. Vries. Location, tilt, and binding: A molecular dynamics study of voltage-sensitive dyes in biomembranes. *J. Phys. Chem. B*, 113(48):15807, 2009.
47. R. Sun, J. F. Dama, J. S. Tan, J. P. Rose, and G. A. Voth. Transition-tempered metadynamics is a promising tool for studying the permeation of drug-like molecules through membranes. *J. Chem. Theory Comput.*, 12(10):5157, 2016.

8

Theories and Algorithms for Molecular Permeation through Membranes

Alfredo E. Cardenas
Institute for Computational Engineering and Sciences, University of Texas at Austin, Austin, TX 78712.

Ron Elber
Institute for Computational Engineering and Sciences, University of Texas at Austin, Austin, TX 78712.
Department of Chemistry, University of Texas at Austin, Austin, TX 78712.

I Introduction

One of the important functions of biological membranes is to prevent unchecked material transport between the cytosol and the environment external to the cell. The membrane barrier must be permeable, however, because the cell needs to remove its waste products and get nutrients. Pumps or channels facilitate the desired transport of specific substances across biological membranes and are controlled by physiological conditions or cell signaling. In particular, pumps are molecular machines that use energy to transport selected molecules against concentration gradients.

Another type of transport through the cell membrane is not assisted by cell machinery and is called *passive* transport. Passive permeation provides the background in which the cellular transport system works and adjusts. For example, proton leaks that follow the concentration gradient across biological membranes may reduce the ability of cells to use this gradient to produce ATP. The cell must pump protons against the thermodynamics drive to keep the ATPases function. Passive transport also enables the penetration of foreign material to cells. The study of translocation of genetic material of viruses or drugs motivates further investigations of passive permeation. This review is focused on computational studies of passive permeation.

The critical barrier for membrane permeation is the hydrophobic moiety between the water layers inside and outside the cell. A molecule that is soluble in water is unlikely to be soluble in the hydrophobic core of the membrane. Indeed, many nonpolar compounds permeate rapidly through the lipid membrane because they tend to partition better inside the hydrophobic region.

This observation means, however, that their solubility in water is low. On the other hand, permeation of more polar, hydrophilic compounds like inorganic ions and amino acids can be very slow. It is therefore of considerable interest to examine compounds that are water-soluble and at the same time permeate membranes efficiently. Such compounds are likely to be good drug carriers.

Living organisms are using soluble *and* permeating compounds to efficiently translocate material into cells; examples are cell-penetrating peptides (CPP) [1] and antimicrobial peptides [2].

Interestingly, some positively charged peptides are able to passively permeate through phospholipid membranes. Negative charges in general are not. Computer simulations provide atomically detailed molecular mechanisms that in conjunction with experimental verification can shed light on permeation mechanisms of interest [3].

Since the pioneering paper of Marrink and Berendsen on water permeation through a DPPC membrane [4], molecular simulations have been used for more than 20 years to study passive permeation mechanisms of solutes across lipid membranes. A challenge in conducting these simulations is of timescales. Straightforward Molecular Dynamics (MD) simulations of kinetics, even when used on high-end computing platforms, are restricted to typical transition times of sub-milliseconds. Many permeation processes take minutes or longer [5]. Hence, straightforward MD simulations are not possible.

Marrink and Berendsen suggested a model based on Smoluchowski equation that exploits atomically detailed modeling and several other assumptions. The solubility–diffusion model developed by Marrink and Berendsen uses the potential of mean force along the membrane axis for the center of mass of the permeant and a position-dependent diffusion coefficient to compute permeation rates. The theory successfully provided a semi-quantitative understanding of permeation of small solutes. However, it is approximate, and these approximations require re-evaluations for larger permeants or more complex permeation processes.

First, Brownian (and Markovian) dynamics is assumed. As we have illustrated recently [6], membrane density fluctuations are frequently non-Markovian, especially close to the membrane center in which the mass density is relatively low. The density fluctuations impact the probability of cavity formation that is likely to assist permeation.

Second, it is assumed that there is only a single dominant coarse variable, namely the center of mass position of the permeant with respect to the membrane normal. We denote this coordinate by Z. The membrane depth is a good reaction coordinate for small permeants that are non-disruptive to the membrane structure. Otherwise, the orientation and other internal degrees of freedom of the permeant molecule, aggregation of permeants or collective transport, formation of water defects, and disruption and re-distribution of the phospholipid heads at the membrane surface are all significant coarse variables that may have important contributions to permeation kinetics.

For example, a terminally blocked tryptophan that we have studied in the past [7] experienced a flip-flop motion or 180 degrees rotation at the membrane center such that the polar part of the molecule points to the closer water phase. This kind of motion is hard to capture in straightforward MD simulations, even if the tryptophan is constrained to the membrane center. The flip-flop is a rare event and a significant waiting time may occur before the actual transition. Since we are constrained by accessible length of MD trajectories, it is necessary to describe the flip-flip transition by another coarse variable. Recent efforts extend the solubility diffusion model to include rotational degrees of freedom besides the center of mass of the permeant [8]. However, the calculations are more time-consuming and difficult to extend to additional degrees of freedom.

Due to these difficulties for straightforward MD simulations, and for one-dimensional permeation models, alternative techniques have been used more recently to extract the potential of mean force. These techniques include adaptive biasing force (ABF [9]) and metadynamics [10]. In ABF, a biasing force is adjusted over time to sample the reaction coordinate. The position-dependent diffusion coefficient is estimated with a Bayesian-inference algorithm. The method was applied recently to the permeation of ethanol across a POPC (1-palmitoyl-2-oleoyl-sn-glycero-3-phosphocholine) membrane. The same membrane system was investigated using bias exchange metadynamics [10]. The free-energy surface was described using seven collective variables. A bin-based kinetic model that was combined with a kinetic Monte–Carlo method to calculate the permeability estimated the diffusion matrix. Both methods provide valuable insights regarding the relevance of additional coordinates, besides the position along the membrane axis for the permeation process but extension for larger systems and the study of time scales and kinetics are far from trivial.

In this review, we will describe our approach to the membrane permeation problem where we have used the Milestoning methodology and also the more traditional solubility–diffusion model. Milestoning allows us to compute kinetics and thermodynamics simultaneously and therefore provides comprehensive picture of the process. In the next section, we define the quantities that are used to describe the permeation, the solubility-diffusion model, and the algorithm of Milestoning. Next, we will describe the application of Milestoning to study the permeation of a blocked tryptophan through a DOPC membrane. Also, we will show how Milestoning can be used to study the dynamics of membrane densities and how this approach provides a different insight to water permeation.

II Passive Permeation: Definition, Measurements, and Modeling

Passive permeation of small molecules through membranes can be thought of as a leakage of material through a biological barrier that determines cell boundaries. If the leakage is harmful and the biological system requires concentration gradient, then membrane pumps must work against the leakage to retain the non-equilibrium state of the cell. Hence, passive permeation can have a negative impact on cell function, which motivates further study.

Moreover, specific biological mechanisms require passive permeation and are designed for it. For example, consider the permeation of CPP through membranes. It is the rapid and efficient passive permeation of water-soluble molecules like CPP, which is particularly interesting. Indeed, in the present review, we consider the efficient permeation of charged amino acids through a neutral membrane.

Permeation is a kinetic observable. It is defined by the ratio of the net flux per unit area through a membrane. Since the net flux at equilibrium must be zero, it is obvious that we must study it at conditions that deviate from equilibrium. The permeability coefficient, P, is used to define the capacity of a particular membrane for permeation. It is defined as:

$$P = \frac{J}{\Delta C} \tag{1}$$

where J is the flux and ΔC is the difference in concentration across the membrane. The flux is the number of molecules that passes through the membrane, per unit time, and per unit area, under stationary flux conditions. The concentration is the number of molecules per unit volume, and the concentration change is the difference of the concentration on one side of the membrane C_i and the concentration at the other side C_f. The change of the concentration is the thermodynamic drive for the process and is given by $\Delta C = C_f - C_i$. The units of $[P] = \frac{N/(t \cdot A)}{N/V} = l \cdot t^{-1}$ are of length per time where N is the number of permeants, t the time, A the surface area, and V is the volume. The permeation coefficient is frequently expressed in centimeter per second. Note that by writing Eq. (1) without explicit time dependence, we assume a stationary flux. However, we did not assume equilibrium or even a Markovian process. In that sense, Eq. (1) is more general than the solubility-diffusion model that assumes overdamped dynamics (see below) and is therefore a more comprehensive starting point to discuss permeation.

Before discussing theory and simulations, it is useful to understand experimental measurements of permeability. Consider a typical experimental setup for measurements of permeation. Vesicles of known volume V, and surface area A, are prepared with a number of permeant molecules

$$N = C \cdot V$$

where C is the concentration of permeants inside the vesicle. The external solution is permeant-free at the beginning of the experiment. If the permeation is slow (minutes to hours are not unreasonable permeation time scales), then the external solution of the vesicles may be periodically removed and the remaining number of molecules in the vesicles, N(t), is determined (for example, using calibrated spectroscopy).

If the number of molecules, N(t), follows a first-order or steady state kinetics for the period in which the population is roughly a constant, we have

$$\frac{dN(t)}{dt} = -kN(t) \tag{2}$$

The explicit solution of Eq. (2) is $N(t) = N(0)\exp(-kt)$. To relate the rate and the permeability coefficients, we write for positive flux, $J = -\frac{1}{A}\frac{dN(t)}{dt} = \frac{k}{A}N(t)$ and $P = \frac{kV}{A}$. For a spherical object with a radius R, we have

$$P = \frac{kR}{3} \tag{3}$$

The permeation coefficient seems to depend on the size of the vesicle. This is, however, incorrect. In the above formulation, the rate coefficient, k, is inversely proportional to R. The number of molecules near the surface of the vesicle that are ready to exit is proportional to the surface area while the total population in the vesicle is proportional to the volume. The surface-to-volume ratio is inverse length. Hence, we expect $dN/N \propto 1/R$ and so is the rate coefficient following Eq. (2). As a result, the permeation coefficient is independent of the vesicle size.

Note that so far we have made no assumption on the type of dynamics that describe the permeation. In the past, the process was frequently described by overdamped dynamics [4]. While the choice of a diffusive model is likely to be appropriate for permeation processes, it is nice to

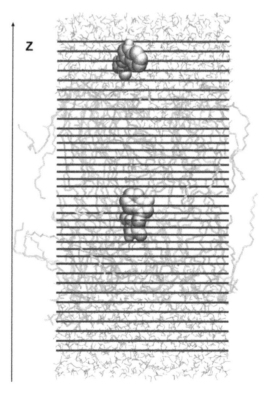

FIGURE 1 An atomically detailed model of a snap shot in time of a permeating trajectory. The Z-axis is normal to the plane and is the coarse coordinate used in the diffusion solubility model. The stick models at the top and bottom of the figure are water molecules. The shaded lines in the center of the figure are phospholipid (DOPC) molecules. The blocked tryptophan molecules (space filling models) and the meaning of the horizontal lines (milestones) are discussed in Section IV. (See color insert for the color version of this figure)

derive permeation from atomically detailed models. Considerable information on molecular mechanisms to guide molecular design is obtained from molecular simulations.

The most straightforward simulation of a permeation coefficient is based on direct calculations of MD trajectories. We simply compute as a function of time positions of the particles as they cross from one side of the membrane to the other. We denote a trajectory by $x(t)$ where x is the phase space position vector (coordinates and velocities) and t is the time. At the beginning of the trajectory ($t = 0$), the coordinates of the permeant are in the aqueous solution at the top of the membrane (Figure 1). At the final time ($t = t_f$) of a transporting trajectory, we find the permeant at the opposite side of the membrane. We estimate the rate coefficient directly, based on the behavior of an ensemble of permeation trajectories.

Let $N(t)$ be the number of trajectories that remain at the same side of the membrane after time t. Then by binning the population at different times and fitting $\log[N(t)]$ vs time, we estimate the rate coefficient from the linear slope assuming that Eq. (2) holds.

The MD approach, as outlined above, mimics the experimental setup and can be used in straightforward estimates of the permeability coefficient. However, it is rare to use straightforward MD simulations to study permeation. Such studies are limited to rapid permeants like oxygen or water molecules [11] which is of order of nanoseconds to microseconds. The time scale of permeation events can be as long as an hour [12, 13] while the basic time step of MD is of a femtosecond, eighteen orders of magnitude shorter. This huge time scale gap forbids direct simulations of polar, charged, or large systems. We should also keep in mind that the algorithm we describe for the calculations of the rate coefficient, k, requires an average over many permeating trajectories. Generating N permeating trajectories increases the computational cost by a factor of N.

The shortcoming of MD motivated the developments of a number of approximate simulation methods that make the permeation study possible. There are numerous assumptions in these approaches and it is important to be aware of them. The first simulation approach we describe is the solubility diffusion model. We then continue to discuss the more recent approach of Milestoning.

III Solubility Diffusion Model

Marrink and Berendsen introduced the solubility diffusion model for simulations of permeation through membranes [4]. The discussion of this model below is, however, similar to the more recent approach of Votapka et al. [14].

The first assumption that it is made in this model is that the permeation can be described along a one-dimensional reaction coordinate, which we denote by Z. It is the normal to the membrane plane (Figure 1).

The second assumption is that the dynamics of permeation is described by the Smoluchowski equation for overdamped processes. Combining the two assumptions, we obtain Eq. (4)

$$\frac{\partial \rho(Z,t)}{\partial t} = -\frac{\partial J(z,t)}{\partial Z} = \frac{\partial}{\partial Z}\left[D(Z)\frac{\partial \rho(Z,t)}{\partial Z} + \beta D(Z)\frac{dW(Z)}{dZ}\rho(Z,t)\right] \quad (4)$$

The probability density of the permeant is $\rho(Z,t)$ at position Z and time t. The flux, $J(Z,t)$, is defined for the same variables. The Boltzmann factor is $\beta \equiv 1/k_B T$. The diffusion coefficient, which may depend on the position, is $D(Z)$ and $W(Z)$ is the potential of mean force. MD simulations are used to determine $D(Z)$ and $W(Z)$ of Eq. (4).

The third assumption or approximation is that the diffusion constant can be extracted from the force–force correlation function of atomically detailed simulations [15]:

$$D(Z) = \frac{(k_B T)^2}{\int\limits_0^\infty \langle \Delta F(Z,t) \cdot \Delta F(Z,0) \rangle_0 dt} \tag{5}$$

The microscopic force fluctuation is given by Eq. (6)

$$\Delta F(Z,t) = F(z,t) - \langle F(z,t) \rangle_t = F(z,t) + \frac{dW(Z)}{dZ} \tag{6}$$

The average $\langle ... \rangle_0$ or $\langle ... \rangle_t$ is conducted by 'sliding' the origin of time. The system is in equilibrium and the time origin should not impact the result. Eq. (5) is a result of the fluctuation dissipation theorem [4] and the relationship between the friction and diffusion coefficient ($D = k_B T/\gamma$). The derivative of the potential of mean force, $dW(Z)/dZ$, that appeared in Eq. (4) is defined in Eq. (6). The potential of mean force for membrane permeation can be computed by a variety of methods, including Widom's particle insertion method [16], Umbrella sampling [4], Adapted Biasing Force [9, 17], Metadynamics [10], and Milestoning [18].

Accepting the assumptions listed above, we continue to solve the Smoluchowski equation for a permeation process. We consider a stationary process. There is no explicit time dependence for the stationary probability and we therefore have $\partial p(Z,t)/\partial t = 0$. We integrate once over the coordinate, Z, to find the value for the stationary flux J_{stat}.

$$J_{stat} = D(Z)\frac{\partial p(Z,t)}{\partial Z} + \beta D(Z)\frac{dW(Z)}{dZ}p(Z,t) \tag{7}$$

We now substitute for $p(Z) \leftarrow \Theta(Z)\exp(-\beta W)$. This substitution is attractive since we expect Boltzmann statistics of the density at equilibrium. In that case, $\Theta(Z)$ will be a constant. The unknown function is now $\Theta(Z)$. We have

$$J_{stat} = D(Z)\exp(-\beta W)\frac{d\Theta(Z)}{dZ}$$

$$\Theta(Z) = J_{stat}\int^z \frac{\exp(\beta W)}{D(Z)}dZ + \rho_a \tag{8}$$

$$p(Z) = J_{stat}\exp(-\beta W(Z))\int^z \frac{\exp(\beta W(Z'))}{D(Z')}dZ' + \rho_a\exp(-\beta W(Z))$$

The two integration coefficients, J_{stat} and ρ_a, are determined from boundary conditions. We set an absorbing boundary condition at one side of the membrane, Z_f. Each molecule that permeates to the other side disappears. This way we do not need to deal with a backward flux. The probability of finding the permeant at Z_f is zero

$$p(Z_f) = 0 = J_{stat}\int^{Z_f} \frac{\exp(\beta W(Z'))}{D(Z')}dZ' + \rho_a \tag{9}$$

$$\rho_a = -J_{stat}\int^{Z_f} \frac{\exp(\beta W(Z'))}{D(Z')}dZ'$$

On the other hand, the probability or the concentration of permeants at the other side of the membrane, Z_i, is a constant, say C_0

$$C_0 = J_{stat} \exp(-\beta W(Z_i)) \int^{Z_i} \frac{\exp(\beta W(Z'))}{D(Z')} dZ' - J_{stat} \exp(-\beta W(Z_i)) \int^{Z_f} \frac{\exp(\beta W(Z'))}{D(Z')} dZ'$$

$$C_0 = J_{stat} \exp(-\beta W(Z_i)) \int_{Z_i}^{Z_f} \frac{\exp(\beta W(Z'))}{D(Z')} dZ'$$

(10)

Taking the initial potential of mean force $W(Z_i)$ to be zero (the mean force potential is determined only up to a constant), we can determine the stationary flux and the permeation coefficient.

$$J_{stat} = \frac{C_0}{\int_{Z_i}^{Z_f} \frac{\exp(\beta W(Z'))}{D(Z')} dZ'}$$

$$P = \frac{J_{stat}}{C_0} = \left[\int_{Z_i}^{Z_f} \frac{\exp(\beta W(Z'))}{D(Z')} dZ' \right]^{-1}$$

(11)

Hence, the solubility diffusion model makes it possible to compute the permeability coefficient, given the diffusion coefficient and the potential of mean force. The last two are computed independently from atomically detailed simulations. Given the simplicity of the model and the assumptions made, it is desirable to have alternative approaches to compute the permeability. In the next section, we discuss one approach of this type, Milestoning.

IV Milestoning

Milestoning was first introduced in 2004 as a theory and an algorithm to compute kinetics in complex systems, given a one-dimensional reaction coordinate [19]. The requirement of a priori reaction coordinate was later dropped and replaced by reaction space following the work by Vanden Eijnden and Venturoli [20]. They use boundaries between Voronoi cells as milestones, an approach that we adopted [21]. The centers of the cells are sampled conformations in a reaction space. We briefly explain the Milestoning methodology below in the context of membrane permeation. The advantages of Milestoning compared to the solubility diffusion model are the following: (i) No assumption on the type of dynamics is made (Newtonian, Langevin, etc.). All classical mechanic models are adequate, (ii) The model is not restricted to one dimension, (iii) The formulation can be made, in principle, exact without the additional approximate calculations of a diffusion constant.

In a Milestoning calculation, we partition the phase space of a system with dividers (milestones) and compute short trajectories between these milestones (Figure 2). The short trajectories are used to estimate local transition probabilities (or kernels) between milestones α and β. The kernel, $K_{\alpha\beta}(x_\alpha, x_\beta)$, is the probability that a trajectory starting at milestone α in phase space point x_α will cross milestone β at position x_β before crossing any other milestone. The trajectories are also used to estimate the life time of a milestone, $t_\alpha(x_\alpha)$, which is the average time it takes a trajectory initiated at x_α to cross for the first time a milestone different from α.

The distribution of initial conditions for the short trajectories is called FHPD (First Hitting Point Distribution). It is written more explicitly as $f_\alpha(x_\alpha)$ and normalized as $\int f_\alpha(x_\alpha)dx_\alpha = 1$. The exact

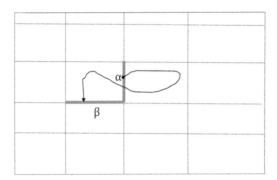

FIGURE 2 A schematic illustration of a Milestoning trajectory (black arrowed curve) that starts at milestone α and end at milestone β (thick line segments). Note that the trajectory is terminated only after 'touching' a milestone different from the milestone it was initiated on. (See color insert for the color version of this figure)

functional form of the FHPD is not known to begin with for a general stationary process; however, it can be computed numerically with iterations [22]. Near equilibrium, the FHPD is estimated directly as $f_\alpha(x_\alpha) \propto \exp[-U(x_\alpha)/k_BT]$. The microscopic potential energy is $U(x)$. The flux of a trajectory that passes milestone α is written as $q_\alpha(x_\alpha) = w_\alpha f_\alpha(x_\alpha)$, where w_α is the milestone weight. Different milestones can, of course, have different weights that are normalized as $\left(\sum_\alpha w_\alpha = 1\right)$, or to the total number of trajectories N. With the trajectory weight at hand, self-consistency equations for a steady flux can be written. The weight of a trajectory initiated at x_α in milestone α that end up in x_β in milestone β is $q_\alpha(x_\alpha)K_{\alpha\beta}(x_\alpha, x_\beta)$. Alternatively, if the number of trajectories initiated at milestone α is n_α and of these trajectories the number that ends up at milestone β is $n_{\alpha\beta}$, we estimate the transition kernel

$$K_{\alpha\beta} \cong n_{\alpha\beta}/n_\alpha \tag{12}$$

It is the probability that a trajectory that starts in milestone α at time zero will cross milestone β at any time t. The transition kernel satisfies a normalization condition $\sum_\beta K_{\alpha\beta} = 1$. Another entity that we extract from the short trajectories is the average lifetime of a milestone. Given that a trajectory is initiated at milestone α, what is the average time that it will take to hit any other milestone. It is computed from the short trajectories as $t_\alpha = \sum_{l=1,\ldots,n_\alpha} t_{l\alpha}$ where $t_{l\alpha}$ is the life time of a trajectory l.

The basic Milestoning formula is a vector-matrix equation for conservation of flux under steady-state conditions [22]. For the membrane system, we consider a flux of permeants from one side of the membrane to the other while retaining the same concentrations on each side of the membrane. This can be achieved if the permeant is found in large quantities on both membrane sides.

$$\mathbf{q}^t = \mathbf{q}^t\mathbf{K} \tag{13}$$

Eq. (13) is an eigenvector problem in which we seek the eigenvector $\mathbf{q}^t$ of the matrix $\mathbf{K}$ with an eigenvalue of one. After we estimate the kernel from short time trajectories (Eq. (12)), it is straightforward to determine the flux vector from Eq. (13). With the flux vector $\mathbf{q}$ at hand, physical observables are readily computed. The free energy of the trajectories that passed last milestone α (assuming that the system is near or at local equilibrium) is given by $F_\alpha = -k_BT \log[q_\alpha \cdot t_\alpha]$. The mean first passage time (MFPT) is the average time it takes a trajectory to hit the product side starting from the reactant, and is of considerable kinetic interest. For the membrane permeation problem, it is the average time it takes a trajectory initiated at one side of the membrane to reach the other side. It is given by $\tau_a = \mathbf{p}_a(\mathbf{I} - \mathbf{K})^{-1}\mathbf{t}$, where p_a is the probability starting at milestone a, $\mathbf{I}$ is the identity matrix, and $\mathbf{t}$ is the vector of milestone lifetimes.

The permeation coefficient in Milestoning can be computed in several ways. Perhaps, the most straightforward approach is to compare Eq. (13) with Eq. (1). The current J in Eq. (1) is the number of molecules that passes through the membrane in unit time and unit area. The flux $\mathbf{q}$ at the last milestone q_f is the number of molecules that make it to the other side of the membrane per unit time. Note that Eq. (13) determines the flux vector only up to a normalization factor.

Hence, if $\mathbf{q}$ is a solution of Eq. (13), so is $\lambda\mathbf{q}$ where λ is a positive scalar. Eq. (13) has a unique solution if we fixed one of the elements of the flux vector. Let the vector element that we fixed be the flux through the initial milestone at the other side of the membrane (Figure 1), q_0. The '0' milestone is set up in solution close to the membrane boundary. We set q_0 be equal to J_0 the initial flux. It is equal to the number of trajectories that crosses the initial milestone per unit time and per unit area and pointing to the membrane. Hence, we can write the set of linear equations:

$$\mathbf{q}^t = \mathbf{q}^t\mathbf{K} \quad \text{subject to}\,(\mathbf{q})_1 = J_0 \quad \text{and then } P = q_L \tag{14}$$

where L is the number of milestones and q_L is the flux at the absorbing milestone of the product side of the membrane. We can estimate J_0 from relatively short MD trajectories of permeants equilibrated in aqueous solution at concentration C. The permeant is diffusing rapidly in solution, making the estimate of the flux for the initial conditions a straightforward task. Examining trajectory configurations, we enumerate the number of times that a permeant crosses milestone 0 (the first milestone) in a particular time interval and per unit area and determine the flux. Let Z be the direction normal to the membrane plane (Figure 1) and let Z_0 be the location of milestone 0. Let the integration step in time be Δt. The equilibrium trajectory is approaching the membrane from top (Figure 1) and is contributing to the flux at time t if between t and $t + \Delta t$ it crosses the first milestone. The flux is determined by the time average of an equilibrium trajectory at one side of the membrane

$$J_0 = \lim_{t' \to \infty} \frac{1}{At'} \int_0^{t'} H(Z(t) - Z_0)H(Z_0 - Z(t + \Delta t))dt \tag{15}$$

where A is the surface area of the membrane and is zero or one for a negative or non-negative arguments, respectively. Here we choose the flux for trajectories that come into the membrane to be positive and hence for these type of trajectories, $Z(t) > Z_0$ and $Z_0 > Z(t + \Delta t)$, which makes J_0 positive.

V Permeation of Blocked Tryptophan with Milestoning

The first permeation system that we discuss is the translocation of *N*-acetyl-L-tryptophanamide (NATA), a neutral, blocked tryptophan through a DOPC membrane [23]. Tryptophan has affinity to membrane/water interface and is frequently seen as an anchor residue. From an experimental perspective, it is frequently used because the indole ring is easy to probe spectroscopically.

Studies using the parallel artificial membrane permeation assay (PAMPA [24]) technique suggested permeation times of hours for NATA through a DOPC membrane.

This slow permeation could be expected for a charged species but not for a small, neutral molecule. To provide a more detailed view of this permeation process, we used Milestoning to determine equilibrium and kinetic properties. The reaction coordinate is the center of mass of the permeant along the membrane axis. To define milestones, we partition the system with planes perpendicular to the membrane axis (Figure 1). The separation between the planes is variable. For planes 14 Å away from the membrane center, the distance separating successive planes is 2 Å and for planes closer to the membrane center it is 1.5 Å. The total number of milestones was 35.

With the positions of the milestones at hand, we sampled configurations of the permeant at each milestone plane using MD with a harmonic restraint. Then, the restraint was removed and the permeant was free to move along the membrane axis. When the permeant crossed a neighboring milestone, the trajectory was stopped and the transition time and identity of the crossed milestone were recorded. Those two pieces of information is what is needed to evaluate the transition kernel and solve the Milestoning equations (Section IV).

Figure 3 shows the free energy profile along the membrane axis. The results obtained by Milestoning are compared with the results computed by the solubility–diffusion theory. The profiles are similar with Milestoning giving a slightly higher barrier at the center of the membrane. Milestoning provides directly permeation time information with overall MFPT of 3.8 ± 3.7 hours. This is an average value of the results for solute permeation from the aqueous phase to the membrane center for the two individual layers of the bilayer membrane. The large error bars suggest the presence of one or more hidden reaction coordinates, other than the center of mass of the permeant, that are not completely sampled during the preparation of configurations in each milestone. One of these possible reaction coordinates is a rotational degree of freedom that change the orientation of NATA so that during its permeation process, the polar blocked termini point to the closer water phase. Due to this orientational preference, when the solute reaches the membrane center, it must flip-flop when it moves from one membrane layer to the other.

The influence of this orientational degree of freedom on the permeation of NATA was later investigated with Milestoning [7]. The position of the solute along the membrane axis (Z) was augmented by the orientation of the backbone of the molecule to create a two-dimensional reaction space. Milestoning enables the determination of kinetics and thermodynamics in multi-dimensions

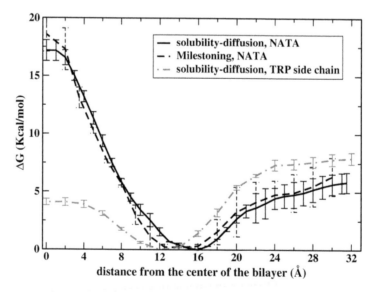

FIGURE 3 The potential of mean force computed both with Milestoning and Umbrella sampling are similar with a large permeation barrier for translocation of NATA in the middle of the membrane and favorable binding in the glycerol region of the phospholipid molecules. (Reprinted with permission from A. E. Cardenas, G. J. Jas, K. Y. DeLeon, W. A. Hegefeld, K. Kuczera, and R. Elber, "Unassisted transport of N-acetyl-L-tryptophanamide through DOPC membrane: Experiment and simulation", J. Phys. Chem. B, 116,2739-2750. Copyright 2012 American Chemical Society, Ref. [23]). (See color insert for the color version of this figure)

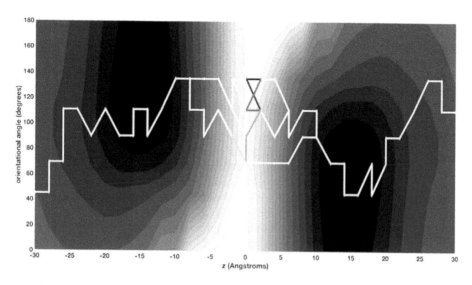

FIGURE 4 The free energy for permeation of NATA as a function of the location of the center of mass of the permeant along the Z axis and its orientation. Changes in the free energy are shown with gray-shaded contour plots: lower free energy is darker and high energy are in white. An orientational preference is clearly observed but the larger changes of free energy occur along the z axis. The figure also shows paths of maximum flux for permeation. The transition state for the paths displayed are colored with different grays (Reproduced with the permission of Taylor and Francis from reference [7]). (See color insert for the color version of this figure)

(e.g., Figure 2). Sampled configurations in the reaction space (that in Milestoning are called anchors) are used as the centers of Voronoi cells. The boundaries of these cells are the milestones. Inclusion of a second reaction coordinate increases the number of potential milestones to 1512. However, the network between milestones is not dense and some milestone-to-milestone transitions are not detected. To adequately describe the flow from reactant to product, 1046 milestones are sufficient.

Figure 4 shows the free-energy estimates for the permeation as a function of the position of NATA along the membrane axis and the orientational angle. Similar to the 1D case, the barrier for permeation (white areas) is mostly a function of the position of the permeant along the membrane axis. The location of the minima shows dependence of the orientational angle. The figure also shows the paths of maximum flux according to the milestone fluxes computed with Milestoning. For the path with maximum flux (shown with the black segment at the center of the bilayer), the orientational flip-flop only occurs after the permeant has crossed through the membrane center.

The permeation time calculated in this 2D milestoning calculation was 1.0 ± 0.9 hours, slightly smaller than the result estimated with 1D milestoning. The magnitude of the errors also decreases. That seems to suggest that the explicit consideration of the orientational motion of the permeant during sampling is important for permeation studies of medium-sized permeants. Still the large uncertainties for the permeation rates indicate that other hidden variables can be important for the determination of the permeation kinetics.

VI Fluctuations of Membrane Density with Milestoning

A factor that can be relevant for the permeation of medium-sized permeants is the membrane atomic density. Motions of one or more phospholipids molecules may be necessary to create empty spaces for the permeant to move into deeper regions of the membrane. However, those motions can be very slow compared with the usual length of MD trajectories used for Milestoning sampling. In a previous work, we studied the fluctuations and dynamic of the membrane packing density in a pure phospholipid bilayer and related these fluctuations to the permeation of small neutral solutes through the membrane. The fluctuations of water density also provided a different view of membrane permeation mechanism of water molecules.

We used Milestoning to compute fluctuations of different atomic components in a bilayer membrane system [25]. A 200 ns trajectory of a bilayer membrane system composed of pure DOPC lipids and water was used in the analysis of the density fluctuations. The simulation box was mapped onto a three-dimensional rectangular grid (Figure 5). In each of the cells in the grid (with a volume of about 64 Å^3), we count the number of atoms of different kinds (for example, lipid atoms, or oxygen water) in every configuration (coordinates were saved every 1 ps). When the number of atoms in a given cell changes from one configuration to the next one, then a density transition has occurred in that cell. For every cell in the simulation box, we keep track of those density changes and store the information of the neighboring cell that exchanges atoms with that cell. We use that information to construct a transition kernel that provides the probability per unit time that a cell will experience a given change of density exchanging particles with a neighboring cell. Many of the cells considered are related by symmetry (for example, a pure bilayer system is homogeneous in the lateral direction of the membrane), and we used that symmetry to accumulate transitions in a single column cell along the membrane axis. At the end, the total number of milestones included when sampling transitions of both lipid and water densities was 5810. The

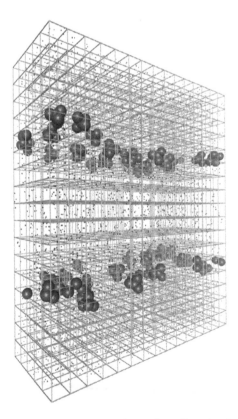

FIGURE 5 Example of the mapping of the simulation box into a three-dimensional cubic grid. The number of atoms in each grid determines local density. The phosphate atoms are represented with larger spheres.

transition matrix is highly sparse and the Milestoning equations can be solved efficiently with standard libraries.

Using this Milestoning analysis of density transitions, we were able to study changes of density and dynamics of water permeating inside the membrane. Figure 6 shows the free energy surface for number density of water molecules at different locations along the membrane axis. At the water phase, there is a valley in the surface where three to four water molecules are more often present in a cell. Moving closer to the membrane center, the valley disappears. In that region, the more prevalent occupation number is zero but still enough water molecules permeate deep in the region so some fluctuation dynamics of water density is observed. The fluctuations enable the calculation of paths for water permeation inside the membrane (also shown in the figure). It is observed that when water is inside the hydrophobic core of the membrane, the paths (black and gray lines) often include multiple water density numbers (two or three). This suggests that a possible mechanism for water permeation could entail a collaborative effort of two or more water molecules going inside the membrane close to each other, to try to reduce the energetic cost of water being inside a low dielectric environment region and the loss of hydrogen bonding. This kind of multiple permeant mechanism cannot be seen by the conventional solubility–diffusion model because in its usual implementation, only one permeant molecule is considered at a time.

Our density analysis can be used to study the dynamics of cavity formation inside the membrane. Figure 7 displays free-energy profiles along the membrane axis for cavity formation for different volumes of the grids used to discretize the simulation box. The probability of observing holes is always larger at the membrane center and it is smaller at the headgroup region. The plot also shows the free-energy profile for a xenon atom permeating inside the same type of membrane.

The result for xenon permeation is similar to the cavity profile for cells that have a volume big enough to contain a single xenon atom. This notable similarity indicates that the permeation of a small neutral molecule is controlled by the formation of membrane cavities.

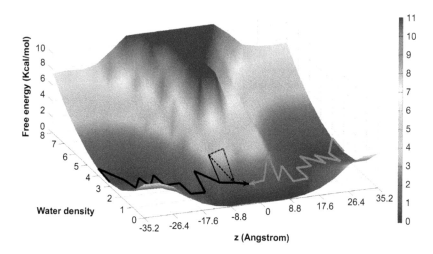

FIGURE 6 Free-energy surface for water permeation as a function of water occupancy in each cell and the position along the membrane axis. The figure also shows maximum flux path for water permeation to the membrane center from the left-side water (in black) and right-side water phases (in gray). The three maximum flux paths are shown at the left (they only differ from each other in the segments shown with dashed lines). These maximum flux paths clearly involve more than one water moving inside the membrane core. (Reproduced from A. E. Cardenas and R. Elber, "Modeling Kinetics and Equilibrium of Membranes with Fields: Milestoning Analysis and Implication to Permeation", J. Chem. Phys. 141,054101(2014) (Ref. 25), with the permission of AIP Publishing). (See color insert for the color version of this figure)

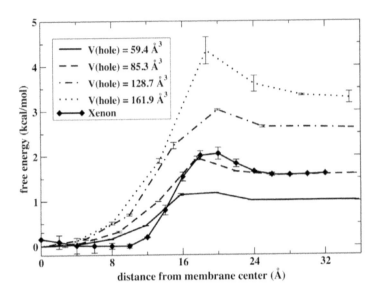

FIGURE 7 The potential of mean force (PMF) for hole formation along the distance from the center of the membrane. The solid and dashed lines show its dependence on the size of the cubic grid used in the density discretization (see Fig. 5). The line with filled diamonds shows the potential of mean force for permeation of xenon atom. The diameter of a xenon atom is 4.3Å, while the size of the cubic grid for the dashed line (the closest to the PMF of the xenon) is 4.4Å (enough to enclose a xenon atom). (Reproduced from reference 25, with the permission of AIP Publishing). (See color insert for the color version of this figure)

VII Difference of Permeation of Positive and Negative Charged Molecules

Previous studies using organic hydrophobic ions with similar structures but different charges have shown than the negative-charged species permeates a lot faster than the positive ion species [26]. This has been explained by the presence of oriented water dipoles at the membrane interface caused by both phosphate and carbonyl groups of the glycerol esters. This positive contribution from the oriented water overcompensates the negative contribution of the P^-N^+ dipole. However, many simulations of monatomic ions have shown that the permeation barrier for translocation is slightly higher for anions than cations. For example, McCammon et al. reported a barrier of 21.9 kcal/mol for Na^+ and 23.6 kcal/mol for Cl^- [27]. Simulation have also shown that the side chain of arginine has a smaller barrier than the side chains of glutamic and aspartic acids [28]. In a recent work, we address this issue by choosing a permeant for which variations of size and molecular flexibility between the cation and anion were minimized in order to focus on the charge effect [3]. We used positively and negatively charged tryptophan and its permeation through a DOPC membrane bilayer in our study. The choice of tryptophan was also motivated by its easy detection with fluorescence measurements. We determined the potential of mean force using the solubility–diffusion theory. Our previous studies comparing Milestoning with the solubility–diffusion results have shown that both provide similar results within the statistical uncertainties of the calculations.

Figure 8 shows the PMF for permeation of positively and negative charged tryptophan, and also for the zwitterionic and neutral molecules. There is a very large barrier for the permeation of the negatively charged (and zwitterionic) tryptophan compared with the positively charged species. The positively charged tryptophan also binds more strongly at the hydrophobic/polar interface of the membrane. The permeability coefficients estimated using the solubility–diffusion model ranges from 9.5×10^{-8} to 4.2×10^{-6} cm/s for the positively charged tryptophan and 1.1×10^{-16} to 1.6×10^{-14} cm/s for the negatively charged tryptophan. These simulation results were in qualitative agreement with our experimental results that measure the permeation of the same tryptophan inside vesicles made of DOPC lipids.

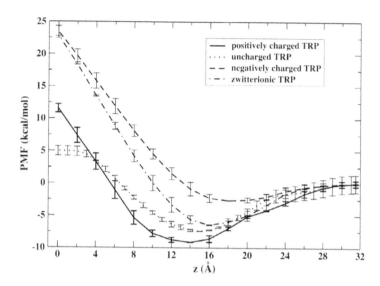

FIGURE 8 Free-energy profile as function of the membrane depth for different charge tryptophan species. The positively charged tryptophan has a notably smaller barrier at the membrane center compared to the negatively charged molecule. (Reprinted with permission from A. E. Cardenas, R. Shrestha, L. J. Webb, and R. Elber, "Membrane Permeation of a Peptide: It is Better to be Positive", J. Phys. Chem. B, 119,6412-6420. Copyright 2015 American Chemical Society. Ref. [3].)

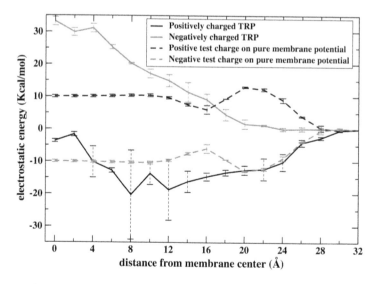

FIGURE 9 Electrostatic potential energy of a permeant charge moving inside the membrane. The solid lines are the energy felt by the positive- (solid black line) and negative-charged permeant (solid gray line). The dashed line is the results obtained by placing a charge probe in a pure lipid membrane. The pure membrane prefers to transport a negatively charged molecule while the membrane perturbed by the presence of the permeant prefers to move a positively charged molecule. (Reprinted with permission from A. E. Cardenas, R. Shrestha, L. J. Webb, and R. Elber, "Membrane Permeation of a Peptide: It is Better to be Positive", J. Phys. Chem. B, 119, 6412-6420. Copyright 2015 American Chemical Society. Ref. [3].) (See color insert for the color version of this figure)

Why the membrane permeation is more difficult for the negatively charged tryptophan compared to the positively charged molecule? A detailed analysis revealed that membrane perturbation is greater when the negatively charged tryptophan is at the hydrophobic core (an average number of 15 water molecules get inside the membrane). For the positively charged tryptophan, the membrane is also distorted but the perturbation is not that large with a smaller amount of water found near the center (on average, six water molecules). Also, in this case, up to two lipids bend over to allow its phosphate and ester oxygens to stabilize the positively charged tryptophan. A similar stabilization is hard to accomplish for the negatively charged tryptophan because in this case terminal choline groups, typically located at the outer region of the membrane, need to get inside producing a larger membrane defect.

The observed membrane distortions induce modifications of the electrostatic interactions of the membrane. Due to those distortions, the permeant does not experience the electrostatics of the unperturbed membrane that favors the anion (Figure 9). Instead, computing the electrostatic energy profile along the membrane axis shows that the opposite is happening, with the positively charged tryptophan interacting favorably with phosphate and the esters of the glycerol group.

When this permeant moves closer to the membrane center, those interactions are more difficult and the energy increases. For the negatively charged tryptophan, the positively charged choline groups cannot move inside the membrane easily and the electrostatic energy of the permeant increases when it moves closer to the membrane center.

Therefore, the specific structure of the phospholipid with phosphate and glycerol groups deeper in the membrane favors the permeation of cations over anions.

VIII Conclusions

We considered and discussed at length the process of small molecule permeation through membranes. The process is critical to life as membranes are the gatekeepers of material entering and leaving biological cells or compartments. As such, membrane permeation deserves careful experimental and modeling attention. The experimental measurements of the permeation coefficient for synthetic vesicles were described and analyzed. The widely used molecular theory of permeation – the solubility-diffusion model – is described in detail and the more recent approach to permeation, based on Milestoning, is discussed as well. Finally, a few examples of peptide permeation (neutral, positively, and negatively charged) are described. We also examine the permeation of water molecules, which is profoundly different from peptide permeation due to the much larger concentration of water that leads to co-operative transport mechanism.

REFERENCES

1. Giulia Guidotti, Liliana Brambilla, and Daniela Rossi, "Cell Penetrating Peptides: From Basic Research to Clinics", *Trends Pharm. Sci.*, 38, 406–424 (2017).
2. Ling-Juan Zhang and Richard L. Gallo, "Antimicrobial Peptides", *Cur. Biol.*, 26, R1–R21 (2016).
3. Alfredo Cardenas, Rebika Shrestha, Lauren Webb, and Ron Elber, "Membrane Permeation of a Peptide: It is Better to be Positive", *J. Phys. Chem. B.*, 119, 6412–6420 (2015).
4. Siewert-Jan Marrink and Hermans J. C. Berendsen, "Simulation of Water Permeation through a Lipid Membrane", *J. Phys. Chem.*, 98, 4155–4168 (1994).
5. Siewart J. Marrink and Herman J. C. Berendsen, "Permeation Process of Small Molecules across Lipid Membranes Studied by Molecular Dynamics Simulations", *J. Phys. Chem.*, 100, 16729–16738 (1996).
6. Alfredo E. Cardenas and Ron Elber, "Markovian and Non-Markovian Modeling of Membrane Dynamics with Milestoning", *J. Phys. Chem. B.*, 120, 8208–8216 (2016).
7. Alfredo E. Cardenas and Ron Elber, "Computational Study of Peptide Permeation through Membrane: Searching for Hidden Slow Variables", *Mol. Phys.*, 111, 3565–3578 (2013).

8. G. Parisio, M. Stocchero, and A. Ferrarini, "Passive Membrane Permeability: Beyond the Standard Solubility-Diffusion Model", *J. Chem. Theory Comput.*, 9, 5236–5246 (2013).

9. Jeffrey Comer, Klaus Schulten, and Christophe Chipot, "Permeability of a Fluid Lipid Bilayer to Short-Chain Alcohols from First Principles", *J. Chem. Theory Comput.*, 13, 2523–2532 (2017).

10. Zhaleh Ghaemi, Domenico Alberga, Paolo Carloni, Alessandro Laio, and Gianluca Lattanzi, "Permeability Coefficients of Lipophilic Compounds Estimated by Computer Simulations", *J. Chem. Theory Comput.*, 12, 4093–4099 (2016).

11. E. Awoonor-Williams and C. N. Rowley, "Molecular Simulation of Nonfaciliated Membrane Permeation", *Biochim. Biophys. Acta.*, 1858, 1672–1687 (2016).

12. S. Paula, A. G. Volkov, A. N. Van Hoek, T. H. Haines, and D. W. Deamer, "Permeation of Protons, Potassium Ions, and Small Polar Molecules through Phospholipid Bilayers as a Function of Membrane Thickness", *Biophys. J.*, 70, 339–348 (1996).

13. A. C. Chakrabarti and D. W. Deamer, "Permeability of Lipid Bilayers to Amino-Acids and Phosphate", *Biochim. Biophys. Acta*, 1111, 171–177 (1992).

14. Lane Votapka, Christopher T. Lee, and Rommie E. Amaro, "Two Relations to Estimate Membrane Permeability using Milestoning", *J. Phys. Chem. B*, 120, 8606–8616 (2016).

15. S. Harris, "Force Correlation Function Representation for the Self-Diffusion Coefficient", *Mol. Phys.*, 23, 861–865 (2006).

16. P. Jedlovszky and M. Mezei, "Calculation of the Free Energy Profile of H_2O, O_2, CO, CO_2, NO, and $CHCl_3$ in a Lipid Bilayer with a Cavity Insertion Variant of the Widom Method", *J. Amer. Chem. Soc.*, 122, 5125–5131 (2000).

17. Eric Darve, D. Rodriguez-Gomez, and Andrew Pohorille, "Adaptive Biasing Force Method for Scalar and Vector Free Energy Calculations", *J. Chem. Phys.*, 128, 1444129 (2008).

18. Anthony M. A. West, Ron Elber, and David Shalloway, "Extending Molecular Dynamics Time-scales with Milestoning: Example of Complex Kinetics in a Solvated Peptide", *J. Chem. Phys.*, 126, 145104 (2007).

19. Anton K. Faradjian and Ron Elber, "Computing Time Scales from Reaction Coordinates by Milestoning", *J. Chem. Phys.*, 120, 10880–10889 (2004).

20. Eric Vanden-Eijnden and Maddalena Venturoli, "Markovian Milestoning with Voronoi Tessellations", *J. Chem. Phys.*, 130, 194101 (2009).

21. Peter Májek and Ron Elber, "Milestoning without a Reaction Coordinate", *J. Chem. Theory Comput.*, 6, 1805–1817 (2010).

22. Juan M. Bello-Rivas and Ron Elber, "Exact Milestoning", *J. Chem. Phys.*, 142, 094102 (2015).

23. Cardenas, A. E.; Jas, G. S.; DeLeon, K. Y.; Hegefeld, W. A.; Kuczera, K.; Elber, R., Unassisted Transport of N-Acetyl-L-tryptophanamide through Membrane: Experiment and Simulation of Kinetics. *J. Phys. Chem. B* **2012**, *116 (9)*, 2739–2750.

24. Manfred Kansy, Frank Senner, and Klaus Gubernator, "Physiochemical High Throughput Screening: Parallel Artificial Membrane Permeation Assay in the Description of Passive Absorption Processes", *J. Med. Chem.*, 41, 1007 (1998).

25. Alfredo E. Cardenas and Ron Elber, "Modeling Kinetics and Equilibrium of Membranes with Fields: Milestoning Analysis and Implication to Permeation", *J. Chem. Phys.*, 141, 054101 (2014).

26. R. J. Clarke, "The Dipole Potential of Phospholipid Membranes and Methods for its Detection", *Adv. Colloid Interface Sci.*, 89, 263–281 (2001).

27. Ilja V. Khavrutskii, Alemayehu A. Gorfe, Benzhuo Lu, and J. Andrew McCammon, "Free Energy for the Permeation of Na^+ and Cl^- Ions and Their Ion-Pair through a Zwitterionic Dimyristoyl Phosphatidylcholine Lipid Bilayer by Umbrella Integration with Harmonic Fourier Beads", *J. Amer. Chem. Soc.*, 131, 1706–1716 (2009).

28. J. L. MacCallum, W. F. Drew Bennett, and D. Peter Tieleman, "Distribution of Amino Acids in a Lipid Bilayer from Computer Simulations", *Biophys. J.*, 94, 3393–3404 (2008).

9

Nanoparticle–Membrane Interactions: Surface Effects

G. Rossi, S. Salassi, F. Simonelli, and A. Bartocci
Department of Physics, University of Genoa, via Dodecaneso 33, 16146 Genoa (Italy)

L. Monticelli
Molecular Microbiology and Structural Biochemistry, UMR 5086, CNRS and Universitè de Lyon, Lyon (France)

1 Introduction

Biological membranes consist mainly of lipids and proteins, with carbohydrates sometimes linked to the polar region of the lipids. They envelop and compartmentalize living cells, they represent the first barrier towards permeation of exogenous materials into cells, and regulate molecular transport (import of nutrients, and export of waste products). The properties and functioning of biological membranes depend on their chemical composition. For example, the lipidic component regulates the functioning of membrane proteins, both through specific interactions (e.g., specific binding sites have been recognized in several membrane proteins) and by modulating the membrane properties, notably the elastic properties. One way to alter cellular functions is by modifying the membrane composition, either in the membrane interior or on the membrane surface. Both alterations can be achieved by exposing membranes to natural or synthetic nanoparticles (NPs), i.e., particles whose size is in the nanometer range. By virtue of their very high surface/volume ratio, NPs interact strongly with biological materials. The possibility to modify biological activity in a controlled way is one of the reasons for the great interest in NP–membrane interactions. A second important reason is the need to reduce NP toxicity, by designing NPs that do not penetrate across cell membranes and/or do not interact strongly with them.

NP permeation and, more generally, NP interaction with membranes depend on the chemical composition and properties of the biological membrane, as well as on a number of properties of the NP, notably the size, the shape, the electrostatic charge, and the arrangement of the ligands on the NP surface. Such dependency on NP properties has been explored using experiments, theory, and simulations, but no consensus has been reached on general principles that allow to predict NP–membrane interactions. In fact, experiments on seemingly similar NPs yielded contrasting results. Moreover, a large body of experiments remain unexplained at the molecular level, and a number of questions remain open.

In the present review, we focus on the effect of NP surface properties on NP–membrane interaction. Surface functionalization is the simplest way to alter the chemical nature of the NPs and their physical properties. It is also extremely versatile, as it can be used to change the electrostatic charge of the NP, its ability to stick to proteins (relevant for NP transport within living organisms) and/or other natural organic matter (relevant for transport in natural waters and the environment), the geometrical patterns of ligands, the size, and the softness of the NP. A large body of literature has shown that

NP surface properties determine NP interaction with biological membranes. Establishing clear relationships between surface properties and NP–membrane interactions is highly desirable, as such relationships would allow the prediction and control of biological effects by manipulating NP surface chemistry. Here we briefly summarize the most relevant experimental results on the effect of surface properties on NP–membrane interaction, and then highlight the contributions from computer simulations, and particularly simulations at the molecular level.

2 Experimental Data in Search for a Molecular-Scale Interpretation

Here we list the main molecular features that certainly contribute to define the NP–membrane interaction: charge, hydrophobicity/hydrophilicity, and surface patterning. One of the most important challenges is to predict which molecular characteristics have the greatest influence on NP–membrane interaction, and whether the effects of different features are cooperative.

2.1 Effects of NP Charge on Membrane Binding and Membrane Damage

Surface charge and hydrophilicity are important factors driving the fate of functionalized NPs within organisms. They influence NP solubility and circulation time in the blood stream; they affect the NP interactions with serum proteins and the stability of the protein corona[1,2]; eventually, they contribute to determine the NP interaction with the cell membrane.[3,4]

When looking at the interaction between charged NPs and model lipid membranes, it is tempting to interpret the experimental data by simple electrostatic arguments. Electrostatic attraction between oppositely charged NPs and bilayers certainly favors the formation of stable NP–lipid complexes. When the NP diameter is below 20 nm, the NPs are expected to adhere to the membrane surface or be embedded in the membrane core without inducing large bilayer deformations. The group of Bothun has investigated the interaction of Ag NPs in this size range, functionalized with neutral and hydrophilic (polyethylene glycol (PEG)) or positively charged (amine) or negatively charged (carboxyl) groups[5], with model vesicles consisting of mixtures of dipalmytoylphosphatidylcholine (DPPC), dipalmitoylphosphatidylglycerol (DPPG), and dipalmitoyltrimethylammonium-propane (DPTAP). They found evidence of strong binding between oppositely charged NPs and vesicles. The interaction between oppositely charged NPs and lipid bilayers can cause transient damage to the membrane, as well. Liposome leakage assays by Goodman *et al.*,[6] for example, reported the disruptive effects of cationic NPs on negatively charged bilayers composed by a mixture of *sn*-2-oleoyl-1-stearoyl-glycero-3-phospho-L-serine (SOPS) and 1-stearoyl-2-oleoyl-*sn*-glycero-3-phosphocholine (SOPC).

The interaction of larger NPs (with a diameter >20 nm) with unilamellar vesicles can lead to the formation of different NP–bilayer complexes, such as bilayer-wrapped NPs. Even these assemblies are stabilized when the NP and the bilayer carry opposite charges.[7] The trend appears to be independent of the nature of the NP core. In addition to metal NPs, also positively and negatively charged oxide NPs (Al_2O_3 and SiO_2) have been shown to damage oppositely charged bilayers.[8] The same reasoning is generally invoked to explain why cationic NPs are better bactericidal agents than anionic NPs, due to the negative charge of bacterial membranes.[9,10]

A large amount of data points to favorable electrostatics-induced NP–bilayer interactions, but there are a number of situations in which the role of NP surface charge is less clear. Electrostatic attraction is not a necessary ingredient to the formation of stable NP–bilayer complexes, nor to toxicity, which can take place also when the NP and the membrane have a ζ potential of the same sign.[8,9,11] Another open question concerns the fate of charged NPs interacting with the surface of membranes exposing neutral lipid head groups, like in the extracellular leaflet of mammalian plasma membranes. A consistent body of experimental work from different groups focused on cationic and anionic gold (Au) NPs with a core in the 2–8 nm range, and showed that they can interact passively with mammalian cell membranes and model lipid bilayers[12–14] (see, for example, the left panel of Fig. 1). The role played by the sign of the NP charge, though, is still debated. Neutron reflectometry studies by Tatur *et al.*[15] suggest that anionic Au NPs could adhere to the bilayer surface without penetrating it, at variance with cationic NPs that would interact with the bilayer in a more disruptive way. According to Goodman *et al.*,[6] Au anionic NPs can induce the largest membrane leakage to pure SOPC bilayers. On the contrary, experiments by Van Lehn *et al.*[13] showed no membrane translocation of the fluorophore in multilamellar 1,2-dioleoyl-*sn*-glycero-3-phosphocholine (DOPC) vesicles in the presence of anionic Au NPs co-localized with the vesicle bilayers.

From the experimental point of view, it is still a challenge to disentangle purely electrostatic effects from other molecular and atomic features of the NP–membrane interface. For this reason, a comprehensive interpretation of this large body of experimental data is still lacking. In Section 3.1, we will discuss to what extent molecular simulations are contributing to the interpretation of experimental results, taking advantage of the possibility to design simulations in which each of these factors is considered separately.

2.2 Effect of NP Charge on Lipid Phase Behavior

In the early 2000s, the group of Steve Granick published a series of papers investigating the interactions of model zwitterionic, phosphatidylcholine (PC) liposomes with NPs bearing anionic or cationic surface charges. Granick's experiments showed that large concentrations of NPs (polystyrene beads with a diameter of 20 nm) functionalized with anionic or cationic groups have different effects on the stabilization of liposome suspensions: while cationic NPs maintain the suspension fluid[16] also at very large liposome concentrations, anionic NPs cause liposome aggregation (but not fusion[17]) and a decrease in the fluidity of the suspension.[16] Later on, the same authors reported a dramatic change in the gel-to-liquid transition of DPPC liposomes upon interaction with anionic and cationic polystyrene NPs: anionic NPs were reported to locally stabilize the DPPC gel phase, and cationic NPs, on the contrary, were reported to fluidize DPPC membranes.[18] These experimental evidences were interpreted as a purely electrostatic effect. According to this interpretation, as an anionic NP approaches the surface of a PC membrane, electrostatic attraction between the anionic surface and the choline groups of the lipid would cause the lipid head group tilt angle to increase, aligning the headgroups to the membrane normal. This, in turn, would favor the stabilization of patches of gel phase over the liquid phase.

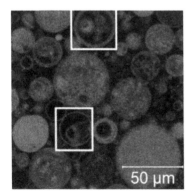
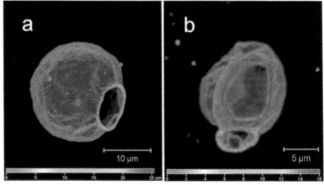

FIGURE 1 Confocal microscopy images of NP-membrane complexes. On the left: bodipy-labeled Au NPs (2 nm diameter), functionalized by a mixture of hydrophobic and anionic ligands, co-localize with the bilayer of multi-lamellar DOPC vesicles. The NPs are shown to be present in outer as well as in inner vesicles, suggesting spontaneous passive translocation (adapted from Van Lehn *et al.*[13]). On the right: SiO$_2$ NPs (18 nm diameter, negative zeta potential) stabilize unusual curvatures and holes in DOPC giant unilamellar vesicles (adapted from Zhang *et al.*[11]). (See color insert for the color version of this figure)

As originally proposed, such an effect should be fairly general and thus applicable to a broad range of negatively charged NPs.

Unfortunately, recent experiments portray a much more variegated scenario, in which this interpretation is questioned. Zhang *et al.*[11] investigated the interaction of DOPC giant unilamellar vesicles with SiO_2 NPs, of the same size as the polystyrene beads used by Granick, bearing a negative charge of 0.13 e/nm^2. The NPs caused the GUV to crinkle and the formation of large holes was observed (Fig. 1). The study of lipid diffusion via Fluorescence Recovery After Photobleaching (FRAP) indicated a slowing down of lipid dynamics. The lipid dynamics was reported to be uniform, with no indication of a coexistence of gel and liquid patches. The slow evolution in time of the lipid mobility also suggested that no sharp liquid-to-gel first-order transition had taken place. Wang and Liu,[19] a few years later, reported that Au NPs (diameter of 15 nm), non-covalently functionalized by citrate, would shift the gel-to-liquid transition of DOPC to higher temperatures (in line with Granick's result), but did not observe this effect when using Au NPs covalently functionalized by mercaptopropionic acid, bearing a similar negative charge. Moreover, the same authors reported that the interaction of citrate-capped Au NPs was stronger with liposomes in the fluid phase than with liposomes in the gel phase.[20] Bhat *et al.*[21] probed the fluidity of dimyristoylphosphatidylcholine (DMPC) large unilamellar vesicles by means of fluorescence spectroscopy, and reported that citrate-capped Au NPs with a diameter of 20 nm induced a fluidization of DMPC.

While it looks unquestionable that negatively charged NPs can stably interact with zwitterionic lipid membranes (see Fig. 1), it seems unlikely that electrostatics alone can account for this variety of experimental observations. NPs with a diameter of 15–20 nm are not expected to passively penetrate the membrane core, nor to be wrapped. Their adsorption on the membrane surface implies a close contact with little patches of lipid and, in this situation, considering the NP surface as a rigid, charged plane seems a quite rough approximation. It is likely that other physical and chemical characteristics of the NP surface should be taken into account, such as its mechanical properties, its propensity to make H-bonds, its roughness, and its protonation state in the membrane vicinity. The contribution of molecular simulations to understanding which NP surface properties can induce lipid phase changes has been very limited so far.

2.3 Effect of the Chemical Nature of the Ligands on the Formation of a Protein Corona

NPs designed to be administered via intravenous routes are prone to interact with serum proteins, which can stably cluster around the NP forming a protein corona.[22–24] The non-specific adsorption of proteins on NPs alters their designed function and can lead to enhanced immune response,[25] influencing the fate of the NPs in the body. The control of protein adsorption[26] and the minimization of early clearance from the bloodstream are crucial to the clinical integration of synthetic NPs.[27] Proteins may also be part of the NP functionalization to drive specific interactions with cell receptors.[28] Thus, it is important to accurately control NP–protein interaction to drive NP–cell interaction. One possible route to act on the NP–protein interaction involves the functionalization of the NPs with proper anti-fouling functional groups.

PEG is known to be a good anti-fouling material.[29] Protein-repellent properties of PEG grafted on surfaces are influenced by PEG chain length,[30] density, and environment temperature,[31,32] and not always the amount of adsorbed proteins is a monotonic function of these parameters.[33] The surface of solid NPs functionalized by PEG chains offers many more degrees of freedom, including curvature, presence of low-affinity coordination sites at surface edges, and surface roughness, which may well influence the anti-fouling properties of PEG itself. The use of PEG as a stealth agent, on the other hand, also has some drawbacks, such as its non-biodegradability, immunogenicity,[34] and its accumulation in membrane-bound organelles.[35]

Zwitterionic functionalization has been used as a strategy alternative to PEGylation. Zwitterionic groups can indeed extend the circulation time of the NPs and increase their ability to effectively penetrate cell membranes.[4] The group of Rotello, for example, recently showed how the protein corona formation on the surface of monolayer-protected gold NPs can be reduced by the use of phosphocholine-terminated ligands.[1]

Considering that anti-fouling groups reduce protein adsorption on NPs, it becomes relevant to investigate how different ligands drive the interaction of the NP with the cell membrane. In Section 3.2, we will discuss computational work on the interaction of cell membranes and proteins with NPs functionalized by PEG or by zwitterionic groups.

2.4 Effect of Ligand Pattern on the Mechanism of NP Internalization

During the synthesis of monolayer-protected NPs, different ligands can spontaneously arrange in ordered patterns on the surface of the NP.[36–38] For NPs functionalized by a mixture of two different ligands, the pattern may correspond to a random, patched, or striped arrangement of the ligands. If the phase separation between the ligands on the NP surface is complete, only two patches form and the NP is usually referred to as a Janus NPs.

From an experimental perspective, the facile engineering of Au NPs functionalized by thiols has made this system a reference model for the study of surface patterning and of its effects on the fate of NPs in the biological environment.[38] Typically, surface patterning on small (< 10 nm) Au NPs can be obtained by mixing purely hydrophobic thiols, such as octanethiol, with charged thiols, in which a hydrophobic chain is terminated by carboxylate or sulfonate groups. On bigger Au NPs (> 10 nm), surface patterning can be obtained using longer immiscible polymer chains. After spontaneous self-assembly, the experimental characterization of the NP surface is a challenging task; both direct and indirect methods have been proposed, followed by heated scientific controversies. Among the different techniques proposed to probe ligand segregation on the surface of NPs, the most common are scanning tunneling microscopy, nuclear magnetic resonance (in particular Overhauser effect spectroscopy), and mass spectrometry. For an excellent presentation of the pros and cons of these experimental approaches, and of their applicability to NPs in different size ranges, the reader is referred to the review by Pengo *et al.*[38]

As for the biological consequences of NP surface patterning, Au NPs with a striped arrangement of hydrophobic and charged ligands show reduced non-specific interactions with proteins, compared to mono-ligand NPs. Moreover, the ligand pattern has been reported to affect the mechanism of cell internalization: striped NPs are able to penetrate the plasma membrane passively, while NPs with a random arrangement of ligands prefer endocytic internalization mechanisms.[14]

Theory and computer simulations have substantially contributed to understanding the physical driving forces for pattern formation. In recent years, molecular simulations have also been used to elucidate how surface patterning can affect NP–membrane interactions. These works will be presented in Section 3.3.

3 What We Understood So Far Using Computational Tools

3.1 Effect of Charge on Membrane Binding, Membrane Damage, and Membrane Permeability

3.1.1 Adsorption of Small Anionic and Cationic NPs at the Surface of Membranes

The experimental data presented in Section 1.1 coherently point to a favorable interaction between charged NPs and neutral, often zwitterionic, membranes. The interaction between small (diameter < 10 nm) ligand-protected NPs and lipid membranes of various composition has been approached via both unbiased and biased Molecular Dynamics (MD) simulations (step 1 in Fig. 2). There is a consensus in the computational community regarding the favorable adsorption of charged NPs at

STEP 1   STEP 2   STEP 3

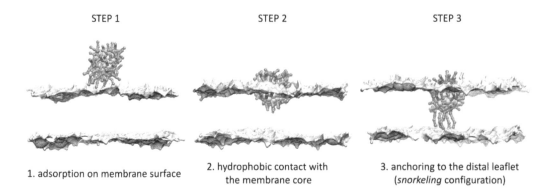

1. adsorption on membrane surface   2. hydrophobic contact with   3. anchoring to the distal leaflet
                                            the membrane core   (*snorkeling* configuration)

FIGURE 2 The 3 steps of NP–membrane interaction, as simulated by MD at CG level. Step 1 corresponds to surface adsorption of the monolayer-protected NP. In Step 2, the NP establishes a hydrophobic contact with the membrane core. In Step 3, the NP is anchored to both membrane leaflets. Water and lipid tails are not shown, lipid headgroups are shown as gray surfaces, the Au NP core is yellow, hydrophobic ligands are green, and anionic ligands are pink. (See color insert for the color version of this figure)

the surface of neutral, in some cases zwitterionic, lipid bilayers.[39–47] The groups of Vattulainen and Akola found, by means of atomistic simulations, that anionic NPs can spontaneously adhere to zwitterionic palmitoyloleoylphosphatidylcholine (POPC) membranes,[40] while cationic NPs can bind to the membrane surface by overcoming of a small energy barrier (12 kJ/mol).[39] The group of Alexander-Katz simulated, with atomistic resolution, the binding of Au NPs functionalized by mixtures of hydrophobic and anionic ligands, obtaining spontaneous binding of the NPs to the surface of PC membranes.[46,47] Similar results, for anionic NPs, were found by others using explicit solvent coarse-grained MD simulations,[41,48–51] and even a DPD approach.[52] For CG force fields, the data concerning cationic NPs are somehow more problematic. Liu *et al.*[50] found, using the Martini CG force field,[53] that adhesion of a cationic NP onto a zwitterionic DPPC vesicle is spontaneous at low (1 e/nm^2) or very high (14.3 e/nm^2) surface charge density, while no attraction is measured in the middle range. Unpublished data from our group using the same force field point to no spontaneous adhesion of cationic NPs on POPC membranes. Adsorption of the cationic NP is even less favorable if the polarizable version[54,55] of the Martini force field is used. The result seems independent on ligand flexibility, as Liu's NPs are rigid and spherical while those used in our group are functionalized with flexible ligands.

3.1.2 Interaction of Small Anionic and Cationic NPs with the Membrane Core

As shown in STEP 2 and 3 of Fig. 2, if the NP size allows for direct translocation across the membrane core, the NP–membrane interaction is a 3-step process. After adsorption on the membrane surface (step 1), the NPs have to find their way to the membrane core. The description of this penetration step depends crucially on the resolution with which the NP is described.

We start by reviewing the simulations in which the charged NPs were modeled as rigid, spherical NPs. These simulations are always performed at coarse-grained level. NPs with realistic surface charge densities (≤ 1 e/nm^2) are usually favorably embedded in the membrane core, provided that the neutral NP surface is sufficiently hydrophobic. The mechanism by which rigid NPs enter the membrane core is simple: the displacement of the lipid headgroups in the vicinity of the adsorbed NP allows for the NP core to establish a favorable contact with the hydrophobic membrane core. As a consequence, the NP progressively approaches the center of the membrane, with rotational adjustments to minimize the contacts between its surface charges and the lipid tails. The process does not always happen spontaneously during unbiased MD simulations, not even at coarse-grained level, due to significant free energy barriers. For this reason, enhanced sampling techniques, most often

Umbrella Sampling,[56] have been used to sample the path leading to the complete NP penetration in the membrane core, and to derive a quantitative estimate of the free energy gain associated to the process. Gkeka *et al.*[57] used the Martini force field to model a rigid NP with a mixed hydrophobic/hydrophilic surface interacting with a PC bilayer, and estimated that the NP penetration corresponded to a free energy gain of a few tens of $k_B T$. Liu *et al.*[50] calculated free energy differences of the same order of magnitude, again using the Martini force field, for anionic NPs interacting with a DPPC vesicle.

The mechanism of membrane penetration by charged NPs has been explored in more detail using NP models in which ligand flexibility and chemistry is explicitly taken into account. The group of Alexander-Katz has used an implicit solvent, implicit bilayer model to look at the energetics of NP embedding. The NP ligands were modeled both as rigid and flexible molecules[13,58,59] with hydrophilic character. The most favorable NP–membrane configuration was found to correspond to the *snorkeling* configuration, in which the NP center of mass is located at the center of the membrane, and the hydrophilic ligand terminals interact with the lipid headgroups of both leaflets (see STEP 3 in Fig. 2). The estimated free energy gains between the water and the snorkeling phase, for NPs with a diameter of 2 nm, are of the order of 160 $k_B T$ for the rigid ligand model, and almost 2 times larger for the flexible ligand model.[58] Enlarging the NP diameter in the 1–10 nm range, as well as increasing the number of charged ligands, implies much larger embedding free energies, up to about 500 $k_B T$.[59]

The mechanisms by which the NP can penetrate the membrane interior were studied in detail by unbiased and biased MD simulations, both by the group of Alexander-Katz and, later, by our group. The main contribution coming from the simulation at high resolution is the identification of the rare events that trigger the NP penetration into the membrane. A first transition from the adsorbed to the embedded state is the protrusion of a lipid tail towards the headgroup region, where the tail can establish a contact with the hydrophobic moieties of the NP ligands. In CG simulations, this event takes place spontaneously on a microsecond time scale,[41] while it does not take place spontaneously on this time scale in atomic-resolution runs.[46,47] Indeed, Van Lehn *et al.*[46] have estimated an energy cost of about 10 $k_B T$ for the lipid tail protrusion.

The progressive embedding of the NP into the membrane core proceeds again via a sequence of rare events corresponding to the translocation of the charged ligands, one by one, from the entrance to the distal leaflet.[41,60] The free energy barriers for the translocation process are due to need for charged ligand terminals to cross the hydrophobic membrane core, often dragging along a solvation shell of water molecules and ions.[42] This process is captured only qualitatively by CG force fields, such as the standard version of the Martini force field,[53] which do not include long-range electrostatic interactions. The refined version of Martini, including long-range electrostatics and polarizable water beads,[55,61] and atomistic models allow for a better estimate of translocation free energy barriers, of about 35 $k_B T$ for each ligand translocation event.[42] We remark that, despite the large free energy barriers separating the adsorbed state from the anchored state, according to high resolution models, the anchored state should be by far more populated at equilibrium,[41] in agreement with the predictions of the implicit solvent-implicit bilayer model of Alexander-Katz.[13]

Unpublished results by our group indicate that Au NPs functionalized by anionic or cationic ligands exhibit the same molecular mechanisms of membrane penetration, and very similar kinetics and thermodynamics. It is thus unlikely that charges alone are responsible for the different behavior of anionic and cationic NPs that is often reported experimentally. Other elements and mechanisms may be invoked to interpret the experimental data. It is possible, for example, that the charge state of titratable groups on the NP surface changes in proximity of the membrane,[62] as a consequence of a shift of the pKa of the ligands in the lipid headgroup regions, similarly to what happens for charged amino acids.[63] Also the possibility to establish hydrogen bonds between the ligands functionalizing the NP surface and the lipids has been invoked to explain the different interactions of charged NPs with model membranes,[8] an hypothesis that has not yet been explored computationally.

3.2 Effect of Surface Properties on Protein Corona Formation

The use of computer simulations for the study of the thermodynamics and kinetics of aggregation of proteins around NPs is still in its infancy. The corona formation is a process that spans time scales of seconds. The relevant NP sizes for biomedical applications range from a few to hundreds of nm, while the thickness of the protein corona on metal or metal oxide NPs varies from 20 to 40 nm.[26] These length and time scales are not accessible by unbiased, explicit solvent molecular simulations, and different approaches need to be considered. Models can be simplified (by coarse-graining), and sampling can be enhanced, using either implicit solvent simulations or enhanced sampling techniques.

Shao *et al.*[64] have investigated the adsorption of two small proteins (Trp-cage and WW domain) on a 10-nm NP by means of implicit solvent Discontinuous Molecular Dynamics[65] (DMD). In Shao's model, the protein residues preserve some chemical specificity, while the NP is modeled as a hard sphere with nonpolar character and strong affinity for nonpolar amino acids, such as alanine. The enhancement of sampling offered by DMD allows to capture the time evolution of the proteins adsorbed on the NP, and use the simulated adsorption isotherms to test the reliability of alternative adsorption models. The lack of a chemically specific description of the NP surface, though, makes quantitative predictions problematic. Other authors have proposed a coarse-grained, but bead-based description of the NP (and of proteins). Feng Ding *et al.*,[66] for example, simulated the surface of an Ag NP as a mixture of hydrophobic and (in small fraction) positively charged beads, to account for the presence of residual Ag ions. A similar approach to the description of the NP surface was adopted by Hong-Ming Ding *et al.*,[67] for an NP with hydrophilic, hydrophobic, or charged surface, and by Yu *et al.*[68] for a hydrophilic SPION NP.

In experiments, it is rare that metal NPs expose their bare surface to the biological environment. A common non-covalent passivating agent for Ag and Au NPs is citrate. Ding *et al.*[66] included citrate in their simulations, and indeed found that citrate can compete with ubiquitin for the NP surface; ubiquitin prevails over citrate only at very high protein concentrations. The role of citrate has been explored at atomistic level, as well, by Brancolini *et al.*[69] These authors considered the docking of ubiquitin on a flat (111) Au surface, with and without citrate, and found that some of the citrate molecule passivating the surface can coexist with the adsorbed proteins, mediating the protein–surface interaction via short, non-electrostatic interactions.

The study of the effects of the interaction with NPs on the secondary structure of proteins is particularly challenging. Coarse-grained models either disregard completely secondary structure,[67] or restrain it to the native configuration.[70] Two atomistic studies concerning the interaction of human serum albumin with fullerenes[71] and a positively charged lysozyme with polymer-coated Ag NPs[72] suggested that these proteins may preserve their secondary structure upon contact with the NP. Different is the case of alpha lactalbumin,[72] which loses its secondary structure when in contact with the oppositely charged, polymer-coated Ag NPs. Such a structural denaturation may be responsible for the formation of a more stable corona.[72] Effects of NPs on protein secondary structure are expected to be chemically specific, i.e., they depend on the nature of the NP, the nature of the protein, and the concentration of any species that can affect NP–protein interaction (e.g., non-covalent passivating agents). In general, computational results reported so far in the literature are too sparse to provide a consistent picture of the role of NP surface composition on protein adsorption.

3.3 Formation of Ligand Patterns and Their Effect on the Mechanism of NP Internalization

The experimental observation of ordered stripe-like patterns of hydrophobic and hydrophilic ligands on the surface of small Au NPs raised an obvious interest for understanding the physical driving forces leading to this peculiar phase segregation. The picture that is nowadays accepted is based on several simulation works by the group of S. Glotzer.[73–75] In these papers, MD simulations at atomistic or coarse-grained resolution were used to study the equilibrium distribution of immiscible surfactants on the surface of Au NPs. The authors concluded that a necessary ingredient to stripe formation is that the two immiscible ligands differ in length, and that stripe formation is entropic in nature[73]: the formation of stripes implies a conformational entropic gain for the longer ligands along the interface with the shorter ones. The thickness of stripes may further depend on other physical parameters, such as the absolute ligand length, the ligand terminal charge, and the degree of immiscibility.

The entropic factor identified by these computational works certainly plays a role in shaping the ligand surface pattern on NPs, but other factors may also contribute. The immiscibility assumption is valid in some cases, such as for hydrogenated and fluorinated ligands,[36,76] but is questionable in others. Octanethiol and mercapto-undecane sulfonate ligands, for example, which are known to give rise to stripe formation,[12] can be hardly described as immiscible, as the hydrophobic stretch of their tails are very similar from a chemical standpoint. It is still unclear to which extent thiols can diffuse on the surface of Au NPs[38]: for very low diffusion rates, ligands may be trapped in an out-of-equilibrium configuration determined by the NP synthesis route. A computational effort in the direction of (a) simulating the synthesis process and (b) modeling more realistic ligand mixtures is thus envisaged.

Molecular simulations offer the possibility to build NP models exhibiting the whole spectrum of possible surface patterns, and to look at how the surface pattern influences the NP interaction with model lipid membranes.

As a first approximation, the NP can be modeled as spherical or faceted cluster with neutral interactions with water and lipids, while reproducing the chemical nature of ligands – hydrophobic or hydrophilic, charged or neutral. Ligands can be represented either modifying the interactions of some NP surface beads, or attaching to the model NP extra beads with the desired properties. This NP representation disregards the details of the NP core surface (e.g., defects, degree of roughness, covalent bonds with the ligands) and the flexibility of ligands, which are modeled as spherical beads. The advantage, though, is to be able to play freely with the placement of ligands on the NP surface, and to use coarse-grained models which allow for the simulation of large NPs (>10 nm), large membrane patches, and even entire liposomes.

The cell uptake of NPs with a diameter larger than 10–20 nm often happens via receptor-mediated endocytosis. Schubertova *et al.*[77] have simulated the receptor-mediated endocytosis of model NPs passivated by ligands with a high affinity to model membrane receptors. A minimum density of ligands is found to allow for membrane wrapping (within the simulated time scale). More interestingly, at intermediate ligand density, the ligand patterning on the NP surface was reported to play an important role modulating membrane wrapping. Homogeneous ligand distributions were the most prone to be uptaken. The result was rationalized taking into account the reorientation of NPs that can take place during the endocytic process.[78,79]

NP rotation, and its interplay with the ligand surface distribution, is important also for smaller NPs (<10 nm) interacting with the membrane via passive translocation into the membrane core. The presence of large patches of hydrophobic vs. hydrophilic ligands on the NP surface affects the NP–membrane interaction at all stages, from the adsorption on the membrane surface to the NP embedding in the membrane core. Implicit solvent coarse-grained[80] simulations show that, in the presence of Janus NPs with only one charged patch,[48] the NP adheres to the membrane surface with the orientation that favors the charged ligand–lipid headgroups interactions. A penetration of the NP in the membrane core would require an NP rotation with partial detachment of the charged patch from the lipid headgroups. This is presumably unfavorable from the energetic standpoint, and indeed it is not observed on the simulation time scale. The insertion–rotation mechanism was observed to take place also in unbiased, implicit solvent DPD simulations by Li *et al.*,[81] who built a coarse-grained model of NPs passivated by a mixture of hydrophobic and hydrophilic ligands. The authors considered three different surface patterning, striped (with different stripe thicknesses), patched, and random. The presence of large stripes or patches can hinder the rotation of the NP at the membrane surface, thus slowing down or impeding the NP embedding into the membrane core. On the contrary, thin stripes or random placement of the ligands allows for the quick and spontaneous incorporation of the NP in the membrane, on timescales of the order of microseconds.

An alternative approach relies on the use of explicit solvent, atomistic, or coarse-grained models for the description of both the membrane and the NP. As discussed in Section 2.1, this resolution allows for a detailed molecular description of the NP–membrane interaction. The simulations performed at coarse-grained[41,42] and atomistic[42] level on small (2 nm diameter) Au NPs functionalized by a mixture of anionic and hydrophobic ligands show again a difference in the way random and striped NPs interact with the membrane. Remarkably, even at this degree of resolution, the NP rotation during the NP–membrane interaction is identified as an important conformational change during the NP–membrane interaction. The anchoring mechanism shown in Fig. 2 (STEPS 2 and 3) requires that the NP rotates to allow for the stable binding of the charged ligands to the distal leaflet. Rotations are more unfavorable if the NP is striped (see Fig. 3), as only one specific NP orientation is able to optimize the hydrophobic contact between the NP and the membrane core. Simulations can also provide quantitative estimates of the free energy cost of each step of the progressive NP penetration into the membrane core. As reported by Simonelli *et al.*,[41] the free energy cost associated to the translocation (or anchoring) of a charged ligand from the entrance leaflet to the distal one increases by at least a factor of 2 when going from the random to the striped arrangement. Such difference in free energy cost is the origin of the much faster kinetics of membrane penetration for NPs with random ligand arrangement.

Overall, simulations at different degrees of resolution and on NPs of different sizes in the 2–20 nm range agree on the identification of an insertion–rotation mechanisms of NP–membrane interaction. It should be noticed that the surface pattern of monolayer-protected NPs depends to some extent on the NP size and ligand flexibility. NPs with extremely thin stripe patterns may present a quite homogeneous surface, while little difference could be observed, in terms of rotational freedom when in contact with the membrane, between NPs with a patchy pattern or with thick stripes. Care must then be taken when comparing the simulation data to the experimental findings. The group of Stellacci has shown that striped NPs can readily and passively penetrate cell membranes in vitro,[12,14,82] while random NPs are more likely internalized via endocytic pathways. These results, which may seem in contrast with the picture emerging from computer simulations, should be interpreted taking into account the stripe thickness on the experimental NPs, which corresponds to 1 or 2 ligands only, and is thus compatible with a rather homogeneous character of the NP surface. Another possibility is that

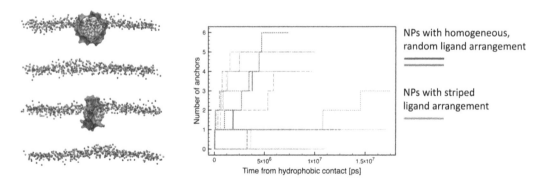

FIGURE 3 Left: Au NPs with a striped arrangement of hydrophobic and anionic ligands interact with a POPC lipid bilayer. Only the hydrophobic ligands are shown in red, while lipid headgroups are represented as blue/gray beads. Lipid tails and water are not shown. The NP inserts in the membrane with a fixed orientation of its hydrophobic stripe. Right: the kinetics of anchoring for NPs with different ligand patterns. The NPs with a random ligand arrangement, corresponding to a homogeneous distribution of charges and hydrophobic groups on their surface, anchor to the membrane with a faster kinetics. (See color insert for the color version of this figure)

the ligands on the NP surface preserve some degree of diffusional mobility, possibly enhanced by the contact with the lipids. Such freedom of the ligands to redistribute on the NP surface may well affect the NP–membrane interaction. This hypothesis has been explored computationally only by Van Lehn *et al.*[83] back in 2011, by means of a model which did not take into account explicit barriers for the diffusion of ligands on the NP surface. Developments in this direction could be envisaged to better link experimental and simulation data.

4 Perspectives

Nanomaterials are nowadays being used in most industrial sectors, from medicine to electronics, and their production is increasing steadily and rapidly. Understanding how NPs interact with living organisms well before they reach the market is paramount to guarantee economic viability. Understanding NP–membrane interaction is one of the first steps towards predictions of NP permeation into cells and cell damage, necessary for the production of safe-by-design nanomaterials. Considering the sheer complexity of biological systems, we expect that experiments and simulations on simplified model systems will continue to play a major role in shaping the way we think about NP interaction with biological systems.

To provide useful predictions on NPs of practical industrial interest, molecular simulations will have to face a number of challenges. The most obvious challenges are common to molecular simulations in general: sampling and accuracy. Sampling relevant time and length scales will remain an important issue in nanomaterial simulations due to the sheer size of the NPs: many of the ones nowadays produced feature sizes of tens of nanometers and larger, and their interaction with membranes takes place over time scales of milliseconds and longer – difficult to tackle with traditional all-atom descriptions, and even with commonly used coarse-grained models. Limitations in sampling are one of the main reasons for the substantial lack of molecular simulations interpreting experiments on the effect of NPs on membrane phase behavior; predicting membrane phase behavior requires simulation at very large time scales, and the NPs shown to produce phase changes in lipid bilayers have relatively large sizes, difficult to reach with atomic-level representations. The challenge regarding simulation accuracy is particularly important for NPs for two reasons: first of all, classical force fields often do not exist for specific NPs and ligands, and even when they exist, in most cases, they are not nearly as thoroughly tested as the ones for biological macromolecules; second, for NPs including metals, it will be beneficial or even necessary to consider the possibility of chemical reactions – for instance, in the case of rearrangements of ligands on the NP surface (which is probably necessary to understand the interaction of striped and random Au NPs with membranes), modifications of surface functionalization, or stimuli-responsive ligands. Such chemical modifications can be taken into account explicitly either by using quantum chemical approaches, or by using reactive force fields. The parameterization of reactive force fields is typically very costly in terms of human and computational resources, but it will allow classical simulations on time scales that will remain out of reach for quantum approaches in the foreseeable future. The high cost and the complexity of force field parameterization will probably be relieved, during the next decade, by the introduction of machine learning approaches, more and more common in this area.

Another important challenge in modeling NP–membrane interactions is related to complexity: not only the very broad complexity of the NP and its ligands, but also the complexity of biological membranes and the biological milieu. One of the reasons for the difficulties encountered by molecular simulations in explaining experiments on NPs–cell interactions is that simulations only deal with simplified model systems, which generally feature simple composition of the membrane (i.e., few different lipid types), no asymmetry between membrane leaflets, and extremely simple composition of the solutions (in terms of buffers, ions, proteins, etc.). Similar difficulties are also found in interpreting the results of experiments on non-biological model systems (e.g., probing NP interaction with lipid vesicles): simulations generally do not take into account the effects of possible modifications of the NP surface (e.g., oxidation) and non-covalent binding with different species used in the preparation of the samples. During the past decade, several researchers have started tackling the complexity challenge: for example, the group of Marrink has proposed, in 2014, a model of plasma membrane that, for the first time, approaches the complexity of the biological target membrane[84]; and several other groups have made significant progress in simulating more and more realistic bacterial membranes.[85] We expect that significant progress in this respect will be made in the near future.

ACKNOWLEDGMENT

LM acknowledges funding from the Institut national de la santé et de la recherche médicale (INSERM), and from the Agence Nationale de la Recherche (ANR, grant NANODROP).

REFERENCES

1. D. F. Moyano, K. Saha, G. Prakash, B. Yan, H. Kong, M. Yazdani and V. M. Rotello, *ACS Nano*, 2014, **8**, 6748–6755.
2. D. F. Moyano and V. M. Rotello, *Langmuir*, DOI: 10.1021/la2004535.
3. R. R. Arvizo, O. R. Miranda, M. A. Thompson, C. M. Pabelick, R. Bhattacharya, J. D. Robertson, V. M. Rotello, Y. S. Prakash and P. Mukherjee, *Nano Lett.*, 2010, **10**, 2543–2548.
4. Y. Jiang, S. Huo, T. Mizuhara, R. Das, Y. Lee, S. Hou, D. F. Moyano, B. Duncan, X. Liang and V. M. Rotello, *ACS Nano*, 2015, **9**, 9986–9993.
5. A. Xi and G. D. Bothun, Analyst, 2014, **139**, 973–981.
6. C. M. Goodman, C. D. McCusker, T. Yilmaz and V. M. Rotello, *Bioconju. Chem.*, 2004, **15**, 897–900.
7. Y. Chen and G. D. Bothun, *Langmuir*, 2011, **27**, 8645–8652.
8. X. Wei, J. Yu, L. Ding, J. Hu and W. Jiang, J. Environ. Sci., 2017, **57**, 221–230.
9. A. L. Neal, *Ecotoxicology*, 2008, **17**, 362–371.
10. Z. V. Feng, I. L. Gunsolus, T. A. Qiu, K. R. Hurley, L. H. Nyberg, H. Frew, K. P. Johnson, A. M. Vartanian, L. M. Jacob, S. E. Lohse, M. D. Torelli, R. J. Hamers, C. J. Murphy and C. L. Haynes, *Chem. Sci.*, 2015, **6**, 5186–5196.
11. S. Zhang, A. Nelson and P. A. Beales, *Langmuir*, 2012, **28**, 12831–12837.
12. A. Verma, O. Uzun, Y. Hu, Y. Hu, H.-S. Han, N. Watson, S. Chen, D. J. Irvine and F. Stellacci, *Nat. Mater.*, 2008, **7**, 588–595.
13. R. C. Van Lehn, P. U. Atukorale, R. P. Carney, Y.-S. Yang, F. Stellacci, D. J. Irvine and A. Alexander-Katz, *Nano Lett.*, 2013, **13**, 4060–4067.
14. S. Sabella, R. P. Carney, V. Brunetti, M. A. Malvindi, N. Al-Juffali, G. Vecchio, S. M. Janes, O. M. Bakr, R. Cingolani, F. Stellacci and P. P. Pompa, *Nanoscale*, 2014, **6**, 7052–7061.
15. S. Tatur, M. Maccarini, R. Barker, A. Nelson and G. Fragneto, *Langmuir*, 2013, **29**, 6606–6614.
16. Y. Yu, S. M. Anthony, L. Zhang, S. C. Bae and S. Granick, *J. Phys. Chem. C*, 2007, **111**, 8233–8236.
17. L. Zhang and S. Granick, Nano Lett., 2006, **6**, 694–698.
18. B. Wang, L. Zhang, S. Chul and S. Granick, PNAS, 2008, **105**, 18171.
19. F. Wang and J. Liu, Nanoscale, 2015, **7**, 15599–15604.
20. F. Wang, D. E. Curry and J. Liu, Langmuir, 2015, **31**, 13271–13274.
21. A. Bhat, L. W. Edwards, X. Fu, D. L. Badman, S. Huo, A. J. Jin and Q. Lu, Appl. Phys. Lett., DOI: 10.1063/1.4972785.
22. T. Cedervall, I. Lynch, S. Lindman, T. Berggård, E. Thulin, H. Nilsson, K. A. Dawson and S. Linse, Proc. Natl. Acad. Sci. U. S. A., 2007, **104**, 2050–2055.
23. I. Lynch, A. Salvati and K. A. Dawson, Nat. Nanotechnol., 2009, **4**, 546–547.
24. P. C. Ke, S. Lin, W. J. Parak, T. P. Davis and F. Caruso, ACS Nano, 2017, **11**, 11773–11776.
25. A. E. Nel, L. Mädler, D. Velegol, T. Xia, E. M. V. Hoek, P. Somasundaran, F. Klaessig, V. Castranova and M. Thompson, Nat. Mater., 2009, **8**, 543–557.
26. C. D. Walkey and W. C. W. Chan, Chem. Soc. Rev., 2012, **41**, 2780–2799.
27. P. Decuzzi, ACS Nano, 2016, **10**, 8133–8138.
28. M. Mahmoudi, S. Sant, B. Wang, S. Laurent and T. Sen, Adv. Drug Deliv. Rev., 2011, **63**, 24–46.
29. K. Knop, R. Hoogenboom, D. Fischer and U. S. Schubert, Angew. Chemie – Int. Ed., 2010, **49**, 6288–6308.

30. W. Norde and D. Gags, Langmuir, 2004, **20**, 4162–4167.
31. R. Gref, M. Lück, P. Quellec, M. Marchand, E. Dellacherie, S. Harnisch, T. Blunk and R. H. Müller, Colloids Surfaces B Biointerfaces, 2000, **18**, 301–313.
32. B. Pelaz, P. Del Pino, P. Maffre, R. Hartmann, M. Gallego, S. Rivera-Fernández, J. M. De La Fuente, G. U. Nienhaus and W. J. Parak, ACS Nano, 2015, **9**, 6996–7008.
33. C. Bernhard, S. J. Roeters, J. Franz, T. Weidner, M. Bonn and G. Gonella, Phys. Chem. Chem. Phys., 2017, **19**, 28182–28188.
34. M. Wang, G. Siddiqui, O. J. R. Gustafsson, A. Käkinen, I. Javed, N. H. Voelcker, D. J. Creek, P. C. Ke and T. P. Davis, Small, 2017, **13**, 1–11.
35. N. J. Butcher, G. M. Mortimer and R. F. Minchin, Nat. Nanotechnol., 2016, **11**, 310–311.
36. M. Şologan, D. Marson, S. Polizzi, P. Pengo, S. Boccardo, S. Pricl, P. Posocco and L. Pasquato, ACS Nano, 2016, **10**, 9316–9325.
37. M. Şologan, C. Cantarutti, S. Bidoggia, S. Polizzi, P. Pengo and L. Pasquato, Faraday Discuss., 2016, **191**, 527–543.
38. P. Pengo, M. Şologan, L. Pasquato, F. Guida, S. Pacor, A. Tossi, F. Stellacci, D. Marson, S. Boccardo, S. Pricl and P. Posocco, Eur. Biophys. J., 2017, **46**, 749–771.
39. E. Heikkila, H. Martinez-Seara, A. A. Gurtovenko, M. Javanainen, I. Vattulainen and J. Akola, J. Phys. Chem. C, 2014, **118**, 11131–11141.
40. E. Heikkilä, H. Martinez-Seara, A. A Gurtovenko, I. Vattulainen and J. Akola, Biochim. Biophys. Acta, 2014, **1838**, 2852–2860.
41. F. Simonelli, D. Bochicchio, R. Ferrando and G. Rossi, J. Phys. Chem. Lett., 2015, **6**, 3175–3179.
42. S. Salassi, F. Simonelli, D. Bochicchio, R. Ferrando and G. Rossi, J. Phys. Chem. C, 2017, **121**, 10927–10935.
43. R. Gupta and B. Rai, Sci. Rep., 2017, **7**, 1–13.
44. P. Gkeka, P. Angelikopoulos, L. Sarkisov and Z. Cournia, PLoS Comput. Biol., 2014, **10**, e1003917.
45. J. P. Prates Ramalho, P. Gkeka and L.Sarkisov, Langmuir, 2011, **27**, 3723–3730.
46. R. C. Van Lehn and A. Alexander-Katz, Soft Matter, 2015, **11**, 3165–3175.
47. R. C. Van Lehn, M. Ricci, P. H. J. Silva, P. Andreozzi, J. Reguera, K. Voïtchovsky, F. Stellacci and A. Alexander-Katz, Nat. Commun., 2014, **5**, 4482.
48. F. Aydin and M. Dutt, J. Phys. Chem. B, 2016, **120**, 6646–6656.
49. X. Liu and A. Basu, J. Am. Chem. Soc., 2009, **131**, 5718–5719.
50. L. Liu, J. Zhang, X. Zhao, Z. Mao, N. Liu, Y. Zhang and Q. H. Liu, Phys. Chem. Chem. Phys., 2016, **18**, 31946–31957.
51. E. L. da Rocha, G. F. Caramori and C. R. Rambo, Phys. Chem. Chem. Phys., 2013, **15**, 2282–2290.
52. P. Chen, Z. Huang, J. Liang, T. Cui, X. Zhang, B. Miao and L. T. Yan, ACS Nano, 2016, **10**, 11541–11547.
53. S. J. Marrink, H. J. Risselada, S. Yefimov, D. P. Tieleman and A. H. de Vries, J. Phys. Chem. B, 2007, **111**, 7812–7824.
54. J. Michalowsky, L. V. Schäfer, C. Holm and J. Smiatek, J. Chem. Phys., 2017, **146**, 054501.
55. S. O. Yesylevskyy, L. V Schäfer, D. Sengupta and S. J. Marrink, PLoS Comput. Biol., 2010, **6**, e1000810.
56. G. M. Torrie and J. P. Valleau, J. Comput. Phys., 1977, **23**, 187–199.
57. P. Gkeka, L. Sarkisov and P. Angelikopoulos, J. Phys. Chem. Lett., 2013, **4**, 1907–1912.
58. R. C. Van Lehn and A. Alexander-Katz, J. Phys. Chem. A, 2014, **118**, 5848–5856.
59. R. C. Van Lehn and A. Alexander-Katz, Soft Matter, 2014, **10**, 648–658.
60. R. C. Van Lehn and A. Alexander-Katz. PLoS One, 2019, **14**, e0209492.
61. J. Michalowsky, L. V Schäfer, C. Holm and J. Smiatek, J. Chem. Phys., 2017, **146**, 054501.
62. J. Koivisto, X. Chen, S. Donnini, T. Lahtinen, H. Häkkinen, G. Groenhof and M. Pettersson, J. Phys. Chem. C, 2016, **120**, 10041–10050.
63. J. L. MacCallum, W. F. D. Bennett and D. P. Tieleman, Biophys. J., 2008, **94**, 3393–3404.
64. Q. Shao and C. K. Hall, J. Phys. Condens. Matter, DOI: 10.1088/0953-8984/28/41/414019.
65. A. V. Smith and C. K. Hall, Proteins Struct. Funct. Genet., 2001, **44**, 344–360.
66. F. Ding, S. Radic, R. Chen, P. Chen, N. K. Geitner, J. M. Brown and P. C. Ke, Nanoscale, 2013, **5**, 9162.
67. H.-M. Ding and Y.-Q. Ma, Biomaterials, 2014, **35**, 8703–8710.

68. S. Yu, A. Perálvarez-Marín, C. Minelli, J. Faraudo, A. Roig and A. Laromaine, Nanoscale, 2016, **8**, 14393–14405.
69. G. Brancolini, D. B. Kokh, L. Calzolai, R. C. Wade and S. Corni, ACS Nano, 2012, **6**, 9863–9878.
70. L. Monticelli, S. K. Kandasamy, X. Periole, R. G. Larson, D. P. Tieleman and S.-J. Marrink, J. Chem. Theory Comput., 2008, **4**, 819–834.
71. G. Leonis, A. Avramopoulos, K. D. Papavasileiou, H. Reis, T. Steinbrecher and M. G. Papadopoulos, J. Phys. Chem. B, 2015, **119**, 14971–14985.
72. B. Wang, S. A. Seabrook, P. Nedumpully-Govindan, P. Chen, H. Yin, L. Waddington, V. C. Epa, D. A. Winkler, J. K. Kirby and P. C. Ke, Phys. Chem. Chem. Phys., 2014, **17**, 1728–1739.
73. C. Singh, P. K. Ghorai, M. A. Horsch, A. M. Jackson, R. G. Larson, F. Stellacci and S. C. Glotzer, Phys. Rev. Lett., 2007, **99**, 226106.
74. P. K. Ghorai and S. C. Glotzer, J. Phys. Chem. C, 2010, **114**, 19182–19187.
75. A. Santos, J. A. Millan and S. C. Glotzer, Nanoscale, 2012, **4**, 2640–2650.
76. P. Posocco, C. Gentilini, S. Bidoggia, A. Pace, P. Franchi, M. Lucarini, M. Fermeglia, S. Pricl and L. Pasquato, ACS Nano, 2012, **6**, 7243–7253.
77. V. Schubertová, F. J. Martinez-Veracoechea and R. Vácha, Soft Matter, 2015, **11**, 2726–2730.
78. S. Dasgupta, T. Auth and G. Gompper, Nano Lett., 2014, **14**, 687–693.
79. C. Huang, Y. Zhang, H. Yuan, H. Gao and S. Zhang, Nano Lett., 2013, **13**, 4546–4550.
80. I. R. Cooke and M. Deserno, J. Chem. Phys., DOI: 10.1063/1.2135785.
81. Y. Li, X. Zhang and D. Cao, Soft Matter, 2014, **10**, 6844–6856.
82. A. Verma and F. Stellacci, Small, 2010, **6**, 12–21.
83. R. C. Van Lehn and A. Alexander-Katz, Soft Matter, 2011, **7**, 11392–11404.
84. H. I. Ingólfsson, M. N. Melo, F. J. van Eerden, C. Arnarez, C. A. Lopez, T. A. Wassenaar, X. Periole, A. H. de Vries, D. P. Tieleman, S. J. Marrink, J. Am. Chem. Soc., 2014, **136**, 14554−14559.
85. A. Boags, P. C. Hsu, F. Samsudin, P. J. Bond, S. Khalid, J. Phys. Chem. Letters, 2017, **11**, 2513–2518.

10

Simulations of Membranes Containing General Anesthetics

Pál Jedlovszky[*]

Department of Chemistry, Eszterházy Károly University, Leányka utca 6, H-3300 Eger, Hungary

1 Introduction

General anesthetics are widely used in the clinical practice since the middle of the 19th century. However, despite their everyday use for such a long time, the mechanism of their action is still largely unexplained. Understanding the molecular mechanism of general anesthesia is hindered by the very wide chemical variety of general anesthetics: they include partially halogenated alkane derivatives (e.g., chloroform, halothane, trichloroethylene), cyclic hydrocarbons (e.g., cyclopropane), ethers (i.e., diethyl ether) and their halogenated derivatives (e.g., enflurane, isoflurane, sevoflurane, methoxyflurane), *n*-alkanols (ranging from ethanol to 1-decanol), aromatic molecules (e.g., ketamine), and also some inorganic compounds (e.g., N_2O) and even noble gases (e.g., xenon). This chemical variety of the general anesthetics is illustrated in Figure 1. Furthermore, a number of compounds that are chemically rather similar to certain known general anesthetics do not have anesthetic property at all. Thus, while chloroform, decanol, and xenon all act as general anesthetics, carbon tetrachloride, dodecanol, and neon are not anesthetics.

The first major step towards the understanding of the molecular mechanism of general anesthesia was made more than a century ago by Meyer (1889) and Overton (1901), who discovered, independently from each other, that the efficiency of general anesthetics depends linearly on the oil:water partition coefficient of these molecules (see Figure 2). More specifically, the effective concentration of the anesthetic molecule in the blood when 50% of the individuals are anesthetized, $[ED_{50}]$, is proportional to the partition coefficient. It should be noted that this correlation is only valid for molecules that can act as general anesthetics, while, evidently, apolar non-anesthetics do not follow it. Nevertheless, the Meyer–Overton rule has several important consequences. First, it reveals that whatever the mechanism of general anesthesia is, the site of its action is located in the hydrocarbon interior of the cell membranes. Second, as it was pointed out by Overton himself (Overton 1901), this correlation as well as the large chemical variety of general anesthetics suggests the existence of a certain physical mechanism behind the phenomenon of general anesthesia.

Another important observation related to the molecular mechanism of general anesthesia is that the anesthetic effect is reverted at high pressure (Johnson and Flagler 1950, Lever et al. 1971, Halsey and Wardley-Smith 1975, Trudell et al. 1975). This pressure reversal was first reported in 1950 by Johnson and Flagler, who observed that tadpoles, anesthetized by 2–5% ethanol, regained their swimming activity when applying a pressure of 200–300 bar (Johnson and Flagler 1950).

---

[*] Electronic mail: jedlovszky.pal@uni-eszterhazy.hu

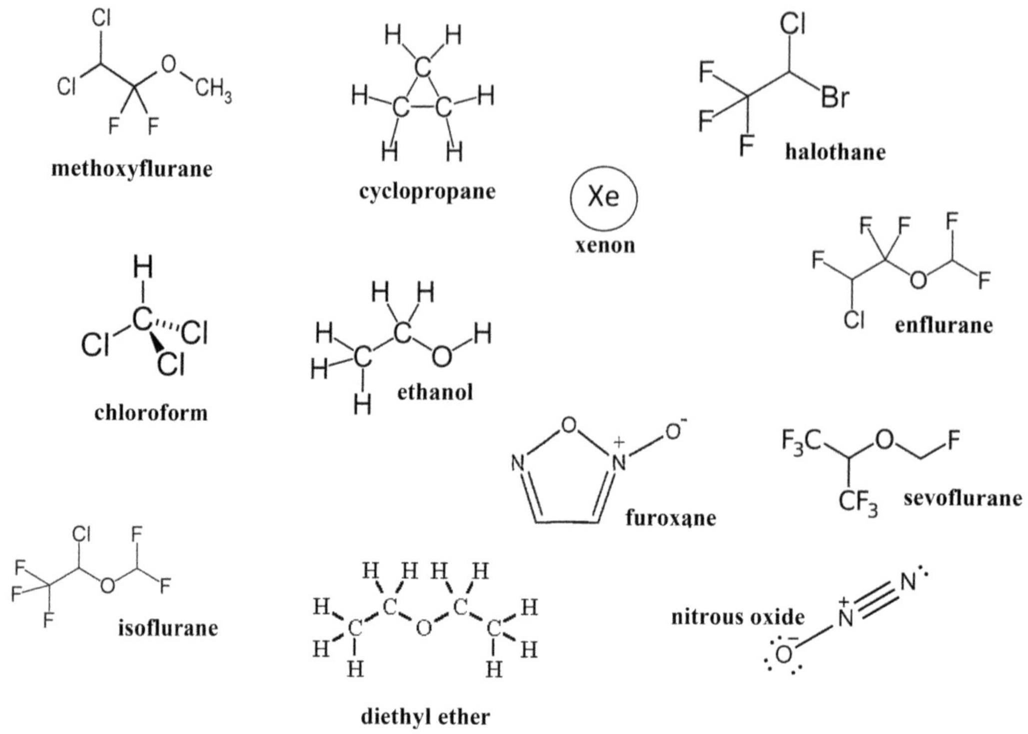

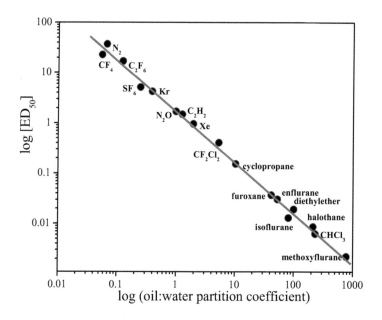

FIGURE 1   Schematic chemical structure of various general anesthetics.

FIGURE 2   Illustration of the Meyer–Overton rule.

Therefore, any reliable explanation of the molecular mechanism of general anesthesia has to account also for its pressure reversal.

For more than half a century, searching for a general physical mechanism that can explain both general anesthesia and its pressure reversal has been in the focus of intensive scientific investigation. These attempts are generally referred to as 'lipid theories.' However, the lack of such a general explanation has also given rise to an alternative view, namely that anesthesia might involve direct interaction of the anesthetic agents with certain membrane-bound proteins, e.g., ion channels or receptors (Franks and Lieb 1978, 1994, Mitchell et al. 1996, Mihic et al. 1997). Due to the wide chemical variety of general anesthetic molecules, however, possible explanations of this kind, called as 'protein theories,' are probably incompatible with the existence of a unique molecular mechanism of general anesthesia. It should be emphasized that lipid theories do not exclude proteins from the molecular mechanism of anesthesia; they just exclude the possibility of a direct, specific, substrate–protein interaction. Instead, the relevant changes of the underlying protein molecules are assumed to be caused by the altered membrane properties due to the presence of anesthetics. Further, a clear distinction should be made in this respect between general anesthesia and local anesthesia, i.e., the induction of the absence of sensation in a specific part of the body. The molecular mechanism of local anesthesia is far better understood than that of general anesthesia, and it certainly involves specific interaction with certain proteins (Lee 1976).

In the frame of the lipid theories, the first important suggestion was made by Mullins, who realized that all anesthetic molecules increase the molar volume of the membrane, an effect that is clearly reverted by the pressure. Based on this observation, he hypothesized that anesthesia occurs if the molar volume of the membrane exceeds a critical value (Mullins 1954). In accordance with this critical volume hypothesis, several general anesthetics were found to increase the thickness of model membranes (Haydon et al. 1977). To explain this increased thickness and rationalize the critical volume hypothesis, it was generally assumed that anesthetics enhance the orientational order of the lipid tails, pushing the two headgroup regions thus farther away from each other. Another possible way of increasing the membrane thickness could be that molecules are accumulated in the middle of the membrane, between the lipid tails, and hence decrease the degree of interdigitation of the lipid molecules of the opposite leaflets. Experimental results, however, did not fully support these hypotheses. Thus, for instance, Trudell et al. (1973) found a decrease in the order of the lipid tails in the presence of halothane and methoxyflurane by performing electron spin resonance (ESR) spectroscopy measurements. Similarly, by performing X-ray and neutron diffraction experiments, Franks and Lieb found the thickness of a mixed lecithin/cholesterol bilayer to be insensitive to the presence of a number of different general anesthetics. Further, at high anesthetic concentration, they found the lipid tails becoming less ordered, while the thickness of the membrane was still unchanged (Franks and Lieb 1979). Hauet et al. (2003) investigated DPPC bilayer in the presence of enflurane, and found that enflurane preferentially interacts with the polar lipid headgroups rather than being located in the middle of the membrane.

Another possible mechanism laying behind the critical volume hypothesis could be that anesthetics decrease the lateral density, i.e., they increase the surface area rather than the thickness of the membrane (Darvas et al. 2012, Fábián et al. 2015). Indeed, in experimental studies, it was often assumed that the surface area of the membrane remains unchanged, and this assumption might led to the observed increase of its thickness in some cases. When no such assumption was made, a clear increase of the membrane surface area was observed upon addition of anesthetics (Ly et al. 2002, Ly and Longo 2004). This explanation assumes that anesthetics are located between the lipid tails rather than in the middle of the membrane, among the tail ends, and it can also easily account for the pressure reversal, as the increase of the pressure evidently increases the lateral density of the membrane.

Another branch of the lipid theories assumes that anesthetics increase the fluidity of the membrane. Thus, Trudell et al. assumed that liquid crystalline (α) and gel (β) phases coexist in the anesthetic-free membrane, and the addition of anesthetics shifts this equilibrium towards the liquid crystalline phase by melting domains of the β phase (Trudell 1977). Shift of the gel–liquid crystalline transition temperature (Janoff and Miller 1982, Forrest and Rodham 1985, Kaminoh et al. 1988,

1992, Sierra-Valdez and Ruiz-Suárez 2013) as well as change of the domain structure in favor of the disordered domains (Turkyilmaz et al. 2011, Weinrich et al. 2012, Weinrich and Worcester 2013) by general anesthetics was also observed by several other authors. Gruner and Shyamsunder assumed that anesthetics alter the lateral pressure profile, which induces changes in the spontaneous curvature of the two individual monolayers (Gruner and Shyamsunder 1991). This view was later criticized by Cantor, pointing out that the existence of the bilayer assumes (nearly) zero total lateral pressure, and hence, in contrast to monolayers, changes in the lateral pressure profile do not induce changes in its curvature (Cantor 1997b). It should also be noted that the assumption that anesthetics act by increasing the fluidity of the membrane are incompatible with the aforementioned suggestion that they enhance the orientational order of the lipid chains. A consistent thermodynamic formalism of the increased membrane fluidity hypothesis was provided by Heimburg and Jackson (2007), which was later extended to local anesthetics as well (Græsbøll et al. 2014). Thus, assuming that anesthetics dissolve ideally in the α, but do not dissolve at all in the β phase, they showed that the observed temperature shift of the gel–liquid crystalline phase transition is analogous with the well-known freezing point depression of simple solutions. Since freezing point depression is a colligative property, its magnitude solely depends on the number of dissolved particles, being independent from their chemical nature. Thus, given that equal amount of different anesthetics are used, the shift of the α–β phase transition temperature they induce depends only on their partition between the aqueous phase and the membrane. This way, the Meyer–Overton relation can simply be accounted for. Further, pressure reversal can also easily be explained, as pressure shifts the gel–liquid crystalline phase transition to higher temperatures. This theory also enables to explain several subtle points of anesthesia, such as the additivity of the effects of local and general anesthetics, or the cutoff effect of *n*-alkanols. This theory was also corroborated by measuring the temperature dependence of the heat capacity of membranes containing different amount of anesthetics (Heimburg and Jackson 2007, Græsbøll et al. 2014).

In 1997, Cantor suggested that general anesthesia might be related to the changes induced by the presence of the anesthetics in the lateral pressure profile across the membrane. Using thermodynamic arguments, he demonstrated that if (i) a protein that is relevant for anesthesia exists in two conformations (i.e., an active and a passive one) in equilibrium in the membrane, anesthetics induce changes in the lateral pressure profile, which alters the cross-sectional area profile of both conformers, and (ii) if this change of the area profile is not uniform along the membrane normal, then the change in the ratio of the passive and active conformers depends exponentially on the amount of anesthetics (Cantor 1997a, 1997b). To support this idea, Cantor also performed calculations on a lattice model of lipid monolayers as well as bilayers both in the absence and presence of *n*-alkanol-like anesthetics (Cantor 1997b). Based on these calculations, neglecting the headgroup interactions, he also predicted the first and second moments of the lateral pressure profile in bilayers consisting of various different acyl chains (Cantor 1999).

Computer simulation methods can provide a very useful tool to investigate the molecular mechanism of general anesthesia. The routinely available computer capacity, however, only reached the level of meaningfully addressing these problems about one and half decades ago. Early simulations of anesthetic containing membranes suffer from very small system size and/or very short simulation time (Huang et al. 1995, Tu et al. 1998, Koubi et al. 2000, 2001, Chanda and Bandyopadhyay 2004, 2006, Chau et al. 2007, 2009). In this paper, we review the simulation works reported in the past one and half decades concerning membranes containing anesthetics, and discuss in detail the general picture that emerges from the results of these studies. Since we are looking for general conclusions that are valid for all anesthetics, studies concerning specific protein–anesthetic interaction in the membrane are not included in this review.

2 Studies Involving Chemically Similar Molecules

In understanding the molecular mechanism of general anesthesia, one of the most puzzling points is that (i) molecules of very different chemical characters can equally act as anesthetics, while (ii) largely similar molecules can have very different anesthetic ability. To shed some light to the origin of this puzzling behavior, it is instructive to systematically study the effect of series of chemically similar compounds, including both anesthetics and non-anesthetics, on the properties of lipid membranes. Two such groups have been studied in detail by computer simulation methods so far, namely *n*-alkanols and noble gases.

2.1 Ethanol and Other *n*-alkanols

Due to its chemical simplicity, ethanol was a prototypical anesthetic molecule in a number of studies. The first such simulation was performed by Feller et al., who studied ethanol in a POPC bilayer at the anesthetic:lipid molar ratio of 1:1. Unlike the vast majority of the subsequent simulations, this study was performed at constant volume rather than at constant pressure. Besides investigating the relaxation dynamics of various molecules also by nuclear Overhauser enhancement spectroscopy (NOESY) experiments, the most important finding of this study was that ethanol is partitioning between the aqueous phase and the lipid membrane, and within the membrane, it prefers to stay at the vicinity of the glycerol segments (Feller et al. 2002). In a clear contrast with this behavior, Bemporad et al. (2004) found that the solvation free energy of methanol increases monotonously when going from the bulk aqueous phase to the middle of the DPPC bilayer. This view was further elaborated in a recent study of Comer et al., who performed very extensive (i.e., several µs long) simulations to determine the free-energy profile of methanol, ethanol, 1-propanol, and 1-butanol in the POPC bilayer. They found that the methanol profile has a small, secondary minimum at the region of the glycerol segments; for ethanol, this minimum already corresponds to a lower solvation free-energy value than what is seen in the bulk aqueous phase (i.e., it becomes the global minimum of the entire profile), and it deepens further with increasing chain length of the alcohol (Comer et al. 2017).

Chanda and Bandyopadhyay simulated ethanol in a DMPC bilayer at the anesthetic:lipid molar ratios of 1:4 (Chanda and Bandyopadhyay 2004) and 1:1 (Chanda and Bandyopadhyay 2006) at atmospheric pressure. In agreement with the earlier results of Feller et al., they also found that ethanol molecules are preferentially located near the glycerol backbone of the lipids, close to the lipid–water interface. With increasing ethanol concentration, they observed a lateral expansion of the membrane, reflected in the increase of the area per lipid value, and also an increasing preference of the headgroup dipole vectors, approximated by the vector pointing from the P to the N atom (PN vector) to point towards the aqueous phase (Chanda and Bandyopadhyay 2004, 2006). As a consequence of the smaller lateral density, the mobility of the lipid molecules was found to be enhanced by ethanol, both in the lateral direction and along the membrane normal (Chanda and Bandyopadhyay 2006).

Performing extensive simulations of the POPE bilayer, Gurtovenko and Anwar studied how the increasing amount of ethanol destroys the bilayer structure. Although the main focus of this study is not relevant in understanding the molecular mechanism of anesthesia, several of their findings, concerning systems in which the bilayer structure is still preserved, are also important in this respect. They found that the bilayer is disintegrated between the ethanol:lipid molar ratios of 6.3:1 and 9.5:1. Increasing ethanol concentration led to a clear increase of the area per lipid, and a clear decrease of the membrane thickness, while the bilayer structure was still intact. This latter effect was explained by the observed increasing interdigitation of the lipid chains of the opposite leaflets with increasing ethanol concentration. Contrary to Chanda and Bandyopadhyay, they did not find any effect of the ethanol concentration on the orientation of the PN vector. Ethanol molecules were found again to prefer staying at the vicinity of the lipid headgroups. Concerning the lipid tails, it was found that ethanol, in general, decreases their order; however, at the smallest ethanol concentration considered, a small increase of this order was observed around the first few carbon atoms of the lipid tails, i.e., where the ethanol molecules are actually located (Gurtovenko and Anwar 2009).

Kaye et al. studied the effect of ethanol on the collective dynamics of the DMPC bilayer both by molecular dynamics simulation and inelastic neutron scattering experiment. They found that ethanol induces a new, low energy, non-dispersive collective transverse mode, which they related to the enhanced permeability of the membrane (Kaye et al. 2011).

Among other molecules, the effect of methanol and ethanol on the structure of the DPPC bilayer was studied by Lee et al. (2005), who found that both alcohols increase the area per lipid, and decrease the thickness of the bilayer, and also decrease the order of the lipid chains. Similar findings were reported by Patra et al. (2006) concerning the effect of alcohols on the properties of both the DPPC and POPC bilayer. They also studied the interplay of the two opposing effects (i.e., lateral expansion and normal contraction) on the molar volume of the membrane, and found that, within the error bars, it remains constant, irrespective of the alcohol concentration. By projecting the lipid centers-of-mass to the bilayer plane and performing Voronoi analysis, they found that both alcohols increase the lateral distance of the lipid molecules without altering their relative arrangement. By calculating the autocorrelation time of the Voronoi areas, they also demonstrated how these alcohols increase the fluidity of the membrane. They also found that both alcohols prefer to stay close to the lipid headgroup region, and this tendency is stronger for ethanol than for methanol. In analyzing the origin of this difference, they found that while 80% of the ethanol molecules are hydrogen bonded to the lipids, methanol always penetrates to the bilayer together with its first hydration shell, and hence does not form any hydrogen bond with the lipids. As a consequence, while the crossing of the ethanol molecules between the two leaflets was observed several times, methanol molecules never penetrated into the region of the hydrocarbon chains. Further, a large fraction of the methanol molecules were found to stay in the bulk aqueous phase, while practically all ethanol molecules penetrated into the bilayer. Concerning the ordering of the various segments of the lipid molecules, they did not find any effect of the presence of these alcohols on the orientation of the PN vector, while both alcohols turned out to slightly increase the order of the lipid tails. In interpreting this result, they hypothesized that this trend is likely to be reverted with increasing alcohol concentration (Patra et al. 2006). This assumption was later fully confirmed, at least for ethanol, by Gurtovenko and Anwar (2009). As a continuation of this study, Terämä et al. calculated the effect of ethanol on the translational and rotational dynamics of the lipid molecules as well as on the lateral pressure profile in these membranes. They found that ethanol enhances both the diffusion and the rotation of the lipids. They attributed these changes to the larger area per lipid, which facilitates lipid rotation, and also to the presence of ethanol–lipid hydrogen bonds, by which ethanol perturbs the interactions between the water and lipid molecules. Concerning the lateral pressure profiles, they found that ethanol smoothes the entire profile, making the amplitude of both the peaks and dips smaller, and hence shifts the pressure from the region of the hydrocarbon chains towards the lipid headgroups (Terama et al. 2008).

The first study concerning the effect of the chain length of alcohols larger than ethanol was reported by Frischknecht and Fink, who performed density functional theory (DFT) calculation of the bilayer of an unspecified lipid, corresponding roughly to DPPC, in the presence of ethanol, 1-butanol, and 1-hexanol using a coarse-grained model of these molecules. Their findings were in line with those of the previous studies: the area per lipid was found to be increased, while the bilayer thickness to be decreased both with increasing alcohol concentration and with increasing chain length. Further, the area compressibility modulus of the bilayer was found to decrease upon addition of alcohols, and this effect was much stronger with increasing chain length. All the three alcohols were found to make the lateral pressure profile smoother by decreasing the amplitude of its peaks and dips (Frischknecht and Fink 2006).

A DPPC bilayer containing ethanol, 1-propanol, and 1-butanol was studied by Dickey and Faller using atomistic simulation. Quite surprisingly, they found a completely different effect of butanol on the membrane properties than that of the two smaller alcohols. Thus, while ethanol and propanol led to the increase of the area per lipid and to the decrease of the rotational relaxation time of the lipid PN vector, opposite changes were observed in the presence of butanol. The lateral diffusion coefficient of the lipid molecules was found to be insensitive to the ethanol concentration, increased upon addition of propanol, while decreased upon addition of butanol to the system. However, the time evolution of the area per lipid showed several sudden jumps during the simulations of the butanol-containing bilayers, which raises some doubt about the proper equilibration of these systems (Dickey and Faller 2007).

In understanding the effect of the alcohol chain length on anesthetic properties, the studies of Griepernau et al. are of great importance (Griepernau et al. 2007, Griepernau and Böckmann 2008). In these works, the effect of ethanol, 1-octanol, 1-decanol, and 1-tetradecanol on the DMPC bilayer was studied in rather long, 30–50 ns simulations. They found that the lipid:water partition coefficient increases with the chain length of the alcohol. Further, long-chain alcohols turned out to have an opposite effect on the membrane properties than ethanol. Namely, they increase the lateral density, the thickness, and the bending rigidity of the membrane, and decrease the mobility of the lipid headgroups, in a clear contrast with what was seen for ethanol. Among these properties, the increase of the pressure led to the increase of the lateral density as well as to the decrease of the membrane thickness and headgroup mobility. Both long-chain alcohols and pressure increased clearly the ordering of the lipid tails. The lateral pressure profile was smoothed by ethanol, but the amplitude of its peaks was clearly increased by long-chain alcohols. Considering that the OH group of all alcohols was found to form hydrogen bonds with the lipid headgroups, and also that the hydrocarbon chain of tetradecanol is already as long as that of DMPC, this ordering effect can simply be understood, as addition of long-chain alcohols (similarly to the increase of the pressure) simply increases the chain density across the entire hydrocarbon region of the membrane. As it can be inferred from the snapshots shown as well as from the tail order parameter and lateral pressure profiles, this increased chain density already induced a liquid crystalline to gel phase transition in systems containing long enough alcohols at high enough concentration or at high enough pressure (Griepernau et al. 2007, Griepernau and Böckmann 2008). The trend seen from these simulations, together with those of Patra et al. (2006) and Frischknecht and Fink (2006), is fully consistent with the well-known cutoff effect of *n*-alcohols, i.e., that their anesthetic potency increases with increasing chain length up to decanol, but above decanol they are no longer anesthetics at all. Thus, methanol is not anesthetic simply because it prefers to stay in the aqueous phase rather than in the membrane. Ethanol already stays preferably in the membrane, and the membrane:water partition coefficient of the alcohols increases with increasing chain length, leading thus also to the increase of their anesthetic ability. This trend continues until alcohols still increase the fluidity of the membrane, i.e., they lower the gel to liquid crystalline phase transition temperature. Once alcohols start increasing this temperature, the cutoff of the anesthetic potency occurs, and alcohols no longer act as anesthetics. This explanation is also fully consistent with the thermodynamic arguments of Heimburg et al. (Heimburg and Jackson 2007, Græsbøll et al. 2014).

2.2 Xenon and Other Noble Gases

Among the general anesthetic molecules exhibiting great chemical diversity, probably the most striking is the anesthetic ability of the noble gas xenon. Indeed, due to the lack of its undesirable side effects, xenon is nowadays increasingly used in clinical narcosis. The fact that a noble gas can act as a rather effective general anesthetic already sets a great challenge to protein theories, as in its frame a specific interaction with a chemically almost completely inactive noble gas needs to be explained. Further, it is rather difficult to imagine any evolutionary mechanism that leads to the development of such a protein whose substrate is not only chemically inactive, but it is only present in the atmosphere in extremely low concentration. The picture is further complicated by the fact that, in contrast with xenon, krypton has only a rather weak anesthetic effect, while smaller noble gases are not anesthetics. Aiming at understanding this puzzling phenomenon, several computer simulation studies of membranes containing noble gases have been performed in the past few years.

The first such study was reported by Stimson et al., who studied xenon in DPPC bilayer in five different concentrations, ranging from the xenon:lipid molar ratio of about 1:5 to 2:1. It should be noted that, similarly to the majority of subsequent simulations of noble gas containing membranes,

Xe atoms were initially placed in the aqueous phase, and they penetrated into the bilayer during the course of the 100 ns long simulations. It was found that, unlike ethanol, xenon prefers to stay in the middle of the bilayer. This different behavior of xenon was explained by the lack of its ability of forming hydrogen bonds with the lipid headgroups. Nevertheless, apart from its peak in the middle of the bilayer, the xenon density profile also exhibited a clear shoulder close to the headgroup region at all concentrations considered. Further, xenon was found to clearly increase both the area per lipid of the membrane and the order of the lipid tails (as seen from the deuterium order parameter value averaged over all C atoms of the hydrocarbon chains) (Stimson et al. 2005).

A more detailed study was later performed by Yamamoto et al., who simulated xenon in a POPE bilayer at the Xe:POPE molar ratio of 1:1. In this study, the effect of the pressure on the system was also investigated, as the simulation was repeated at several high pressures up to 500 bar. It was found that both the area and volume per lipid of the membrane were clearly increased in the presence of Xe, and was decreased at high pressure. The thickness of the membrane turned out to be insensitive to the presence of Xe up to 100 bar, but it was increased by Xe at higher pressures. Also, the thickness of the Xe-containing membrane was increased by increasing pressure. Xe was again found to have two preferred positions along the membrane normal, i.e., besides the main preference for staying in the middle of the bilayer, a secondary preference for positions close to the headgroup region was also observed. The increase of the pressure was found to enhance the main preference by pushing Xe atoms to the middle of the membrane. Interestingly, in spite of this finding, no change was observed in the gap between the chain terminal C atoms of the two bilayer leaflets with increasing pressure. Therefore, the unexpected thickening of the membrane with pressure was attributed to the increased order of the lipid tails. On the other hand, this order was found to be decreased in the presence of xenon. The observed dependence of the tail order on the presence of Xe and on the pressure was attributed to the observed diffusion of the Xe atoms, which collide to the lipid tails and thus decrease their order, while this xenon diffusion was slowed down at high pressures. The motion of the Xe atoms and their effect of swelling the membrane led to an increased diffusivity of the lipid molecules, an effect that is clearly reverted by increasing pressure (Yamamoto et al. 2012).

Rather different findings were reported by Booker and Sum, who studied Xe in the DOPC bilayer in high concentrations, at Xe:lipid molar ratios ranging from 1:1 to 3:1, both at atmospheric pressure and at high pressures ranging up to 200 bar. Similarly to other studies, they also found that the vast majority (i.e., 97%) of the Xe atoms penetrate into the bilayer, and they prefer two distinct positions along the membrane normal, although the outer of these preferred positions was found to be somewhat closer to the membrane interior, i.e., between the double bonds and chain terminal C atoms, than in the other studies. Further, the preference for this outer position was found to be stronger than that for being in the middle of the membrane, again in a clear contrast with the findings of other studies of similar systems. Although the area per lipid of the membrane was found to be increased by Xe, no effect of the pressure on it was reported here. The thickness of the membrane was found to be clearly increased with increasing Xe concentration and, contrary to the aforementioned results of Yamamoto et al., 75–80% of this thickening was attributed to the increase in the gap between the chain terminal C atoms. It was also found that Xe increases the order of the lipid tails beyond the double bond, leading to decreased fluidity of the membrane. This finding seems to be contradicted by the results of differential scanning calorimetry (DSC) measurements, reported in the same paper, which show that the gel-to-liquid crystalline phase transition temperature is lowered by increasing Xe concentration. The lateral pressure profile of the membrane was also reported at atmospheric pressure. Up to the Xe:DOPC molar ratio of 2:1, no clear effect of the presence of Xe on this profile was seen, while above this concentration, similarly to ethanol, xenon also smoothed the profile by reducing the amplitude of its peaks and dips (Booker and Sum 2013).

The effect of the size of the noble gas atoms on their anesthetic properties was investigated in detail by Moskovitz and Yang (2015) and by Chen et al. (2015). Moskovitz and Yang simulated Ne, Ar, and Xe in a DOPC bilayer at high concentrations, corresponding to the noble gas:lipid molar ratio of 3:1, both at atmospheric pressure and at high pressures up to 1000 bar. Similarly to other authors, they also found a dual preference of the noble gas atoms along the membrane normal, as they accumulated close to the headgroup region and also in the middle of the membrane, between its two leaflets. The former preference, which became more prominent with increasing atomic size, was found to be responsible for the swelling (i.e., lateral expansion), while the latter one for the inflation (i.e., thickening) of the membrane in the presence of the noble gases. Pressure was found to decrease both the area per lipid and the thickness of the membrane. Neither the presence of noble gases nor the pressure was found to influence the orientation of the PN vectors of the lipid headgroups. The water coordination number around the lipid headgroup atoms was found to be increased both by Xe and Ne, and decreased by increasing pressure. The diffusivity of the noble gases turned out to decrease with increasing size, while the order of the lipid tails, especially close to the tail ends, was found to be increased by noble gases, and decreased at high pressures (Moskovitz and Yang 2015).

Finally, Chen et al. studied a POPE bilayer in the presence of four noble gases, i.e., Ne, Ar, Kr, and Xe, at high concentrations ranging from 1:1 to 3:1 noble gas:lipid molar ratios, at atmospheric pressure. They found that the membrane:water partition coefficient increases with increasing size, thus, while only 77% of the Ne atoms penetrated into the membrane, for Xe this value is 96%. They also found dual preference of all noble gases along the membrane normal, with the fraction of atoms located close to the headgroup region being considerably higher for Xe than for the other, non-anesthetic noble gases. Thus, at the noble gas:lipid molar ratio of 3:1, about 90% of the Ne atoms accumulated in the middle of the membrane, i.e., in the interspace between the two leaflets, while 40% of the Xe atoms were located close to the headgroup region. Similarly to Moskovitz and Yang, they also found that noble gas atoms located in the middle of the membrane push the two lipid layers farther apart from each other, increasing thus the thickness of the membrane, while those located near the headgroup region push the lipid molecules farther away from each other laterally, increasing thus the area per headgroup of the membrane. Interestingly, while both the area per headgroup and molar volume of the membrane was found to increase with increasing noble gas concentration, no clear change of the membrane thickness was observed. This finding was attributed to the fact that, besides the size of the interlayer gap, the membrane thickness is also influenced by the ordering of the lipid tails, and this order was found to be decreased by noble gases, larger atoms having larger effect in this respect (Chen et al. 2015).

Although the results of existing simulations of noble gas containing lipid membranes are rather controversial in several respects, there are at least some points at which these studies led to similar conclusions. Thus, while the thickness of the membrane as well as the orientational order of the lipid tails seems to be sensitive to the details of the simulations, a clear lateral expansion of the membrane was found in every case. Considering that this swelling is caused by the atoms located close to the headgroup region, where ethanol, a chemically completely different general anesthetic, also prefers to stay, and also that the fraction of noble gas atoms located in this region is considerably higher for the general anesthetic xenon than for other non-anesthetic noble gases, and pressure decreases the fraction of these atoms, it seems plausible to assume that anesthetic effect is related to the atoms located in this region, and to their effect of swelling the membrane. In understanding the origin of the preference of the bulky and apolar Xe atoms for staying close to the crowded region of the polar headgroups, one has to consider the fact that Xe atoms are rather polarizable, in contrast with the smaller noble gas atoms.

3 Position of the Anesthetics in the Membrane

In understanding the molecular mechanism of general anesthesia, one of the key points is to find out how the anesthetic molecules are distributed within the membrane, more specifically, in which part of the membrane are they accumulated. The early studies led to various results, some of which are incompatible with each other, in this respect. Among several other small molecules, Jedlovszky and Mezei calculated the solvation free-energy profile of chloroform across the neat DMPC membrane (Jedlovszky and Mezei 2000) as well as DMPC/cholesterol mixed membranes of various compositions (Jedlovszky and Mezei 2003). They found that in membranes of low cholesterol content, chloroform prefers to stay in the middle of the membrane, but in membranes of high enough cholesterol content, another preferred position emerges at the vicinity of the polar headgroups, although the global free-energy minimum still corresponds to the middle of the membrane (Jedlovszky and Mezei 2003). Koubi et al. (2000) found that halothane stays close to the headgroup region in a DPPC bilayer, whereas Chau et al. (2007, 2009) claimed that halothane stays in the middle of the DMPC membrane. However, all these simulations were far too short (i.e., 1–2 ns) to converge. Furthermore, in the simulations of Chau et al. (2007, 2009), halothane molecules were initially placed in the middle of the membrane, i.e., where they were found to stay during the simulation. Therefore, these contradictions can simply be attributed to insufficient equilibration. In the somewhat longer coarse-grained simulation of Pickholz et al. (2005) and atomistic simulation of Oh and Klein (2009), halothane was found to prefer staying at the vicinity of the headgroup region, although its concentration did not drop to zero in the hydrocarbon region of the membrane either. Similar distribution along the membrane normal axis was found for isoflurane (Wieteska et al. 2015), whereas ketamine (Jerabek et al. 2010) as well as ethanol (Feller et al. 2002, Chanda and Bandyopadhyay 2004, 2006, Patra et al. 2006, Dickey and Faller 2007, Griepernau et al. 2007, Griepernau and Böckmann 2008) was always found to stay almost exclusively close to the headgroup region. Similar conclusion was drawn from the solvation free-energy profile calculation of ethanol in a POPC membrane (Jerabek et al. 2010). The strong preference of ethanol and the somewhat less strong, but still main preference of halothane for staying close to the lipid headgroups was explained by their ability to form strong, O–H donated (Patra et al. 2006) and weak, C–H donated hydrogen bonds (Darvas et al. 2012), respectively, with the ester O atoms of the lipid molecules.

On the other hand, two preferred positions along the membrane normal was found for xenon and other noble gases: the main preference corresponds to the middle of the bilayer, while the position close to the headgroup region corresponds to a lower maximum of the density profile (Stimson et al. 2005, Yamamoto et al. 2012, Booker and Sum 2013, Chen et al. 2015, Moskovitz and Yang 2015). It was also shown that this secondary preference becomes progressively weaker with decreasing size (and thus with decreasing anesthetic potency) of the noble gases (Chen et al. 2015, Moskovitz and Yang 2015). Similar, dual preference was found for chloroform in neat DOPC as well as mixed DSPC/cholesterol membranes (Reigada 2011) as well as for chloroform, halothane, diethyl ether, and enflurane both in the gel (Darvas et al. 2012) and liquid crystalline phases (Fábián et al. 2015) of the DPPC bilayer. Fábián et al. (2015) also showed that the anesthetic density profile can always be very well fitted by the sum of three Gaussian functions, two of which are mirror images of each other, while the third one is centered at the middle of the bilayer. The picture that emerges from these studies is that most of the anesthetics prefer two distinct positions along the membrane normal axis. One of these positions is close to the polar headgroup region, at the boundary of the hydrocarbon phase, while the other one is in the middle of the membrane. The relative strengths of these two preferences can be markedly different for different anesthetics, and possibly also for different membranes; moreover, in some cases, the second of these preferences (i.e., for staying in the middle of the membrane) may even vanish completely, but the preference for staying close to the polar headgroup region is always present. Considering also the fact that general anesthetics are either weakly polar or, at least, strongly polarizable particles, the existing results suggest that the phenomenon of general anesthesia might be related to those anesthetic molecules that are accumulated at the vicinity of the headgroup region.

In contrast with their distribution along the membrane normal axis, the lateral distribution of general anesthetics has, to our knowledge, only been studied once. Thus, Reigada (2013) investigated the lateral distribution of chloroform both in an α phase DOPC membrane containing a β phase mixed raft of DSPC and cholesterol, and also in a β phase DSPC/cholesterol mixed membrane containing an α phase raft of DOPC. He found the boundary of these nanodomains to remain stable even after a 200 ns long trajectory. However, considering that the diffusion coefficient of the lipid molecules is in the order of 1 μm²/s, no dissolution of the nanodomains can be expected even after such a long simulation. More importantly, he found that chloroform is always accumulated in the disordered (α) phase (due to their much larger diffusion coefficient, unlike lipids, the chloroform molecules could indeed explore the membrane plane during the course of the simulation) (Reigada 2013). Since real biological membranes are always rafts of α and β phase domains, this result suggests that the disordered, α phase domains and membrane proteins embedded in such domains are related to the molecular mechanism of general anesthesia.

4 Change of Various Membrane Properties

In this chapter, we review how various membrane properties that are thought to be possibly related to general anesthesia are seen to be influenced by the presence of various general anesthetics in computer simulations. When summarizing recent results in this respect, it should be pointed out that the change of only such membrane properties might be related to the molecular mechanism of anesthesia that changes in the same direction upon addition of any general anesthetics.

4.1 System Dimensions and Lipid Ordering

There is a clear consensus in the literature that general anesthetics increase the area per headgroup, and hence decrease the lateral density of the membrane. Furthermore, Griepernau and Böckmann (2008) also showed that long enough chain 1-alkanols, which are no longer anesthetics, behave in the opposite way in this respect. Patra et al. demonstrated, by performing Voronoi analysis, that this lateral swelling does not change the relative arrangement of the lipids, but just pushes them somewhat farther away from each other (Patra et al. 2006). It has also been shown several times that the area compressibility modulus of the membrane is decreased by ethanol, but it is increased by long-chain non-anesthetic 1-alkanols (Frischknecht and Fink 2006, Griepernau et al. 2007). No such consensus is reached, on the other hand, concerning the change of the membrane thickness due to the presence of general anesthetics. Thus, while ethanol was generally found to decrease the membrane thickness (Lee et al. 2005, Frischknecht and Fink 2006, Patra et al. 2006, Griepernau and Böckmann 2008, Gurtovenko and Anwar 2009), xenon clearly increases it (Booker and Sum 2013, Moskovitz and Yang 2015). Further, no clear effect of halothane (Oh and Klein 2009, Darvas et al. 2012, Fábián et al. 2015), ketamine (Jerabek et al. 2010), chloroform, diethyl ether, and enflurane (Darvas et al. 2012, Fábián et al. 2015) on various membranes was found in this respect. Due to the increase of the area per lipid, the molar volume of the membrane was also typically found to be increased by several anesthetics (Darvas et al. 2012, Yamamoto et al. 2012, Chen et al. 2015, Fábián et al. 2015), even if they also led to the thinning of the membrane (Koubi et al. 2000, Griepernau et al. 2007). These findings suggest that the critical volume hypothesis (Mullins 1954) might well be relevant in explaining general anesthesia; however, it originates in the increase of the area per lipid, and hence a critical area hypothesis would be closer to the molecular origin of this phenomenon (Darvas et al. 2012, Fábián et al. 2015).

The above findings are also in accordance with the results concerning the distribution of the anesthetics along the membrane normal. Thus, it is generally accepted that anesthetics located close to the headgroup region cause the swelling, while those being in the middle of the bilayer cause the thickening of the membrane, by pushing the lipid molecules farther away from each other laterally, and along the membrane normal, respectively (Moskovitz and Yang 2015). The facts that (i) all anesthetics prefer positions close to the headgroup region, even if together with other positions in some cases, and (ii) they all induce swelling of the membrane make the decrease of the lateral density a good candidate as the relevant change of the membrane laying behind the molecular mechanism of general anesthesia.

On the other hand, the change of the membrane thickness depends, besides the interdigitation of the two layers, also on the effect of anesthetics on the lipid tail order. Not surprisingly, existing simulation results do not show any clear trend in this respect. In several cases, the same anesthetic molecule was found to have opposite effects in different studies in this respect. Thus, xenon was reported to increase (Booker and Sum 2013, Moskovitz and Yang 2015) as well as to decrease (Yamamoto et al. 2012, Chen et al. 2015) the lipid tail order. The effect of anesthetics on the local order of the lipid tails seems to depend also on the anesthetic concentration (Gurtovenko and Anwar 2009), and can be different for different segments of the lipid tail (Koubi 2000, Chanda and Bandyopadhyay 2004, 2006, Booker and Sum 2013). Further, as it was shown by Reigada, this effect might be different in different phases of the membrane, i.e., the same anesthetic can increase the order of the lipid tails in the disordered (α) phase, and decrease it in the ordered (β) phase (Reigada 2011, 2013). On the other hand, recent simulations agree that the orientation of the PN vector relative to the bilayer plane is not altered by the presence of general anesthetics (Patra et al. 2006, Gurtovenko and Anwar 2009, Oh and Klein 2009, Darvas et al. 2012, Fábián et al. 2015, Moskovitz and Yang 2015).

4.2 Membrane Fluidity

Several simulation studies addressed also the effect of general anesthetics on the fluidity of the membrane. Although the fluidity of a condensed phase can be accessed through various different physical quantities, the vast majority of the recent studies agree that the fluidity of the membrane is enhanced by general anesthetics. The most typical approach of this problem is the calculation of the lateral diffusion coefficient of the lipid molecules, which was found to be noticeably increased in the presence of ethanol (Chanda and Bandyopadhyay 2004, 2006, Dickey and Faller 2007, Terama et al. 2008), halothane (Pickholz et al. 2005), xenon (Yamamoto et al. 2012), as well as chloroform (Reigada 2011). Similar conclusion was drawn by Patra et al. (2006), who projected the lipid centers-of-mass to the membrane plane, calculated the autocorrelation time of the area of their Voronoi polygons, and found a clear decrease of it in the presence of ethanol. Chanda and Bandyopadhyay (2006) showed that, besides their lateral diffusivity, the diffusion of the lipid molecules along the membrane normal axis is also enhanced by ethanol.

The change of the dynamics of various parts of the lipid molecule in the presence of general anesthetics was also studied. Thus, Griepernau et al. (2007) found that the diffusion coefficient of the lipid headgroups increases in the presence of ethanol. Terämä et al. studied the rotational autocorrelation function of the PN vector as well as the vectors pointing along the glycerol group, and from the first C atom of one of the lipid tails to that of the other tail in different membranes, and found that the correlation time is systematically decreased by ethanol in every case (Terama et al. 2008). Dickey and Faller (2007) also found that the rotational relaxation time of the PN vector of the DPPC headgroup is decreased by ethanol and propanol.

The only statement that is in contradiction with all the above findings was made by Booker and Sum, who claimed that xenon decreases the mobility of the lipid tails. However, this statement was simply based on the observed enhanced order of the lipid tail ends rather than on the calculation of any dynamical property (Booker and Sum 2013).

Although the vast majority of the recent simulation studies claim that general anesthetics increase the membrane fluidity, the corresponding results almost always simply concern the mobility of the individual lipid molecules or certain parts of them. This enhanced lipid mobility can easily be explained by the anesthetic induced lateral expansion of the membrane, which creates additional space for lipid motion. Results that concern membrane fluidity and go beyond the dynamics of the individual molecules are scarce. In accordance with earlier experimental results (Ly et al. 2002, Ly and Longo 2004), Griepernau et al. (2007) found that ethanol decreases the bending modulus of the DMPC membrane, making it thus more flexible. Kaye et al. (2011) found that ethanol induces a new, non-dispersive collective mode, related to the transverse dynamics of the hydrophobic core of the DMPC membrane. We are not aware, however, of any simulation study concerning the dependence of the gel-to-liquid crystalline phase transition temperature of the membrane in the presence of general anesthetics, although earlier experimental data (Janoff and Miller 1982, Forrest and Rodham 1985, Kaminoh et al. 1988, 1992, Sierra-Valdez and Ruiz-Suárez 2013) as well as theoretical claims (Heimburg and Jackson 2007, Græsbøll et al. 2014) concerning the anesthetic-induced membrane fluidity as a possible change related to the molecular mechanism of general anesthesia are typically based on this interpretation of the membrane fluidity.

4.3 Lateral Pressure Profile

Although anesthetic-induced changes in the lateral pressure profile across the membrane represent a possible origin of the molecular mechanism of general anesthesia, detailed investigation of this issue is hindered by several technical difficulties. In fact, the original suggestion of Cantor was solely based on thermodynamic arguments and a few reasonable assumptions (Cantor 1997a, 1997b). Experimental test of this hypothesis seems to be almost impossible, as it would require the measurement of the lateral pressure with Angström resolution. Certainly, we are not aware of any such study, at all. Although computer simulation methods seem to be particularly suitable to address this problem, the calculation of the lateral pressure profile in a simulation is not a straightforward task, either. The problem stems from the fact that the pressure contribution coming from the interaction of a particle pair can be calculated as an integral over an open path, and thus, the resulting profile depends also on the choice of the path. This uncertainty reflects the fact that, by calculating the lateral pressure profile, an inherently non-local quantity (i.e., pressure) is tried to be localized. It has been shown, however, that the use of several reasonably chosen integration paths, such as the Irwing–Kirkwood (Irving and Kirkwood 1950) and Harasima (Harasima 1958) contours, leads to qualitatively similar results (Sonne et al. 2005). The main drawback of using the Irwing–Kirkwood path is that this way non-pairwise additive interaction terms, such as the reciprocal space contribution of the Ewald summation, cannot be taken into account (Sonne et al. 2005). As a consequence, in the vast majority of studies, lateral pressure profile was calculated *a posteriori*, without applying long-range correction of the electrostatic interaction but using as large interaction cutoff as possible. Although it is generally believed that such a change in the potential between the simulation and the analysis does not lead to substantial systematic error as long as the cutoff value used is large enough (Sonne et al. 2005), it was also shown that in certain cases (e.g., two phases of markedly different densities, not sufficiently large cutoff), this systematic error can reach the order of several hundred bar (Sega et al. 2016). Very recently, Sega et al. developed a method for calculating the lateral pressure profile on the fly, without changing the potential, in which the Harasima path is employed and the particle mesh Ewald (PME) correction (Essman et al. 1995) is fully taken into account (Sega et al. 2016).

Due to the above difficulties, lateral pressure profile calculation in simulated membranes containing anesthetics has only been reported a handful of times. One of the first such calculations was published by Terämä et al., who calculated the lateral pressure profile across a DPPC and a PDPC bilayer both in the absence and presence of ethanol. The obtained profiles showed a positive peak at the membrane–water interface, followed by a marked 400–1000 bar deep minimum at the region of the glycerol backbones. These features were interpreted as consequences of the steric repulsion in the crowded headgroup region and attraction between water and the ester groups, respectively. This latter minimum, corresponding to a force that laterally contracts the bilayer, is located at the boundary of the polar and apolar membrane regions, and gives a large contribution to the surface tension of the system. The addition of ethanol to the system smoothes the entire profile, making the peak at the water–lipid interface lower and the dip at the region of the glycerol backbones, i.e., where the ethanol molecules are also located, less deep. This latter change reflects the weakening of the contracting forces in this region, and it is in a good agreement with the ethanol induced lateral expansion of the membrane (Terama et al. 2008). Similar smoothing effect of ethanol on the lateral pressure profile was earlier reported by Frischknecht and Fink (2006) on the basis of DFT calculations.

The lateral pressure profile obtained by Griepernau and Böckmann in the DMPC bilayer showed several different features. Thus, here the peak at the water–lipid interface was very small, and was followed by a secondary minimum in the headgroup region before the main dip in the region of the glycerol backbones. These features in the headgroup part of the lateral pressure profile might well be sensitive to the force field used, as they are determined by the delicate interplay of the steric repulsion and electrostatic attraction occurring in this region. In the hydrocarbon phase, three marked peaks were seen, only some traces of which were earlier detected by Terämä et al. Ethanol was again found to smooth this profile. Similar smoothing effect was observed by the increase of the pressure, although, evidently, it also shifts the entire lateral pressure profile to higher values. In this study, the first and second moments of the lateral pressure profile were also estimated in the neat bilayer. The obtained values turned out to be an order of magnitude smaller than what was predicted by Cantor (1999). This difference was attributed to the neglect of the headgroup repulsion in the original calculation of Cantor (Griepernau and Böckmann 2008).

Qualitatively similar pressure profile was obtained by Oh and Klein in the DMPC bilayer. Although they did not find a clear smoothing effect of halothane on this profile (e.g., halothane increased the peaks in the hydrocarbon region), they observed a decrease of the amplitude of the main dip in the glycerol backbone region, in accordance with the earlier studies (Oh and Klein 2009). Booker and Sum reported a very structured lateral pressure profile in the DOPC bilayer, which, however, exhibits the same general features as the previous ones. They found little effect of xenon on this profile; nevertheless, 100–200 bar decrease of the amplitude of the minimum beneath the headgroup region was observed also here (Booker and Sum 2013). Jerabek et al. reported calculation of the lateral pressure profile in the POPC bilayer in the presence and absence of ketamine. Interestingly, they did not find any marked effect of ketamine on this profile, apart from that it lowered the outer peaks in the hydrocarbon region, at the position where ketamine was accumulated (Jerabek et al. 2010).

To the best of our knowledge, so far the only lateral pressure profile calculation in anesthetic-containing membranes that was performed without changing the potential between the simulation and pressure profile calculation was reported recently by Fábián et al., who studied chloroform, halothane, diethyl ether, and enflurane in the DPPC bilayer. Although the profiles showed similar features than the previously reported ones, they found a different effect of the aesthetics on these profiles. Namely, instead of smoothing the profile, anesthetic generated an inward shift of the pressure around the region of the glycerol backbones. Thus, lateral pressure was decreased at 13–18 Å from the bilayer center, and increased closer to it by the anesthetics. The lateral pressure profile calculation was also complemented by a free volume analysis. It turned out that all anesthetics increase the free volume in the outer edge of the hydrocarbon phase, at 8–16 Å from the bilayer center, where (apart from the middle of the membrane) anesthetic molecules are preferentially located. The accumulation of the anesthetics in this region was found to lead to a lateral swelling of the membrane as they push the lipid molecules farther away from each other. However, this swelling leads to a decrease of the lateral pressure in the nearby region of the ester groups and glycerol backbones, into which the anesthetic molecules do not penetrate. In this region, the additional space created by the above pushing apart of the lipid molecules is still rather large, but it is no longer accompanied by the accumulation of the anesthetic molecules (Fábián et al. 2017).

5 Pressure Reversal

Since the anesthetic effect is known to vanish with increasing pressure, any possible explanation of the mechanism of anesthesia has to account also for its pressure reversal. Concerning lipid theories, this requirement implies that the membrane property, the change of which is responsible for the anesthetic effect, must not only be such that it changes in the same direction upon addition of any general anesthetics, but it should also change in the opposite direction upon increasing the pressure, at least in the membrane containing anesthetics.

The effect of high pressure on anesthetic-containing membranes has only been studied a handful of times. In one of the first of such studies, Chau et al. reported self-association of halothane in DMPC membrane at pressures of 200 bar and 400 bar, while no such self-association was observed at atmospheric pressure (Chau et al. 2007, 2009). However, these simulations were far too short (i.e., below 1 ns) to properly equilibrate the systems (as seen also from the fact that halothane molecules were still accumulated in the middle of the membrane, i.e., at their initial position). Performing orders of magnitude longer simulations, several authors of these contributions showed later, independently from each other, by analyzing anesthetic cluster size distribution (Darvas et al. 2012, Fábián et al. 2015) and nearest neighbor coordination number (Wieteska et al. 2015) of various general anesthetics, that pressure does not induce their noticeable self-association in the membrane.

Among the above-discussed membrane properties, the area per lipid showed a clear pressure reversal in all cases, i.e., its anesthetic-induced increase was always reverted by the pressure-induced decrease (Griepernau and Böckmann 2008, Darvas et al. 2012, Yamamoto et al. 2012, Fábián et al. 2015, Moskovitz and Yang 2015). A decreasing trend of it with increasing pressure was also reported by Booker and Sum (2013), although in their study this change remained within error bars. However, in this study, the pressure was not increased beyond the moderate value of 200 bar.

Similar effect, i.e., pressure-induced decrease reverting the anesthetic-induced increase was observed also for the molar volume of the membrane (Darvas et al. 2012, Yamamoto et al. 2012, Fábián et al. 2015). Interestingly, results concerning not only the addition of anesthetics but also the increase of the pressure turned out to be, on the other hand, rather controversial in respect of the membrane thickness. Thus, in some studies, both the addition of anesthetics and the increase of the pressure were found to decrease (Griepernau and Böckmann 2008) or increase the membrane thickness (Yamamoto et al. 2012), whereas in some other studies, none of them led to clear changes of membrane thickness (Darvas et al. 2012, Fábián et al. 2015). These results all suggest that although the critical volume hypothesis (Mullins 1954) might well be relevant in respect of the molecular mechanism of general anesthesia, it clearly originates in the changes of the lateral density of the membrane, and hence it should be replaced by a critical area hypothesis (Darvas et al. 2012, Fábián et al. 2015).

Consistently with the observed pressure reversal of the area per lipid value, the diffusion coefficient of the entire lipid molecules (Yamamoto et al. 2012) as well as their headgroups (Griepernau and Böckmann 2008) showed an opposite (i.e., decreasing) trend with increasing pressure than upon addition of anesthetics. Interestingly, while the diffusivity of xenon was found to clearly decrease with increasing pressure by Yamamoto et al. (2012), it turned out to be totally insensitive to the pressure in the study of Moskovitz and Yang (2015).

Similarly to the addition of anesthetics, the increase of the pressure also turned out to have no clear effect on the ordering of the lipid tails. In several studies, pressure was found to induce additional order of the lipid tails (Griepernau and Böckmann 2008, Darvas et al. 2012, Yamamoto et al. 2012, Fábián et al. 2015); however, Moskovitz and Yang (2015) observed an opposite effect, while Booker and Sum (2013) found lipid tail ordering to be insensitive to the pressure. All studies agree, on the other hand, that the orientation of the headgroup PN vector does not depend on the pressure at all (Darvas et al. 2012, Fábián et al. 2015, Moskovitz and Yang 2015). Results concerning the changes in the distribution of anesthetics along the membrane normal upon increasing pressure are also controversial: while xenon was found to be pushed to the middle of the POPE membrane by increasing pressure (Yamamoto et al. 2012), an opposite trend was observed in the DOPC membrane (Moskovitz and Yang 2015) as well as for several other anesthetics in the DPPC membrane (Fábián et al. 2015).

Finally, we are only aware of two studies concerning the pressure reversal of the lateral pressure profile. Griepernau and Böckmann studied ethanol in the DMPC bilayer, and found that both ethanol and pressure make the profile smoother by decreasing the amplitude of its local minima and maxima, although this effect of the pressure was much weaker in the anesthetic-containing membrane than in the neat membrane. Although here the effect of the anesthetic and pressure is seemingly the same, it should not be forgotten that the increase of the pressure shifts the entire lateral pressure profile upwards (i.e., it increases, evidently, also its lateral component). Thus, at any part of the profile where anesthetics decrease the lateral pressure (i.e., typically in the region of the peaks), this change is naturally reverted by the increase of the global pressure. Such a change occurs, among others, in the region of the glycerol backbones (Griepernau and Böckmann 2008). Similar decrease of the lateral pressure was observed by Fábián et al. close to the headgroup region, at the hydrophobic side of the hydrocarbon–polar interface of the DPPC bilayer in the presence of four other anesthetics, i.e., chloroform, halothane, diethyl ether, and enflurane. This decrease, attributed to the anesthetic-induced lateral pushing away of the lipid molecules, occurring next to this region, was again clearly reverted by the increase of the global pressure (Fábián et al. 2017).

All these results corroborate the conclusions drawn in the previous sub-section, suggesting that, if general anesthesia can indeed be explained in the frame of the lipid theories, (i) the anesthetic-induced lateral swelling of the membrane as well as (ii) the resulting increased mobility of the lipid molecules (which is often thought to be a sign of the increased membrane fluidity), and (iii) lateral pressure decrease either in the outer preferred region of the anesthetics or just next to it along the membrane normal is a good candidate for being the relevant membrane properties in this respect. All of these changes seem to be such that they are caused by any general anesthetics, and are reverted by increasing pressure.

6 Concluding Remarks

In this contribution, the computer simulation studies of lipid membranes containing general anesthetics, reported in the past 15 years, have been reviewed. Although there are still plenty of unresolved points concerning the molecular mechanism of general anesthesia, the intensive investigation of the problem, enabled by the continuous, rapid increase of the routinely available computing power in the past decades, certainly brought us closer to the molecular-level understanding of this phenomenon.

In the frame of the lipid theories, general anesthesia and its pressure reversal can only be related to such membrane properties that (i) change in the same way upon addition of any general anesthetic, (ii) do not change in the same way upon addition of molecules having no anesthetic effect, even if they are chemically similar to some of the general anesthetics, and (iii) change in the opposite way upon increasing the pressure. Although there are many controversies in the results of the various studies in this respect concerning a number of membrane properties, a growing consensus seems to be reached at least in the following points. First, general anesthetics usually prefer two positions along the membrane normal: close to the glycerol backbones, and in the middle of the bilayer. While the latter of these two preferences is certainly due to the larger free volume available for the anesthetics between the two membrane leaflets than among the lipid chains (Marrink and Berendsen 1994, 1996, Marrink et al. 1996, Jedlovszky and Mezei 2000, 2003, Alinchenko et al. 2004, 2005, Falck et al. 2004, Rabinovich et al. 2005), the former preference is related to the fact that all general anesthetics are either mildly polar, or, at least, strongly polarizable molecules, and hence here they can favorably interact with the polar part of the membrane without entering into the crowded headgroup region. The relative strengths of these preferences vary from anesthetic to anesthetic, and, in certain cases (e.g., for ethanol), the latter preference might be completely missing; however, the former preference has been observed for all anesthetics in all membranes so far. The lack of the preference for staying in the middle of the membrane in some cases suggests that the anesthetic effect can probably be attributed to the molecules staying in the other preferred position, i.e., close to the glycerol backbones. It is also clear that these molecules induce a lateral expansion of the membrane by pushing the lipid molecules laterally farther away from each other, while the ones located in the middle of the bilayer increase the membrane thickness by pushing the two membrane leaflets farther away from each other. The fact that the lateral density of the membrane is increased considerably by any general anesthetics, while there is no such a clear trend concerning the membrane thickness, corroborates the aforementioned assumption that anesthetic effect is probably related to the anesthetics located close to the headgroup region. Furthermore, the lateral membrane density is not only found to decrease upon addition of any general anesthetics, but it also satisfies the other two of the above criteria, namely it increases with increasing pressure, and increases (in the case of long-chain n-alcohols) or only slightly decreases (in the case of small noble gases) when adding non-anesthetics that are chemically similar to certain anesthetics to the membrane.

Besides the decrease of the lateral density, there are at least two more anesthetic-induced changes in the membrane properties that seem to satisfy the above conditions, namely the increased lateral mobility of the lipid molecules and the decrease of the lateral pressure close to the outer preferred position of the anesthetics. These anesthetic-induced changes are clearly related to each other: both the decrease of the lateral pressure, reflecting weaker local interactions, and the increased lateral mobility of the lipid molecules are caused by the anesthetic-induced lateral pushing away of the lipid chains, which creates additional space between the lipid molecules in the nearby region.

This picture is also compatible with several earlier hypotheses concerning the molecular mechanism of general anesthesia in the frame of the lipid theories. Thus, the anesthetic-induced lateral expansion of the membrane increases also its molar volume, hence, it explains also the critical volume hypothesis (Mullins 1954). The increased lipid mobility is often interpreted as a sign of the increased membrane fluidity, while the decrease of the lateral pressure close to the outer preferred position of the anesthetics can specify the changes of the lateral pressure profile that were, in general, assumed to occur by Cantor (Cantor 1997a, 1997b). It was also found that anesthetics prefer to be dissolved in the disordered domains of membranes containing rafts of the α and β phases (Reigada 2013). Thus, in accordance with the thermodynamic model of Heimburg et al. (Heimburg and Jackson 2007, Græsbøll et al. 2014), they lower the transition temperature between the two phases, and hence increase the fraction of the disordered (α) domains, increasing thus also the fluidity of the membrane.

Finally, it should be recalled that the lack of success of finding the relevant membrane property, the change of which is behind the molecular mechanism of general anesthesia, in spite of the intensive research carried out in this direction in the 20th century led to the rise of the view that maybe there is no such membrane property, instead, anesthesia is due to some specific interaction with certain membrane-bound proteins. Needless to say that, due to the large chemical variety of general anesthetics, such a protein theory would exclude the existence of a unique mechanism of general anesthesia. In the light of the recent results, reviewed in this paper, we are now much closer to find the relevant membrane property, and hence it seems that there is no need to give up the idea that a unique mechanism lays behind the phenomenon of general anesthesia. In other words, at least the principle of Occam's razor supports the lipid theory against the protein theory.

ACKNOWLEDGEMENT

This work has been supported by the Hungarian NKFIH Foundation under Project No. 119732.

REFERENCES

Alinchenko, M. G., A. V. Anikeenko, N. N. Medvedev, V. P. Voloshin, M. Mezei, and P. Jedlovszky. 2004. Morphology of Voids in Molecular Systems. A Voronoi-Delaunay Analysis of a Simulated DMPC Membrane. *J. Phys. Chem. B* 108: 19056–19067.

Alinchenko, M. G., V. P. Voloshin, N. N. Medvedev, M. Mezei, L. Pártay, and P. Jedlovszky. 2005. Effect of Cholesterol on the Properties of Phospholipid Membranes. 4. Interatomic Voids. *J. Phys. Chem. B* 109: 16490–16502.

Bemporad, D., J. W. Essex, and C. Luttmann. 2004. Permeation of Small Molecules through a Lipid Bilayer: A Computer Simulation Study. *J. Phys. Chem. B* 108: 4875–4884.

Booker, R. D. and A. K. Sum. 2013. Biophysical Changes Induced by Xenon on Phospholipid Bilayers. *Biochim. Biophys. Acta* 1828: 1347–1356.

Cantor R. S. 1997a. Lateral Pressures in Cell Membranes: A Mechanism for Modulation of Protein Function. *J. Phys. Chem. B* 101: 1723–1725.

Cantor R. S. 1997b. The Lateral Pressure Profile in Membranes: A Physical Mechanism of General Anesthesia. *Biochemistry* 36: 2339–2344.

Cantor R. S. 1999. The Influence of Membrane Lateral Pressures on Simple Geometric Models of Protein Conformational Equilibria. *Chem. Phys. Lipids* 101: 45–56.

Chanda, J. and S. Bandyopadhyay. 2004. Distribution of Ethanol in a Model Membrane: A Computer Simulation Study. *Chem. Phys. Letters* 392: 249–254.

Chanda, J. and S. Bandyopadhyay. 2006. Perturbation of Phospholipid Bilayer Properties by Ethanol at a High Concentration. *Langmuir* 22: 3775–3781.

Chau, P. L., P. N. M. Hoang, S. Picaud, and P. Jedlovszky. 2007. A Possible Mechanism for Pressure Reversal of General Anaesthetics from Molecular Simulations. *Chem. Phys. Letters* 438: 294–297.

Chau, P. L., P. Jedlovszky, P. N. M. Hoang, and S. Picaud. 2009. Pressure Reversal of General Anaesthetics: A Possible Mechanism from Molecular Dynamics Simulations. *J. Mol. Liquids* 147: 128–134.

Chen, J., L. Chen, Y. Wang, X. Wang, and S. Zeng. 2015. Exploring the Effects on Lipid Bilayer Induced by Noble Gases via Molecular Dynamics Simulations. *Sci. Rep.* 5: 17235-1-6.

Comer, J., K. Schulten, and C. Chipot. 2017. Permeability of a Fluid Lipid Bilayer to Short-Chain Alcohols from First Principles. *J. Chem. Theor. Comput.* 13: 2523–2532.

Darvas, M., P. M. N. Hoang, S. Picaud, M. Sega, and P. Jedlovszky. 2012. Anesthetic Molecules Embedded in a Lipid Membrane: A Computer Simulation Study. *Phys. Chem. Chem. Phys.* 14: 12956–12969.

Dickey, A. N. and R. Faller. 2007. How Alcohol Chain-Length and Concentration Modulate Hydrogen Bond Formation in a Lipid Bilayer. *Biophys. J.* 92: 2366–2376.

Essman, U., L. Perera, M. L. Berkowitz, T. Darden, H. Lee, and L. G. Pedersen. 1995. A Smooth Particle Mesh Ewald Method. *J. Chem. Phys.* 103: 8577–8594.

Fábián, B., M. Darvas, S. Picaud, M. Sega, and P. Jedlovszky. 2015. The Effect of Anesthetics on the Properties of a Lipid Membrane in the Biologically Relevant Phase: A Computer Simulation Study. *Phys. Chem. Chem. Phys.* 17: 14750–14760.

Fábián, B., M. Sega, V. P. Voloshin, N. N. Medvedev, and P. Jedlovszky. 2017. Lateral Pressure Profile and Free Volume Properties of Phospholipid Membranes Containing Anesthetics. *J. Phys. Chem. B* 121: 2814–2824.

Falck, E., M. Patra, M. Karttunen, M. T. Hyvönen, and I. Vattulainen. 2004. Impact of Cholesterol on Voids in Phospholipid Membranes. *J. Chem. Phys.* 121: 12676–12689.

Feller, S. E., C. A. Brown, D. T. Nizza, and K. Gawrisch. 2002. Nuclear Overhauser Enhancement Spectroscopy Cross-Relaxation Rates and Ethanol Distribution across Membranes. *Biophys. J.* 82: 1396–1404.

Forrest, B. J. and D. K. Rodham. 1985. An Anaesthetic-Induced Phosphatidylcholine Hexagonal Phase. *Biochim. Biophys. Acta* 814: 281–288.

Franks, N. P. and W. R. Lieb. 1978. Where Do General Anaesthetics Act? *Nature* 274: 339–342.

Franks, N. P. and W. R. Lieb. 1979. The Structure of Lipid Bilayers and the Effects of General Anaesthetics. *J. Mol. Biol.* 133: 469–500.

Franks, N. P. and W. R. Lieb. 1994. Molecular and Cellular Mechanisms of General Anaesthesia. *Nature* 367: 607–614.

Frischknecht, A. L. and L. J. D. Fink. 2006. Alcohols Reduce Lateral Membrane Pressures: Predictions from Molecular Theory. *Biophys. J.* 91: 4081–4090.

Græsbøll, K., H. Sasse-Middelhoff, and T. Heimburg. 2014. The Thermodynamics of General and Local Anesthesia. *Biophys. J.* 106: 2143–2156.

Griepernau, B. and R. A. Böckmann. 2008. The Influence of 1-Alkanols and External Pressure on the Lateral Pressure Profiles of Lipid Bilayers. *Biophys. J.* 95: 5766–5778.

Griepernau, B., S. Leis, M. F. Schneider, M. Sikor, D. Steppich, and R. A. Böckmann. 2007. 1-Alkanols and Membranes: A Story of Attraction. *Biochim. Biophys. Acta* 1768: 2899–2913.

Gruner, S. M. and E. Shyamsunder. 1991. Is the Mechanism of General Anesthesia Related to Lipid Membrane Spontaneous Curvature? *Ann. N. Y. Acad. Sci.* 625: 685–697.

Gurtovenko, A. A. and J. Anwar. 2009. Interaction of Ethanol with Biological Membranes: The Formation of Non-bilayer Structures within the Membrane Interior and their Significance. *J. Phys. Chem. B* 113: 1983–1992.

Halsey, M. J. and B. Wardley-Smith. 1975. Pressure Reversal of Narcosis Produced by Anaesthetics, Narcotics and Tranquillisers. *Nature* 257: 811–813.

Harasima, A. 1958. Molecular Theory of Surface Tension. *Adv. Chem. Phys.* 1: 203–237.

Hauet, N, F. Artzner, F. Boucher, et al. 2003. Interaction between Artificial Membranes and Enflurane, a General Volatile Anesthetic: DPPC-Enflurane Interaction. *Biophys. J.* 84: 3123–3137.

Haydon, D. A., B. M. Hendry, S. R. Levinson, and J. Requena. 1977. The Molecular Mechanisms of Anaesthesia. *Nature* 268: 356–358.

Heimburg, T. and A. D. Jackson. 2007. The Thermodynamics of General Anesthesia. *Biophys. J.* 92: 3159–3165.

Huang, P., E. Bertaccini, and G. H. Loew. 1995. Molecular Dynamics Simulation of Anesthetic-Phospholipid Bilayer Interactions. *J. Biomol. Struct. Dyn.* 12: 725–754.

Irving, J. H. and J. G. Kirkwood. 1950. The Statistical Mechanical Theory of Transport Processes. IV. The Equations of Hydrodynamics. *J. Chem. Phys.* 18: 817–829.

Janoff, A. S. and K. W. Miller. 1982. *Biological Membranes*, ed D. Chapman, 417–476. London: Academic Press.

Jedlovszky, P. and M. Mezei. 2000. Calculation of the Free Energy Profile of H_2O, O_2, CO, CO_2, NO, and $CHCl_3$ in a Lipid Bilayer with a Cavity Insertion Variant of the Widom Method. *J. Am. Chem. Soc.* 122: 5125–5131.

Jedlovszky, P. and M. Mezei. 2003. Effect of Cholesterol on the Properties of Phospholipid Membranes. 2. Free Energy Profile of Small Molecules. *J. Phys. Chem. B* 107: 5322–5332.

Jerabek, H., G. Pabst, M. Rappolt, and T. Stockner. 2010. Membrane-Mediated Effect on Ion Channels Induced by the Anesthetic Drug Ketmine. *J. Am. Chem. Soc.* 132: 7990–7997.

Johnson, F. H. and E. A. Flagler. 1950. Hydrostatic Pressure Reversal of Narcosis in Tadpoles. *Science* 112: 91–92.

Kaminoh, Y., S. Nishimura, H. Kamaya and I. Ueda. 1992. Alcohol Interaction with High Entropy States of Macromolecules: Critical Temperature Hypothesis for Anesthesia Cutoff. *Biochim. Biophys. Acta* 1106: 335–343.

Kaminoh, Y., C. Tashiro, H. Kamaya and I. Ueda. 1988. Depression of Phase-Transition Temperature by Anesthetics: Nonzero Solid Membrane Binding. *Biochim. Biophys. Acta* 946: 215–220.

Kaye, M. D., K. Schmalzl, V. Conti-Nibali, M. Tarek, and M. C. Rheinstädter. 2011. Ethanol Enhances Collective Dynamics of Lipid Membranes. *Phys. Rev. E* 83: 050907(R)-1-4.

Koubi, L., M. Tarek, S. Bandyopadhyay, M. L. Klein, and D. Scharf. 2001. Membrane Structural Perturbations Caused by Anesthetics and Nonimmobilizers: A Molecular Dynamics Investigation. *Biophys. J.* 81: 3339–3345.

Koubi, L., M. Tarek, M. L. Klein, and D. Scharf. 2000. Distribution of Halothane in a Dipalmitoylphosphatidylcholine Bilayer from Molecular Dynamics Calculations. *Biophys. J.* 78: 800–811.

Lee, A. G. 1976. Model for Action of Local Anesthetics. *Nature* 262: 545–548.

Lee, B. W., R. Faller, A. K. Sum, I. Vattulainen, M. Patra, and M. Karttunen. 2005. Structural Effects of Small Molecules on Phospholipid Bilayers Investigated by Molecular Simulation. *Fluid Phase Equilibria* 228–229: 135–140.

Lever, M. J., K. W. Miller, W. D. M. Paton, and E. B. Smith. 1971. Pressure Reversal of Anaesthesia. *Nature* 231: 368–371.

Ly, H. V., D. E. Block, and M. L. Longo. 2002. Interfacial Tension Effect of Ethanol on Lipid Bilayer Rigidity, Stability, and Area/Molecule: A Micropipet Aspiration Approach. *Langmuir* 18: 8988–8995.

Ly, H. V. and M. L. Longo. 2004. The Influence of Short-Chain Alcohols on Interfacial Tension, Mechanical Properties, Area/ Molecule,and Permeability of Fluid Lipid Bilayers. *Biophys. J.* 87: 1013–1033.

Marrink, S. J. and H. J. C. Berendsen. 1994. Simulation of Water Transport through a Lipid Membrane. *J. Phys. Chem.* 98: 4155–4168.

Marrink, S. J. and H. J. C. Berendsen. 1996. Permeation Process of Small Molecules across Lipid Membranes Studied by Molecular Dynamics Simulations. *J. Phys. Chem.* 100: 16729–16738.

Marrink, S. J., R. M. Sok, and H. J. C. Berendsen. 1996. Free Volume Properties of a Simulated Lipid Membrane. *J. Chem. Phys.* 104: 9090–9099.

Meyer, H. 1889. Zur Theorie der Alkoholnarkose. *Naunyn-Schmiedebergers Archiv für Experimentelle Pathologie und Pharmakologie* 42: 109–118.

Mihic, S. J.; Q. Ye, M. J. Wick, et al. 1997. Sites of Alcohol and Volatile Anaesthetic Action on GABA$_A$ and Glycine Receptors. *Nature* 389: 385–389.

Mitchell, D. C., J. T. R. Lawrence, and B. J. Litman. 1996. Primary Alcohols Modulate the Activation of the G Protein-Coupled Receptor Rhodopsin by a Lipid-Mediated Mechanism. *J. Biol. Chem.* 271: 19033–19036.

Moskovitz, Y. and H. Yang. 2015. Modelling of Noble Anaesthetic Gases and High Hydrostatic Pressure Effects in Lipid Bilayers. *Soft Matter* 11: 2125–2138.

Mullins, L. J. 1954. Some Physical Mechanisms in Narcosis. *Chem. Rev.* 54: 289–323.

Oh, K. J. and M. L. Klein. 2009. Effects of Halothane on Dimyristoylphosphatidylcholine Lipid Bilayer Structure: A Molecular Dynamics Simulation Study. *Bull. Korean Chem. Soc.* 30: 2087–2092.

Overton, E. 1901. *Studien über die Narkose zugleich ein Beitrag zur allgemeinen Pharmakologie.* Jena: Gustav Fischer Verlag.

Patra, M., E. Salonen, E. Terama, et al. 2006. Under the Influence of Alcohol: The Effect of Ethanol and Methanol on Lipid Bilayers. *Biophys. J.* 90: 1121–1135.

Pickholz, M., L. Saiz, and M. L. Klein. 2005. Concentration Effects of Volatile Anesthetics on the Properties of Model Membranes: A Coarse-Grain Approach. *Biophys. J.* 88: 1524–1534.

Rabinovich, A. L., N. K. Balabaev, M. G. Alinchenko, V. P. Voloshin, N. N. Medvedev, and P. Jedlovszky. 2005. Computer Simulation Study of Intermolecular Voids in Unsaturated Phosphatidylcholine Lipid Bilayers. *J. Chem. Phys.* 122: 084906-1-12.

Reigada, R. 2011. Influence of Chloroform in Liquid-Ordered and Liquid-Disordered Phases in Lipid Membranes. *J. Phys. Chem. B* 115: 2527–2535.

Reigada, R. 2013. Atomistic Study of Lipid Membranes Containing Chloroform: Looking for a Lipid-Mediated Mechanism of Anesthesia. *Plos One* 8: e52631-1–10.

Sega, M., B. Fábián, and P. Jedlovszky. 2016. Pressure Profile Calculation with Mesh Ewald Methods. *J. Chem. Theory Comput.* 12: 4509–4515.

Sierra-Valdez, F. J. and J. C. Ruiz-Suárez. 2013. Noble Gases in Pure Lipid Membranes. *J. Phys. Chem. B* 117: 3167–3172.

Sonne, J., F. Y. Hansen, and G. H. Peters. 2005. Methodological Problems in Pressure Profile Calculations for Lipid Bilayers. *J. Chem. Phys.* 122: 124903-1-9.

Stimson, L. M., Vattulainen, I., Róg, T., and M. Karttunen. 2005. Exploring the Effect of Xenon on Biomembranes. *Cell. Mol. Biol. Letters* 10: 563–569.

Terama, E., O. H. S. Ollila, E. Salonen, et al. 2008. Influence of Ethanol on Lipid Membranes: From Lateral Pressure Profiles to Dynamics and Partitioning. *J. Phys. Chem. B* 112: 4131–4139.

Trudell, J. R. 1977. A Unitary Theory of Anesthesia Based on Lateral Phase Separations in Nerve Membranes. *Anesthesiology* 46: 5–10.

Trudell, J. R., W. L. Hubbell, and E. N. Cohen. 1973. The Effect of Two Inhalation Anesthetics on the Order of Spin-Labelled Phospholipid Vesicles. *Biochim. Biophys. Acta* 291: 321–327.

Trudell, J. R., D. G. Payan, J. H. Chin, and E. N. Cohen. 1975. The Antagonistic Effect of an Inhalation Anesthetic and High Pressure on the Phase Diagram of Mixed Dipalmitoyl-Dimyristoylphosphatidylcholine Bilayers. *Proc. Natl. Acad. Sci. USA* 72: 210–213.

Tu, K., M. Tarek, M. L. Klein, and D. Scharf. 1998. Effects of Anaesthetics on the Structure of a Phospholipid Bilayer: Molecular Dynamics Investigation of Halothane in the Hydrated Liquid Crystal Phase of Dipalmitoylphosphatidylcholine. *Biophys. J.* 75: 2123–2134.

Turkyilmaz, S., P. F. Almeida, and S. L. Regen. 2011. Effects of Isoflurane, Halothane, and Chloroform on the Interactions and Lateral Organization of Lipids in the Liquid-Ordered Phase. *Langmuir* 27: 14380–14385.

Weinrich, M., H. Nanda, D. L. Worcester, C. F. Majkrzak, B. B. Maranville, and S. M. Bezrukov. 2012. Halothane Changes the Domain Structure of a Binary Lipid Membrane. *Langmuir* 28: 4723–4728.

Weinrich, M. and D. L.Worcester. 2013. Xenon and Other Volatile Anesthetics Change Domain Structure in Model Lipid Raft Membranes. *J. Phys. Chem. B* 117: 16141–16147.

Wieteska, J. R., P. R. L. Welche, K. M. Tu, et al. 2015. Isoflurane does not Aggregate Inside POPC Bilayers at High Pressure: Implications for Pressure Reversal of General Anaesthesia. *Chem. Phys. Letters* 638: 116–121.

Yamamoto, E., T. Akimoto, H. Shimizu, Y. Hirano, M. Yasui, and K. Yasuoka. 2012. Diffusive Nature of Xenon Anesthetic Changes Properties of a Lipid Bilayer: Molecular Dynamics Simulations. *J. Phys. Chem. B* 116: 8989–8995.

11

Cation-Mediated Nanodomain Formation in Mixed Lipid Bilayers

Sai J. Ganesan, Hongcheng Xu, and Silvina Matysiak

1 Introduction

Cellular membranes are inhomogeneous in nature and are composed of various kinds of lipids and proteins in interaction with their environments. These interactions give rise to many interesting patterns which not only have unique physicochemical properties, but also biological significance, that we are only beginning to understand [1]. Studies in the last few decades have suggested that cellular membranes are composed of different fluid–fluid phase-separated lipid micro- and nano-domains, formed by both lipid–lipid interaction [2,3] and lipid–ion interaction [4]. These lipid domains give rise to membrane structure regulation. Since distinct cell types have different lipid compositions, investigating dynamic membrane properties of multiple lipid compositions can help us understand many key biological processes including nerve cell signaling, membrane reorganization, and remodeling [5–7].

Ion interactions with biological membranes not only affect structure, dynamics, and stability of bilayers, but also affect protein binding, folding, aggregation, and insertion in a lipid bilayer environment. Most of the biological processes mentioned above are poorly understood at this time. As is widely known, ion interaction with phospholipids is driven mainly by Coulombic forces near the headgroup region and is therefore more pronounced for anionic lipids [8]. Interactions of ions, especially cations with phospholipid membranes, are known to agree with the Hofmeister series based on theoretical and computational predictions, as well as experiments [9–12]. In general, multivalent cations are known to interact more strongly than most monovalent cations, with the exception of lithium ions [13]. In other words, among monovalent metal ions of biological significance, K^+ and Na^+ are known to cause only subtle effects on bilayer structure, stability, and chain packing [13,14]. However, even at low concentrations, Li^+ is found to influence the state of a bilayer composed of anionic lipids phosphatidylserine (PS), by increasing its melting temperature, thus inducing the transition from a liquid-disordered state to a gel state [14–16]. Among multivalent cations, Ca^{2+} is known to have a higher binding affinity to both anionic lipids and saturated zwitterionic lipids than Mg^{2+} and Zn^{2+} [17,18]. Membranes in a more ordered state or gel-like phase are also known to bind more divalent cations than membranes in liquid-disordered state [19]. Li^+, Mg^{2+}, and Ca^{2+} are known to form dehydrated and more ordered ion–PS complexes [16,20], but are hypothesized to differ in lipid binding modes [21].

Results from computational studies on ion–lipid interactions are inconsistent, especially on the ion–lipid binding modes, and are dependent on the Molecular Dynamics (MD) force field used. For example, studies using GROMOS force field with mixed PC and PS bilayers found Na^+ to interact with carboxyl ester groups of POPS lipids [22], and with serine carboxyl and phosphate groups when pure PS lipids were used [23]. A study using AMBER force field for DOPC lipids [24] and other ion-specific parameters for Na^+ and K^+ [25] found the ions to interact preferentially with the phosphate region [12]. Addressing these fundamental discrepancies is a central part of an open-science collaboration project, the NMRlipids project [26]. Since typical atomistic MD force fields use fixed partial charges parameterized in a certain environment and hence are not transferable to other environments, transferable polarizable force fields that capture induced molecular polarization are being developed by various groups [27–30]

Studies in the last few decades reveal that monovalent and divalent ions, along with pH and lipid composition, can influence the phase of lipid membranes [31–33]. In biological systems, where temperature is usually maintained remarkably constant, the lipid phase transition can be initiated in membranes by various factors other than the temperature change. Membrane phase changes induced by lipid–lipid interaction is reasonably well studied compared to phase changes and local/global structural changes induced by ion–lipid interactions. Cholesterol is known to associate with multiple saturated lipids to form liquid-ordered (L_o) domains and induce nano- and micro-domain formation in lipid mixtures, which is largely known to be driven by a tail-induced ordering effect [34–38]. In contrast, the changes induced by ion–lipid interactions occurred in headgroup regions. The re-alignment of the lipid headgroups is precipitated by the binding of ions, which leads to the compression of the lipid tails by lateral pressure, resulting in a more ordered state [39–42]. Consequently, the number and the type of ions present in the environment can influence structural changes in lipid membranes. For example, Ca^{2+} is known to induce phase separation of acidic phospholipids from neutral ones by (a) the presence of individual endothermic peaks in differential scanning calorimetry results [43], (b) changes in mobility of probes in the bilayers [44] and (c) shifts in NMR signals [43,45]. Studies suggest that the presence of Ca^{2+} in mixtures with various saturated zwitterionic lipids can induce crystal-line, dehydrated PS–Ca^{2+} domains [45,46]. These domains have also been structurally charac-terized by techniques ranging from X-ray diffraction to freeze-fracture electron microscopy in unilamellar and multilamellar vesicles [46].

Molecular simulations can provide a mechanistic view of macroscopic experimental observables. Coarse-grained (CG) models, which allows us to explore longer time scales and larger systems, have yielded significant results in our understanding of lipid–lipid interaction or tail-induced phase separation in lipid mixtures [37,47–52]. CG models have also been used to explore the effect of pH on lipid charge, shape, and interactions [53,54]. However, there are not many CG models that are suitable for exploring ion-induced lipid phase separation processes. Nevertheless, further simplistic Monte–Carlo (MC) based on a two-dimensional triangular binary lipid lattice model has been used to show the formation of PC–PS domains as a function of mixing energy, without the physical use of ions or electrostatic interactions in the model [55].

In this chapter, we present a generic Water-Explicit Polarizable MEMbrane (WEPMEM) lipid model for anionic and zwitterionic lipids with the inclusion of dipolar interactions in the lipid headgroup region [56]. The addition of these dipolar particles improves the bilayer potential and dielectric gradient at membrane–water interfaces [56]. Because of these properties of WEPMEM, we have been able to investigate the role of monovalent ions of different sizes and ionic radii on lipid bilayers of different compositions. Our studies, as explained further in this chapter, suggest that cations of particular sizes have the ability to induce nanodomains in lipid bilayers, which can significantly affect structural and organizational properties of the bilayer and hence regulate function of cellular membranes.

2 Methods

To investigate cation-induced lipid domain formation, we use our recently developed WEPMEM model. WEPMEM uses a 4:1 mapping scheme (i.e., four heavy atoms are mapped into one CG bead) consistent with the MARTINI model for POPC (1-palmitoyl-2-oleoyl-*sn*-glycero-3-phosphocholine) and POPS (1-palmitoyl-2-oleoyl-*sn*-glycero-3-phospho-L-serine) [51], as shown in Figure 1a. An environment-sensitive dipole is added into polar CG beads to make the bead polarizable (Figure 1b).

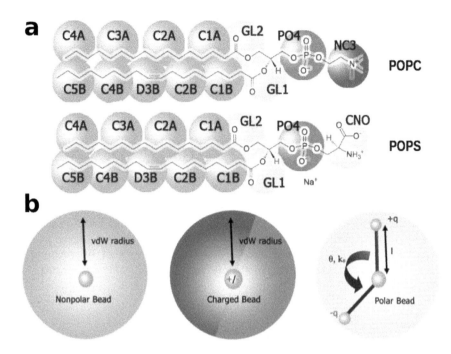

FIGURE 1 Coarse-grained (CG) mapping scheme and bead types in WEPMEM (Water-Explicit Polarizable MEM-brane) model. (a) CG mapping scheme of POPC (upper) and POPS (lower) lipid. POPC lipid consists of choline (NC3), phosphate (PO4), and glycerol (GL1/GL2) groups as headgroups. POPS lipid consists of serine (CNO), phosphate (PO4), and glycerol (GL1/GL2) groups as headgroups. Both lipids have a palmitoyl tail (C1A, C2A, C3A, C4A) and an oleoyl tail (C1B, C2B, D3B, C4B, C5B). (b) The color of beads represents different bead types. Yellow beads are polarizable, green beads are hydrophobic, red beads are negatively charged, and blue beads are positively charged. Reproduced from Ganesan, S.J., Xu, H. and Matysiak, S., Effect of lipid headgroup interactions on membrane properties and membrane-induced cationic β-hairpin folding, Phys. Chem. Chem. Phys., 18, 17836–17850, 2017. With permission from the PCCP Owner Societies. (See color insert for the color version of this figure)

The polarizable beads consist of a pair of oppositely charged dummy particles in addition to a main CG bead. These dummy particles are used to mimic dipoles that can be influenced by an external electric field. A phospholipid molecule is mapped onto a structure consisting of 13 CG sites made of 3 different bead types. Polarizable (Pol), hydrophobic (H), and charged (C) bead types are used to mark the differences in polarity and polarizability of the corresponding chemical groups. The MARTINI CG polarizable water [57] is used with WEPMEM model as explicit solvent. The CG ions are modeled as hydrated ions.

The force field (see Eq. (1)) used in WEPMEM model consists of bonded (harmonic bond and angular potential) and non-bonded terms (Lennard–Jones (LJ) and Coulombic potentials):

$$U_{\text{total}} = U_{\text{bonded}} + U_{\text{non–bonded}}, \tag{1a}$$

$$U_{\text{bonded}} = U_{\text{bonds}} + U_{\text{angles}}, \tag{1b}$$

$$U_{\text{non–bonded}} = U_{\text{vdW}} + U_{\text{electrostatics}}, \tag{1c}$$

$$U_{\text{bonds}} = \frac{1}{2} k_l (l - l_0)^2, \tag{1d}$$

$$U_{\text{angles}} = \frac{1}{2} k_\theta (\theta - \theta_0)^2, \tag{1e}$$

$$U_{\text{vdW}} = 4\varepsilon_{ij} \left[\left(\frac{\sigma_{ij}}{r_{ij}} \right)^{12} - \left(\frac{\sigma_{ij}}{r_{ij}} \right)^{6} \right], \tag{1f}$$

$$U_{\text{electrostatics}} = \frac{q_i q_j}{4\pi\varepsilon_0 \varepsilon_r r_{ij}}, \tag{1g}$$

where U denotes potential energy, k_l and k_θ are the bond and angle strength, l and θ are bond length and angle, l_0 and θ_0 are the equilibrium bond length and angle, ε_{ij} denotes the strength of LJ interaction between two beads, σ_{ij} is the effective size between two beads, and q denotes the charge of bead.

The non-dipolar bonded interactions are based on the MARTINI force field [51,52], and non-bonded interactions are fine-tuned using MARTINI's LJ parameter scale as a baseline. All parameter modifications have been validated with calculations of free energy data and compared to experimental observables. For details regarding the setup of WEPMEM model, refer to our previous papers [56,58].

3 Model Validation

The WEPMEM model improves both structural and dielectric properties of a lipid bilayer in comparison to the MARTINI model. The density distribution of each chemical group in WEPMEM model shows good agreement with the distribution obtained in an atomistic model, as shown in Figure 2. In POPC density distribution, the choline groups have a broader distribution than the distribution of phosphate groups, and are located more outward from the bilayer surface. Although both serine groups in POPS lipids and choline groups in POPC lipids are located at the head of corresponding lipids, serine groups in POPS bend to approach phosphate groups and therefore stay relatively buried in the bilayer surface. In comparison, choline groups in POPC are less bent toward bilayer but rather are more immersed in the solvent. The difference between serine groups and choline groups is in the formation of hydrogen bonds in lipid headgroup region. In the atomistic GROMOS model, serine groups form strong hydrogen bonds with phosphate groups and other serine groups, while choline groups do not. Similarly, in WEPMEM model, serine beads form strong dipole–dipole or dipole–charge interactions with other serine beads and phosphate beads to form a dipole network. The dipole network also condenses the bilayer by around 5 Å^2 compared to MARTINI models. The condensation effect is in agreement with the experimental observation that area per lipid in POPS bilayers is lower than that of POPC bilayers [56]. In summary, the dipole–dipole interactions in lipid headgroup region forms a network to improve lipid bilayer structural properties in WEPMEM model, and the WEPMEM model is more consistent with structural properties in the atomistic model.

In addition, the dielectric properties in the lipid headgroup region are better characterized with the introduction of polarizable dipoles. It has been difficult to capture dielectric properties in the bilayer interfacial region in non-polarizable models such as MARTINI model [51]. MARTINI model is not able to model dipoles that are susceptible to an external electric field in the bilayer interfacial region and therefore leads to an artificial dielectric gradient. To assess the improvement in bilayer dielectric property, susceptibility profile of bilayers along the normal direction was computed on WEPMEM model, MARTINI model, and atomistic GROMOS model. In this analysis, bilayer was divided in slabs of 5Å and the variance of dipoles in polarizable beads was computed for each slab as indicated in Figure 3. In each slab, the computed dielectric constant was underestimated due to its limited sampling size. Nevertheless, using a larger slab means averaging dielectric constant in a broader lipid headgroup region, and inevitably lowers the spatial precision in local dielectric calculation. In

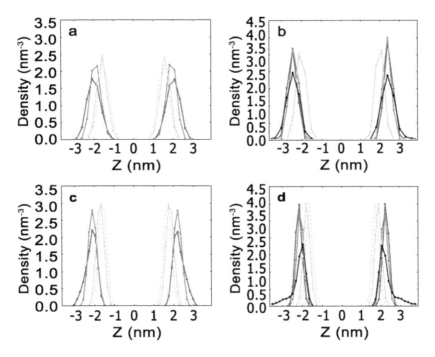

FIGURE 2 Density distribution of POPC and POPS lipids with atomistic GROMOS force field and WEPMEM model: (a) atomistic POPC lipid; (b) WEPMEM POPC lipid; (c) atomistic POPS lipid; (d) WEPMEM POPS lipid. Red: choline/serine group, blue: phosphate group, solid yellow: glycerol group in oleoyl tail, dashed yellow: glycerol group in palmitoyl tail, black: counterion. The center of the bilayer is located at $z = 0$. Reproduced from Ganesan, S.J., Xu, H. and Matysiak, S., Effect of lipid headgroup interactions on membrane properties and membrane-induced cationic β-hairpin folding, Phys. Chem. Chem. Phys., 18, 17836–17850, 2017. With permission from the PCCP Owner Societies. (See color insert for the color version of this figure)

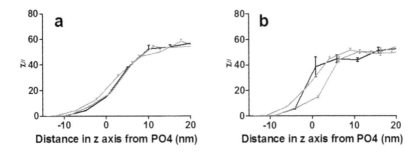

FIGURE 3 Susceptibility profile of (a) POPC and (b) POPS bilayer. Black curves are calculated from WEPMEM model, gray curves are from MARTINI model, and red curves are from GROMOS atomistic force field. Reproduced from Ganesan, S.J., Xu, H. and Matysiak, S., Effect of lipid headgroup interactions on membrane properties and membrane-induced cationic β-hairpin folding, Phys. Chem. Chem. Phys., 18, 17836–17850, 2017. With permission from the PCCP Owner Societies. (See color insert for the color version of this figure)

addition, further analysis indicates the dielectric constant of bulk water eventually converges to experimentally observed value (~80) as we expand slab size in the analysis. A gradient of dielectric constant from the bilayer center to bulk water was found in both POPC and POPS lipid bilayers. In the POPS bilayer, a region with high dielectric constant in the membrane–water interface was observed with WEPMEM model, as in good agreement with atomistic model. To the contrary, the MARTINI POPS has a much lower dielectric constant in membrane–water interface (Figure 3b). The presence of Drude-like dipoles susceptible to external electric field gives rise to the dielectric constant in the interfacial region. In experiments, the presence of a highly susceptible interfacial region in POPS was also observed in scattering experiments by Cheng *et al.* [59]. These authors found that water molecules near POPS multilamellar vesicles are more polarized in the interfacial region than those near POPC vesicles. In addition, incorporating dipoles into lipid headgroups also improves the agreement of bilayer dipole potential between WEPMEM model and atomistic model [56].

4 Results and Discussions

4.1 Effect of Monovalent Cations on Nanodomain Formation

Experimentally, Li^+ was found to influence the state of PS bilayer similarly as Ca^{2+} and Mg^{2+} [14–16]. However, other monovalent cations such as Na^+ and K^+ only cause minor changes to bilayers even with high concentrations [14]. It is not clear why these monovalent cations (Li^+, Na^+, K^+) have so distinctive effects on lipid bilayers. To investigate the effect of monovalent ion size on bilayers consisting of anionic lipid PS, we performed simulations of 1:1 mixed 240 POPC/POPS lipids bilayer, and 1:1 mixed 960 POPC/POPS lipids bilayer with monovalent cation sizes of 1.0σ, 0.9σ, 0.8σ, and 0.7σ. The size σ, defined as 4.7Å, was used for all the other CG beads in WEPMEM model. To quantify the formation of POPS lipid nanodomains, we first performed a cluster analysis to these lipids in the small system with 240 lipids. The time evolution of the largest cluster size was then computed to show the growth of clusters over time in Figure 4. Most

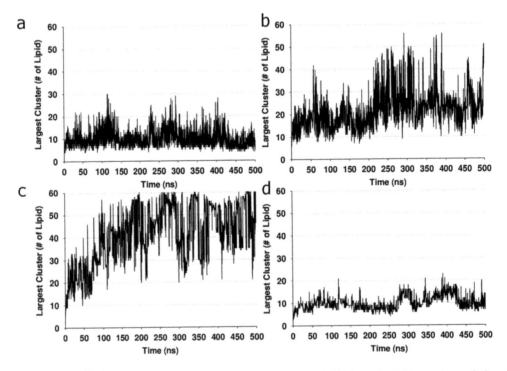

FIGURE 4 Effect of 0.4 M M^+ ion size on the 240–lipid 1:1 POPC/POPS lipid bilayer. (a–d) Cluster time evolution of POPS lipid clusters with 1.0 σ, 0.9σ, 0.8σ, and 0.7σ ions, respectively. Data obtained from the clustering method based on ion–lipid interaction cutoff distances. Reprinted with permission from Ganesan, S.J., Xu, H., and Matysiak, S., Influence of monovalent cation size on nanodomain formation in anionic—zwitterionic mixed bilayers, *J. Phys. Chem. B*, 121, 787–799. Copyright 2017 American Chemical Society.

clusters cease growing at a size of 20 lipids when ion size is either too large (1.0σ) or too small (0.7σ). Larger domains of lipids were only observed with ion sizes of 0.8σ or 0.9σ. Also, the cluster growth is the most evident and fastest in the presence of 0.8σ ions. The size of largest lipid nanodomain also fluctuates considerably due to the dynamics nature of clusters, which is attributed to the re-alignment of ion–lipid interactions. The POPS lipids in the mixed bilayer form nanodomains through direct interactions between headgroups, and indirect interactions mediated by monovalent cations. The finite size effect in the simulation of lipid nanodomain formation was also investigated by comparing lipid bilayer simulations with the same lipid composition but 4 times the number of lipids. In the larger bilayer system of 960 lipids, most POPS lipid cluster sizes are under 10 lipids when the cation size is 1.0σ. However, when the cation size is 0.8σ, the lipid nanodomains grow up to a size of 60 lipids and the growth keeps evolving over time. Evidently, the cation size of 0.8σ induces the formation of larger and more stable POPS lipid nanodomains than the other cation sizes.

4.2 Membrane Curvature Induced by Nanodomain Formation

Apart from the lipid aggregation in planar bilayer surface induced by cation binding, we also observed structural changes of lipid bilayer in a lateral perspective. In Figure 5, with the ion size of 0.8σ, the binding of ions with POPS lipid headgroups induces a significant curvature. A distinct curvature is not evident when the ion size is 1.0σ. Moreover, in Figure 5b, the POPS-rich domains (red region) always localize to regions with a negative curvature, while the POPC-rich domains always adopt a positive curvature. There is also compositional asymmetry of lipid distribution across the bilayer as evident from Figure 5b. It is more likely to find POPC-rich domain in the opposite leaflet when there is a POPS-rich domain at the same location, and *vice versa*. In Figure 5b, most cations (represented as yellow beads) are located in POPS-rich domain, indicating a significant role played by cations in the POPS lipid nanodomain formation.

To quantify the bilayer curvature and lipid compositional asymmetry in the bilayer with ion size of 0.8σ, we performed Voronoi tessellation to all the lipids in each bilayer leaflet. Each Voronoi cell corresponds to one lipid and reflects the planar area each corresponding lipid

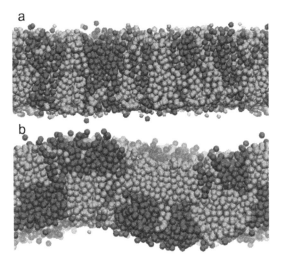

FIGURE 5 Side view snapshots of 1:1 mixed 960 POPC/POPS lipids bilayer demonstrating the difference in bilayer curvature. (a) The bilayer with ion size of $1.0\ \sigma$; (b) The bilayer with ion size of 0.8σ. Blue regions represent POPC lipids, red regions represent POPS lipids, and small yellow beads are monovalent ions. Reprinted with permission from Ganesan, S.J., Xu, H., and Matysiak, S., Influence of monovalent cation size on nanodomain formation in anionic—zwitterionic mixed bilayers, *J. Phys. Chem. B*, 121, 787–799. Copyright 2017 American Chemical Society. (See color insert for the color version of this figure)

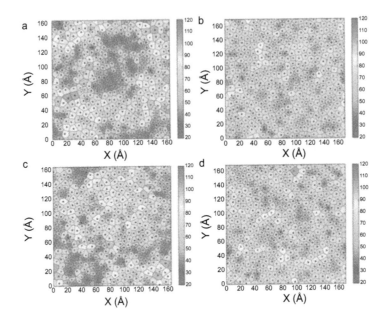

FIGURE 6 Voronoi diagram of 1:1 mixed 960 POPC/POPS lipids bilayer showing the area of each lipid. (a) The lipid bilayer upper leaflet with ion size of 0.8 σ; (b) The lipid bilayer lower leaflet with ion size of 0.8σ; (c) The lipid bilayer upper leaflet with ion size of 1.0σ; (d) The lipid bilayer lower leaflet with ion size of 1.0σ. Each Voronoi cell corresponds to a single lipid shown in red (POPS lipid) or blue (POPC lipid) dots. Color bar represents the area of each lipid in Å^2. Reprinted with permission from Ganesan, S.J., Xu, H., and Matysiak, S., Influence of monovalent cation size on nanodomain formation in anionic–zwitterionic mixed bilayers, *J. Phys. Chem. B*, 121, 787–799. Copyright 2017 American Chemical Society. (See color insert for the color version of this figure)

occupies in the leaflet. By comparing the opposite leaflets in the same bilayer (Figures 6a and 6c), it is evident that POPS-rich domain is located opposite to POPC-rich domain in the other leaflet. Our correlation analysis indicates that the presence of PS lipid in one leaflet is negatively correlated with the presence of PS lipid at the same location in the opposite leaflet (Pearson correlation coefficient $r = -0.46 \pm 0.07$). In addition to the bilayer curvature and compositional asymmetry, there is also a condensation in the area of PS lipid induced by small monovalent cations (0.8σ). The condensation effect is always associated with POPS-rich domain and appears complementary in both leaflets. In POPC-rich domains, the area of each POPC lipid is usually greater than average area per lipid in the bilayer.

A typical snapshot of cation–POPS complex is shown in Figure 7 to demonstrate its cone-like geometry induced by ion binding. The headgroup region of the complex is condensed by strong interactions mediated by cations. The interactions are between ions and serine/phosphate headgroups. The tails of the complex are tilted outwards. The cone-like geometry induced by cations gives rise to the formation of negative curvature in the bilayer with small cations (0.8σ).

Schematic images in Figure 8 are proposed to explain the alignment of positive and negative curvature in both leaflets across the bilayer. The presence of negative curvature at the same location in both leaflets will result in the formation of voids, which is entropically unfavorable. A better alignment is formed for lipids by having positive curvature on one side and negative curvature on the other side.

4.3 Thermodynamics of Nanodomain Formation

To understand the molecular mechanism behind the cation-induced POPS lipid domain formation, we investigated ion coordination number around POPS lipid headgroups, ion partition, and pairwise interactions between ion and lipid headgroups. As shown in Figure 9a, b, cations

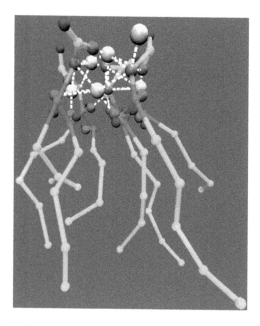

FIGURE 7 A typical snapshot of cation–PS complex demonstrating interactions in POPS lipid headgroup. Yellow beads are monovalent cations, tan beads are phosphate groups, and magenta beads are glycerol groups. Drude-like dummy particles with positive charges are shown in red, and with negative charges shown in blue. The white dashed lines indicate the binding of monovalent cations to lipid headgroups. Reprinted with permission from Ganesan, S.J., Xu, H., and Matysiak, S., Influence of monovalent cation size on nanodomain formation in anionic—zwitterionic mixed bilayers, *J. Phys. Chem. B*, 121, 787–799. Copyright 2017 American Chemical Society. (See color insert for the color version of this figure)

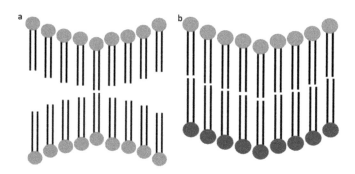

FIGURE 8 Schematic images demonstrating the necessity of adopting opposite curvature in both leaflets of lipid bilayer. (a) Lipid bilayer with both leaflets in negative curvature; (b) Lipid bilayer with upper leaflet in negative curvature and lower leaflet in positive curvature. Lipids in leaflet with positive curvature are shown in black, and in leaflet with negative curvature shown in red. Reprinted with permission from Ganesan, S.J., Xu, H., and Matysiak, S., Influence of monovalent cation size on nanodomain formation in anionic—zwitterionic mixed bilayers, *J. Phys. Chem. B*, 121, 787–799. Copyright 2017 American Chemical Society.

with the size of 0.8σ have the highest coordination number around phosphate and serine groups. To the contrary, cations with the size of 1.0σ have the lowest coordination number. Figure 9c shows the transfer free energy of cations from bulk water to the bilayer. Ions with a size of 0.8σ have the lowest free energy of partitioning into the bilayer, while the partition of 1.0σ ions is not as favorable.

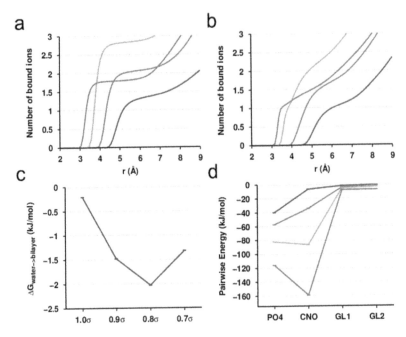

FIGURE 9 Ion coordination number near lipid headgroups, ion transfer free energy between water and bilayer interface, and ion-to-lipid headgroup beads pairwise interaction energy. (a) Ion coordination number near lipid phosphate group; (b) ion coordination number near lipid serine group; (c) ion transfer free energy between water and bilayer interface; (d) ion-to-lipid headgroup beads (PO4: phosphate, CNO: serine, GL1/GL2: glycerol groups) pairwise interaction energy. Color scheme in a, b, and d: blue curves are for ion size of 1.0 σ, red curves for ion size of 0.9σ, green curves for ion size of 0.8σ, and magenta curves for ion size of 0.7σ. Reprinted with permission from Ganesan, S. J., Xu, H., and Matysiak, S., Influence of monovalent cation size on nanodomain formation in anionic—zwitterionic mixed bilayers, *J. Phys. Chem. B*, 121, 787–799. Copyright 2017 American Chemical Society. (See color insert for the color version of this figure)

Ions of size 0.7σ and 0.9σ have similar lipid coordination number and transfer free energy. However, ions with size 0.7σ give rise to smaller clusters in comparison with ions of size 0.9σ. The distinction between 0.7σ cations and 0.9σ cations can be explained by Figure 9d. Cations of size 0.7σ (magenta curve) overwhelmingly interact with serine groups, thus hindering lipid–lipid interactions mediated both by serine–serine and serine–phosphate interactions. On the other hand, ion–phosphate and ion–serine interactions are proportional in the bilayer system with 0.8σ, and therefore the ions are able to bridge more lipids together using two distinct, equally stabilizing binding modes.

5 Conclusion

With our recently developed WEPMEM model [56], we were able to capture many membrane properties with improved accuracy and efficiency. The density distribution of lipid bilayers resembles that of all-atom simulations, and the dielectric profile along the bilayer normal axis is consistent with all-atom simulations. With a CG representation, larger length- and time-scales are possible, and enabled the exploration of lipid nanodomain formation. The addition of a dipole moment into polar beads adds structural polarization to our model. These dipoles interact with charged beads and other dipoles to mimic many dipole–dipole interactions such as hydrogen bonds. The dipole network formed by POPS lipid headgroups was found to condense the bilayer and is in a good agreement with atomistic model in structural properties.

The importance of having dipoles in a CG lipid model was clearly demonstrated as it facilitates in capturing the size effect of monovalent cations. The interactions between lipid headgroups and ions are mostly electrostatic. Therefore, the explicit dipoles in polar serine beads are more realistic than modeling ion–lipid interaction only by van der Waals interactions. As a result, we successfully captured the size-dependent effect of monovalent cations in inducing the growth of anionic lipid nanodomains. We have found that anionic lipid nanodomain formation was hindered when the cations size is either too large or too small. When cations of appropriate size were added into the anionic/zwitterionic mixed bilayer, they not only induce anionic lipid aggregation in a planar surface, but also give rise to spontaneous lateral curvature. In addition, we analyzed the thermodynamics of cations with different sizes in interacting with lipid headgroups. We found that there was a window of sizes where the ions had a balanced interaction with both serine and phosphate groups. If the size of ions is larger, they interact more with phosphate group, and if the size of ions is smaller, they interact more with serine groups. When the cation size decreases, the overall electrostatic interactions increase between cations and lipid headgroups. Moreover, the interactions between cations and lipid headgroup are maximized when the balance between serine–cation interactions and phosphate–cation interactions is achieved. When the size of cations is 0.7σ, the cations interact too strongly with serine and when the size of cations is 1.0σ, the interactions between cations and lipid headgroups are too weak. The inserted Drude-like dipole in lipid headgroups is the key to capturing the cation size-dependent binding preference.

REFERENCES

[1] Donald M. Engelman. Membranes are more mosaic than fluid. *Nature*, 438(7068):578–580, 2005.

[2] Saame Raza Shaikh, Alfred C. Dumaual, Laura J. Jenski, and William Stillwell. Lipid phase separation in phospholipid bilayers and monolayers modeling the plasma membrane. *Biochim. Biophys. Acta, Biomembr.*, 1512(2):317–328, 2001.

[3] Elliot L. Elson, Eliot Fried, John E. Dolbow, and Guy M. Genin. Phase separation in biological membranes: integration of theory and experiment. *Annu. Rev. Biophys.*, 39:207, 2010.

[4] Joachim Seeligt, Renate Lehrmann, and Evelyne Terzi. Domain formation induced by lipid-ion and lipid-peptide interactions. *Mol. Membr. Biol.*, 12(1):51–57, 1995. doi: 10.3109/09687689509038495.

[5] S. Sonnino and A. Prinetti. Membrane domains and the "lipid raft"concept. *Curr. Med. Chem.*, 20(1):4–21, 2013.

[6] Helgi I. Ingolfsson, Peter Tieleman, and Siewert Marrink. Lipid organization of the plasma membrane. *Biophys. J.*, 108(2):358a, 2015.

[7] Mary L. Kraft. Plasma membrane organization and function: moving past lipid rafts. *Mol. Biol. Cell*, 24(18):2765–2768, 2013.

[8] Hans Binder and Olaf Zschörnig. The effect of metal cations on the phase behavior and hydration characteristics of phospholipid membranes. *Chem. Phys. Lipids*, 115(1):39–61, 2002.

[9] Benjamin Klasczyk, Volker Knecht, Reinhard Lipowsky, and Rumiana Dimova. Interactions of alkali metal chlorides with phosphatidylcholine vesicles. *Langmuir*, 26(24):18951–18958, 2010. doi: 10.1021/la103631y. PMID: 21114263.

[10] Frédéric F. Harb and Bernard Tinland. Effect of ionic strength on dynamics of supported phosphatidylcholine lipid bilayer revealed by frapp and langmuir—blodgett transfer ratios. *Langmuir*, 29(18):5540–5546, 2013. doi: 10.1021/la304962n. PMID: 23581462.

[11] Georg Pabst, Aden Hodzic, Janez Štrancar, Sabine Danner, Michael Rappolt, and Peter Laggner. Rigidification of neutral lipid bilayers in the presence of salts. *Biophys. J.*, 93(8):2688–2696, 2007. ISSN 0006-3495. doi: https://doi.org/10.1529/biophysj.107.112615.

[12] Robert Vajcha, Shirley W. I. Siu, Michal Petrov, Rainer A. Böckmann, Justyna Barucha-Kraszewska, Piotr Jurkiewicz, Martin Hof, Max L. Berkowitz, and Pavel Jungwirth. Effects of alkali cations and halide anions on the dopc lipid membrane. *J. Phys. Chem. A*, 113(26):7235–7243, 2009. doi: 10.1021/jp809974e. PMID: 19290591.

[13] Andrea Catte, Mykhailo Girych, Matti Javanainen, Claire Loison, Josef Melcr, Markus S. Miettinen, Luca Monticelli, Jukka Määttä, Vasily S. Oganesyan, OH Samuli Ollila, et al. Molecular electrometer and binding of cations to phospholipid bilayers. *Phys. Chem. Chem. Phys.*, 18(47):32560–32569, 2016.

[14] Helmut Hauser and G. Graham Shipley. Interactions of monovalent cations with phosphatidylserine bilayer membranes. *Biochemistry*, 22(9):2171–2178, 1983.

[15] H. Hauser and G. Graham Shipley. Crystallization of phosphatidylserine bilayers induced by lithium. *J. Biol. Chem.*, 256(22):11377–11380, 1981.

[16] HL Casal, HH Mantsch, and H. Hauser. Infrared studies of fully hydrated saturated phosphatidylserine bilayers. Effect of lithium and calcium. *Biochemistry*, 26(14):4408–4416, 1987.

[17] Or Szekely, Ariel Steiner, Pablo Szekely, Einav Amit, Roi Asor, Carmen Tamburu, and Uri Raviv. The structure of ions and zwitterionic lipids regulates the charge of dipolar membranes. *Langmuir*, 27(12):7419–7438, 2011.

[18] Daniel Huster, Klaus Arnold, and Klaus Gawrisch. Strength of Ca^{2+} binding to retinal lipid membranes: consequences for lipid organization. *Biophys. J.*, 78(6):3011–3018, 2000.

[19] Koichi Satoh. Determination of binding constants of Ca^{2+}, Na^{+}, and Cl^{-} ions to liposomal membranes of dipalmitoylphosphatidylcholine at gel phase by particle electrophoresis. *Biochim. Biophys. Acta, Biomembr.*, 1239(2):239–248, 1995.

[20] Jairajh Mattai, Helmut Hauser, Rudy A. Demel, and G. Graham Shipley. Interactions of metal ions with phosphatidylserine bilayer membranes: effect of hydrocarbon chain unsaturation. *Biochemistry*, 28(5):2322–2330, 1989.

[21] Michel Roux and Myer Bloom. Calcium, magnesium, lithium, sodium, and potassium distributions in the headgroup region of binary membranes of phosphatidylcholine and phosphatidylserine as seen by deuterium nmr. *Biochemistry*, 29(30):7077–7089, 1990.

[22] Piotr Jurkiewicz, Lukasz Cwiklik, Alžběta Vojtíšková, Pavel Jungwirth, and Martin Hof. Structure, dynamics, and hydration of POPC/POPS bilayers suspended in NaCl, KCl, and CsCl solutions. *Biochim. Biophys. Acta, Biomembr.*, 1818(3):609–616, 2012. ISSN 00052736. doi: 10.1016/j.bbamem.2011.11.033.

[23] Anton A. Polyansky, Pavel E. Volynsky, Dmitry E. Nolde, Alexander S. Arseniev, and Roman G. Efremov. Role of lipid charge in organization of water/lipid bilayer interface: insights via computer simulations. *J. Phys. Chem. B*, 109(31):15052–15059, 2005.

[24] Shirley WI Siu, Robert Vácha, Pavel Jungwirth, and Rainer A. Böckmann. Biomolecular simulations of membranes: physical properties from different force fields. *J. Chem. Phys.*, 128(12):125103, 2008.

[25] Liem X. Dang, Gregory K. Schenter, Vassiliki-Alexandra Glezakou, and John L. Fulton. Molecular simulation analysis and X-ray absorption measurement of Ca^{2+}, K^{+} and Cl^{-} ions in solution. *J. Phys. Chem. B*, 110(47):23644–23654, 2006.

[26] The NMRlipids project. URL http://nmrlipids.blogspot.com/. accessed Dec 10, 2016.

[27] Guillaume Lamoureux and Benoît Roux. Absolute hydration free energy scale for alkali and halide ions established from simulations with a polarizable force field. *J. Phys. Chem. B*, 110(7):3308–3322, 2006. ISSN 15206106. doi: 10.1021/jp056043p.

[28] Denis Bucher, Leonardo Guidoni, Patrick Maurer, and Ursula Rothlisberger. Developing improved charge sets for the modeling of the KcsA K^{+} channel using QM/MM electrostatic potentials. *J. Chem. Theory Comput.*, 5(8):2173–2179, 2009. ISSN 15499618. doi: 10.1021/ct9001619.

[29] Toby W. Allen, Olaf S. Andersen, and Benoit Roux. Ion permeation through a narrow channel: using gramicidin to ascertain all-atom molecular dynamics potential of mean force methodology and biomolecular force fields. *Biophys. J.*, 90(10):3447–3468, 2006. ISSN 0006-3495. doi: 10.1529/biophysj.105.077073.

[30] Janamejaya Chowdhary, Edward Harder, Pedro EM Lopes, Lei Huang, Alexander D. MacKerell Jr, and Beñiot Roux. A polarizable force field of dipalmitoylphosphatidylcholine based on the classical drude model for molecular dynamics simulations of lipids. *J. Phys. Chem. B*, 117(31):9142–9160, 2013.

[31] Hermann Träuble and Hansjörg Eibl. Electrostatic effects on lipid phase transitions: membrane structure and ionic environment. *Proceedings of the National Academy of Sciences*, 71(1):214–219, 1974.

[32] Paulo FF Almeida, Winchil LC Vaz, and TE Thompson. Lateral diffusion in the liquid phases of dimyristoylphosphatidylcholine/cholesterol lipid bilayers: a free volume analysis. *Biochemistry*, 31(29):6739–6747, 1992.

[33] Derek Marsh. Cholesterol-induced fluid membrane domains: a compendium of lipid-raft ternary phase diagrams. *Biochim. Biophys. Acta, Biomembr.*, 1788(10):2114–2123, 2009.

[34] Erwin London. Insights into lipid raft structure and formation from experiments in model membranes. *Curr. Opin. Struct. Biol.*, 12(4):480–486, 2002.

[35] H. Jelger Risselada and Siewert J. Marrink. The molecular face of lipid rafts in model membranes. *Proc. Natl. Acad. Sci. U. S. A.*, 105(45):17367–17372, 2008.

[36] Jiang Zhao, Jing Wu, Frederick A. Heberle, Thalia T. Mills, Paul Klawitter, Grace Huang, Greg Costanza, and Gerald W. Feigenson. Phase studies of model biomembranes: complex behavior of DSPC/DOPC/cholesterol. *Biochim. Biophys. Acta, Biomembr.*, 1768(11):2764–2776, 2007.

[37] Jason D. Perlmutter and Jonathan N. Sachs. Interleaflet interaction and asymmetry in phase separated lipid bilayers: molecular dynamics simulations. *J. Am. Chem. Soc.*, 133(17):6563–6577, 2011.

[38] Davit Hakobyan and Andreas Heuer. Phase separation in a lipid/cholesterol system: comparison of coarse-grained and united-atom simulations. *J. Phys. Chem. B*, 117(14):3841–3851, 2013.

[39] Hideo Akutsu and Joachim Seelig. Interaction of metal ions with phosphatidylcholine bilayer membranes. *Biochemistry*, 20(26):7366–7373, 1981.

[40] Shuji Aruga, Ryoichi Kataoka, and Shigeki Mitaku. Interaction between Ca^{2+} and dipalmitoylphosphatidylcholine membranes: I. Transition anomalies of ultrasonic properties. *Biophys. chem.*, 21 (3–4):265–275, 1985.

[41] Jan Westman and LE Göran Eriksson. The interaction of various lanthanide ions and some anions with phosphatidylcholine vesicle membranes a 31p nmr study of the surface potential effects. *Biochim. Biophys. Acta, Biomembr.*, 557(1):62–78, 1979.

[42] Ryoichi Kataoka, Shuji Aruga, Shigeki Mitaku, Kazuhiko Kinosita, and Akira Ikegami. Interaction between Ca^{2+} and dipalmitoylphosphatidylcholine membranes: II. Fluorescence anisotropy study. *Biophys. chem.*, 21(3–4):277–284, 1985.

[43] SW Hui, LT Boni, TP Stewart, and T. Isac. Identification of phosphatidylserine and phosphatidylcholine in calcium-induced phase separated domains. *Biochemistry*, 22(14):3511–3516, 1983.

[44] K. Florinecasteel and GW Feigenson. Partition of headgroup-labeled fluorescent lipids between coexisting gel and liquid-crystal phases in multilamellar phospholipid-vesicles. *Biophys. J.*, 53: A256–A256, 1988.

[45] CPS Tilcock, MB Bally, SB Farren, PR Cullis, and SM Gruner. Cation-dependent segregation phenomena and phase behavior in model membrane systems containing phosphatidylserine: influence of cholesterol and acyl chain composition. *Biochemistry*, 23(12):2696–2703, 1984.

[46] Jens R. Coorssen and R. Peter Rand. Structural effects of neutral lipids on divalent cation-induced interactions of phosphatidylserine-containing bilayers. *Biophys. J.*, 68(3):1009–1018, 1995.

[47] Mark J. Stevens. Coarse-grained simulations of lipid bilayers. *J. Chem. Phys.*, 121(23):11942–11948, 2004. ISSN 00219606. doi: 10.1063/1.1814058.

[48] Zun-Jing Wang and Markus Deserno. Systematic implicit solvent coarse-graining of bilayer membranes: lipid and phase transferability of the force field. *New J. Phys.*, 12(9):095004, 2010.

[49] Emily M. Curtis and Carol K. Hall. Molecular dynamics simulations of dppc bilayers using "lime", a new coarse-grained model. *J. Phys. Chem. B*, 117(17):5019–5030, 2013.

[50] Carlos F. Lopez, Steven O. Nielsen, Goundla Srinivas, William F. DeGrado, and Michael L. Klein. Probing membrane insertion activity of antimicrobial polymers via coarse-grain molecular dynamics. *J. Chem. Theory Comput.*, 2(3):649–655, 2006. ISSN 15499618. doi: 10.1021/ct050298p.

[51] Siewert J. Marrink, H. Jelger Risselada, Serge Yefimov, D. Peter Tieleman, and Alex H. De Vries. The MARTINI force field: coarse grained model for biomolecular simulations. *J. Phys. Chem. B*, 111(27):7812–7824, 2007. ISSN 15206106. doi: 10.1021/jp071097f.

[52] Luca Monticelli, Senthil K. Kandasamy, Xavier Periole, Ronald G. Larson, D. Peter Tieleman, and Siewart-Jan Marrink. The MARTINI coarse grained force field: extension to proteins. *J. Chem. Theory Comput.*, 4:819–834, 2008.

[53] Emily M. Curtis, Xingqing Xiao, Stavroula Sofou, and Carol K Hall. Phase separation behavior of mixed lipid systems at neutral and low pH: coarse-grained simulations with DMD/LIME. *Langmuir*, 31(3):1086–1094, 2015.

[54] Martin Dahlberg and Arnold Maliniak. Mechanical properties of coarse-grained bilayers formed by cardiolipin and zwitterionic lipids. *J. Chem. Theory Comput.*, 6(5):1638–1649, 2010.

[55] Juyang Huang and Gerald W Feigenson. Monte carlo simulation of lipid mixtures: finding phase separation. *Biophys. J.*, 65(5):1788, 1993.

[56] Sai J. Ganesan, Hongcheng Xu, and Silvina Matysiak. Effect of lipid head group interactions on membrane properties and membrane-induced cationic β-hairpin folding. *Phys. Chem. Chem. Phys.*, 18(27):17836–17850, 2016.

[57] Semen O. Yesylevskyy, Lars V. Schäfer, Durba Sengupta, and Siewert J. Marrink. Polarizable water model for the coarse-grained MARTINI force field. *PLoS Comput. Biol.*, 6(6):e1000810, June 2010. ISSN 1553-7358. doi: 10.1371/journal.pcbi.1000810.

[58] Sai J. Ganesan, Hongcheng Xu, and Silvina Matysiak. Influence of monovalent cation size on nanodomain formation in anionic—zwitterionic mixed bilayers. *J. Phys. Chem. B*, 121(4):787–799, 2017.

[59] Ji-Xin Cheng, Sophie Pautot, David A. Weitz, and X. Sunney Xie. Ordering of water molecules between phospholipid bilayers visualized by coherent anti—stokes raman scattering microscopy. *Proc. Natl. Acad. Sci. U. S. A.*, 100(17):9826–9830, 2003.

12

Molecular Dynamics Simulations of Gram-Negative Bacterial Membranes

Syma Khalid, Graham Saunders, and Taylor Haynes

All bacterial cells are surrounded by cell membranes. In Gram-negative bacteria, the cell envelope comprises two membranes separated by the periplasmic space which contains a thin cell wall (Figure 1). In contrast, Gram-positive bacteria are surrounded by a single cell membrane which is separated from the external environment by a thicker cell wall. The membranes provide protection by controlling the movement of chemicals between the cell and the external environment and by providing a physical barrier that mechanically encloses the internal contents of the cells from the environment.[1] Proteins that are embedded within cell membranes perform a range of functions including cell signaling, enabling passive and active movement of molecules into and out of the cell, catalysis, adhesion, and provision of mechanical strength. While the importance of the heterogeneity in the topology, size, and chemistry of the different proteins is straightforward to rationalize, they perform different functions, and thus are adapted accordingly; the heterogeneity of the lipid components of the membranes is more difficult to understand. Indeed, for many years, lipids were considered to be inert bystanders that simply provide a hydrophobic environment in which the proteins can function. Over recent years, it has become apparent that the lipids within the membranes have important roles in modulating (i) the activity of membrane proteins, (ii) membrane curvature, and (iii) membrane thickness. Consequently, building accurate computational models of these lipids is imperative if we are to gain a molecular-level understanding of the biological importance of the chemical composition of lipids that comprise biological membranes. In the following chapter, we review progress in molecular simulations of the outer membranes of Gram-negative bacteria over the past ten years or so, with a focus on asymmetric membrane models.

The outer membrane of Gram-negative bacteria is the most complex of the bacterial membranes in terms of structure and composition. The lipid composition is asymmetric; the outer leaflet is composed of lipopolysaccharide (LPS) molecules whereas the inner leaflet contains a mixture of phospholipids. Given this chemical complexity, and the generally accepted view within the membrane biophysics community for many years that the specific molecular details of the lipids within these membranes was not so important, it is not surprising that these membranes were for many years modeled as a simple bilayer of a single phospholipid species. The specific chemical structure of LPS is dependent upon the bacterial species, strain, and even stage of the lifecycle. In general, it consists of lipid A molecule, a core oligosaccharide with an inner and outer core region, and an O-specific polysaccharide antigen chain. The lipid A component comprises a glucosamine dimer bound to lipid acyl chains that are non-covalently anchored into the hydrophobic region of the outer membrane. Covalently linked to lipid A is the core portion of the molecule, which can be further divided into the inner (Re) and outer (Ra) cores. The inner core is directly attached to lipid A and contains a high proportion of rare sugars such as 2-keto-3-deoxyoctulosonate (Kdo) and L-glycero-D-manno-heptose.[2] In contrast, the outer core is composed of sugars such as hexoses and hexosamines, which are more generally ubiquitous in biology. One of the first atomistic models of LPS to be used in simulation studies was reported by Lins and Straatsma.[3] This was the rough LPS from *P. aeruginosa*, consisting of lipid A (five hydrocarbon tails) and core region. Although these simulations were rather

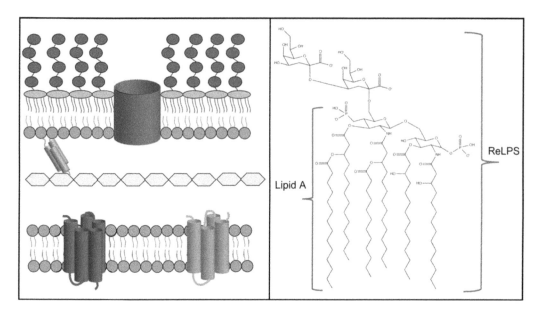

FIGURE 1 A schematic diagram of the cell envelope of Gram-negative bacteria (left). The LPS molecules are purple, blue, and dark green, phospholipids are paler green, peptidoglycan is yellow, outer membrane protein is magenta, inner membrane proteins are orange and dark yellow, and Braun's lipoprotein is blue. The chemical structure of Re-LPS from E. coli (right). (See color insert for the color version of this figure)

short (1 ns), they did nevertheless yield insights into the importance of divalent cations. The simulations were performed using 116 Ca^{2+} ions to neutralize the charge on the 16 LPS molecules. The ions were found to be confined to a ~2-nm-thick layer in the inner core region of the LPS membrane where they interacted predominately with the phosphorylated sugar moieties. The authors commented that phosphate groups are able to trap the ions, and this observation was later also reported by much longer simulations using different force-fields, showing the validity of this initial finding. This model was subsequently used in simulations of a homology model of the protein OprF, eight years later.[4] Indeed, interestingly, it was some years after the Lins and Straatsma model was reported that interest in LPS was rekindled from a simulation perspective. Perhaps in part due to the healthcare problems associated with bacterial resistance to antibiotics becoming the focus of much media attention at this time, in 2011, the first atomistic simulations of *E.coli* LPS were reported by Khalid and co-workers.[5] The asymmetric membrane model was described by the GROMOS53A6 force field, was composed entirely of Rd-LPS molecules in the outer layer, which comprises lipid A (six hydrocarbon tails) and the inner core of the LPS, and a mixture of phosphatidylethanolamine (90%), phosphatidylglycerol (5%), and cardiolipin (5%) lipids in the inner leaflet.[6] A CHARMM version of the *E. coli* force field was reported by Im and co-workers in 2013; this model had a similar inner leaflet composition to the Khalid model, but also contained the O-antigen region in the outer leaflet.[7] Given the widespread use of the GROMOS and CHARMM force-fields within the biomolecular simulation community and their compatibility with freely available simulation codes, the models of Khalid and Im brought LPS-containing membranes into the mainstream of membrane simulations. Since 2013, a few more variants of LPS models have been reported.[8] Coarse-grain models of LPS have also emerged in the last five years or so.[9,10] As with all coarse-grain models, the fewer interaction sites of these models in comparison to their atomistic/united atom counterparts enables simulations of larger systems on longer timescales, at the expense of some fine-grained detail. In the remainder of this chapter, rather than a historical survey of the development of various models, we focus on key findings from simulations of complex bacterial outer membranes.

Slow Diffusion and Importance of Divalent Cations

The binding of divalent cations to the phosphorylated sugars of LPS was first reported from simulations by Lins & Straatsma.[3] Since then similar observations have been made from longer simulations for example by the Soares, Khalid and Im groups using an adapted GLYCAM06 (Kirschner et al . Chem. Theory Comput., 2012, 8 (11), pp 4719–4731) parameter set, GROMOS53A6[5] and CHARMM36 respectively[7]. In particular, in a combined experimental and simulation study, Clifton et al. showed that replacing divalent cations with the appropriate number of monovalent cations in a bilayer composed of LPS in both leaflets leads to breakdown of the bilayer (Clifton et al, Angewandte Chemie, 2015, 127(41):12120-12123). The slow diffusion rate of LPS compared to phospholipids has been observed by both experimental and computational methods; indeed, the former diffuse an order of magnitude slower than the latter. Simulations of the antimicrobial peptide polymyxin B1 showed insertion into a symmetric inner membrane model within 100 ns, but no penetration into an LPS-containing model from wildtype *E. coli*, largely due to the slow diffusion of LPS and the strong electrostatic interactions between the LPS phosphate and hydroxyl groups and the amino groups of polymyxin (Figure 2).[12] Similarly, bilayers composed of five LPS chemotypes, representing different resistance-susceptibility behavior to polymyxin B1, were studied by Santos *et al.*[11] The study revealed differences in diffusion properties of the different chemotypes. Binding of the peptide to the membrane surface leads to an increase in average membrane thickness, which is more pronounced for the antibiotic-susceptible chemotypes. Once again while structural changes were observed in the membrane surface upon binding of the peptide, it was not observed to fully penetrate into the membrane. In other words, the lipophilic tail does not reach the membrane core. Given the mechanism of action of polymyxin B1 is still debated, it is unclear whether the lack of penetration in the simulations is an equilibrium property of polymyxin and LPS-containing simulations, or an artifact of short simulations. This poses a bit of a conundrum for simulation studies; we have better, more

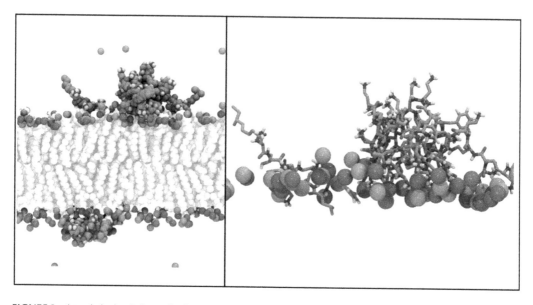

FIGURE 2 Atomistic simulations of polymyxin B1 in a bilayer of LPS did not show any penetration of the lipopeptide into the bilayer.[12] Phosphorus atoms of the LPS molecules are shown as gold spheres, Ca²⁺ ions are pink spheres, LPS tails are white spheres, and polymyxin is shown in stick representation. (See color insert for the color version of this figure)

chemically representative models of the outer membrane when we incorporate LPS, but the slower dynamics mean we have to run much longer simulations. This point was clearly demonstrated by Soares and co-workers when developing their GLYCAM06-based parameters set for LPS; their study showed equilibration of LPS-containing membranes requires >500 ns of simulation (GLYCAM06 (Kirschner et al. Chem. Theory Comput., 2012, 8 (11), pp 4719–4731)). To partly overcome some of these timescale limitations, coarse-grain (CG) force-fields of LPS have been developed. These include models based on the popular MARTINI force-field for membrane simulations (D.H. de Jong et al, JCTC, 9:687–697, 2013). A MARTINI-based model was employed to study the interaction of polymyxin B1 with LPS-containing membranes, to complement the earlier mentioned atomistic study from the same group.[10] Encouragingly, the CG simulations showed strong membrane-binding of polymyxin B1. Furthermore, localized glass-like behavior of the membrane upon the peptide binding was observed, which agrees qualitatively with the atomistic simulations of Santos *et al*, in which some of the LPS-containing membranes thickened upon binding LPS, suggesting increased acyl-chain order. In the CG simulations, insertion of polymyxin into the lipid core of the Re-LPS membrane was observed albeit only once as the benzyl and isobutyl groups of the peptide penetrated the phosphate interface of the bilayer that was cross-linked by Ca^{2+} ions.

Different Dynamics, Not Just Slower

Simulations of transmembrane proteins native to the outer membrane have shown that the dynamics, especially of the long loops on the outer leaflet side of the membrane, show not only slower, but also different conformational dynamics in LPS compared to phospholipids. Comparative simulations of the TonB-dependent transporter FecA in an asymmetric LPS-containing membrane versus a symmetric phospholipid bilayer revealed LPS impacts the loops dynamics via two distinct mechanisms.[13] Short-lived, but frequently formed non-specific hydrogen-bonding interactions alter the local fluctuations in loop movement, whereas the bulky LPS molecules provide steric resistance to large-scale conformational rearrangements of the loops. Similarly, altered loop dynamics in phospholipids compared to LPS were reported for another TonB-dependent transporter, BtuB, by Balusek and Gumbart.[14] Furthermore, this study showed that not only are protein–LPS interactions important in terms of the loops dynamics, but so too are the divalent cations located within the lipid A headgroup region. Calcium binding to an extracellular loop was shown to promote a conformational shift in the loop as well induce secondary-structure formation. Calcium binding had previously been shown from experimental studies, to be essential for the transport for cobalamin through BtuB, but the details of the structural consequences of the binding on the protein were revealed from the simulations of BtuB in a biologically representative membrane environment. There are a number of studies in the literature now in which native proteins have been studied in LPS-containing membranes including OmpA,[15,16] Hia,[17] POTRA domains of BamA,[18] OmpLA,[19] and OprH.[20] The latter study compared the dynamics of OprH outer membranes of *P. aeruginosa* and *E. coli*, which is important consideration given membrane proteins are often expressed and studied in species to which they are not native. The simulations revealed that despite the *E. coli* membrane being thicker than the *P. aeruginosa membrane*, in the vicinity of the protein, the width of both membranes adjusts to match the hydrophobic surface of the protein barrel, similar to observations from simulations of outer membrane proteins (OMPs) in simple phospholipid bilayers. Thus, simulation studies have shown that despite exhibiting slower dynamics compared to phospholipids, LPS molecules can adjust to their conformation to avoid hydrophobic mismatch with nearby proteins.

Specific Binding

A number of atomistic and CG simulation studies have focussed on OmpF, the trimeric porin from *E. coli*, in complex outer membrane models. Patel *et al.* simulated OmpF in different outer membrane models: rough LPS, core LPS, and core LPS with five repeating units of O6-antigen and a combination of these three levels of LPS within the same leaflet.[21] The inner leaflet composition was the same in all three models. These simulations revealed the core sugars of LPS play a key role in shielding the large extracellular loops of OmpF surface epitopes from being recognized by antibodies. This shielding effect is reinforced by O-antigen polysaccharides, which further reduce the access to extracellular loops. Simulations in which a mixture of LPS molecules were considered within the same leaflet, showed that the local lipidic environment of OmpF is of fundamental importance to the ability of the outer membrane to occlude access to recognition sites on OmpF, thereby evading detection by antibodies (Figure 3).

AFM and electron microscopy studies have previously shown that the outer membrane is likely a non-fluid, densely packed structure with LPS being highly immobile.[22–24] Interestingly, slower lipid diffusion up to 6 nm from OmpF in both flat lipid bilayers and spherical vesicles has also been reported from CG simulation studies, even when only phospholipids are considered in the membrane model.[25,26] For a number of years, there has been experimental evidence to suggest that there are one or more LPS-binding sites on the surface of OmpF. More recently, a combination of biophysical studies and a high-resolution X-ray structure from the Lakey and van den Berg groups provided further compelling evidence for specific LPS-binding regions on the surface of OmpF.[27] The protein

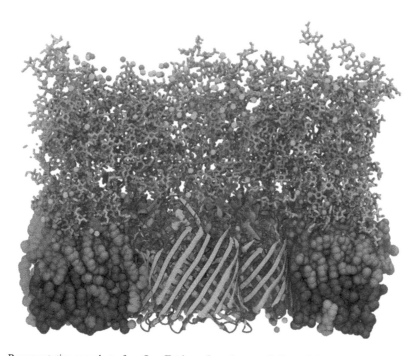

FIGURE 3 Representative snapshot of an OmpF trimer (barrel, green; helix, red; loop and turn, blue) embedded in Outer Membrane (OM) of E. coli LPS with R1 core and five repeating units of O6-antigen. The outer leaflet contains Lipid A, represented as pink spheres, core sugars as orange stick model, and O-antigen polysaccharides as gray stick model. The inner leaflet contains 1-palmitoyl(16:0)-2-palmitoleoyl(16:1 cis-9)-phosphatidylethanolamine (blue spheres), 1-palmitoyl(16:0)-2-vacenoyl(18:1 cis-11)-phosphatidylglycerol (orange spheres), and 1,10-palmitoyl-2,20-vacenoyl cardiolipin with a net charge of −2e (magenta spheres). Ca^{2+} ions are represented as cyan small spheres, K^+ ions as green small spheres, and Cl^- ions as magenta small spheres. For clarity, some portions of the system have been removed.[21] (See color insert for the color version of this figure)

was crystallized in its trimeric form in complex with Re-LPS, while the mutational studies showed a reduction in LPS-binding when charged residues on the OmpF surface were mutated. Subsequently, simulations from Bond and co-workers reported umbrella sampling simulations in which LPS was extracted from the membrane Huber et al. Structure. 2018 Aug 7;26(8): 1151–1161) The free-energy cost of removing the LPS from the 'bulk' membrane was ~9.5 kcal/mol higher than removing it from a membrane that does not contain OmpF proteins. In summary, in terms of OmpF-LPS binding, the X-ray structure showed the binding site, the mutational studies showed what happens when the binding site is modified, and the computational studies provided the energetics. This is clear demonstration of structural biology, biophysical measurements, and computational methods each providing complementary data to advance the understanding of a specific OMP–lipid interaction. It is highly unlikely that OmpF is the only outer membrane protein with specific binding sites for LPS; for example, one of the X-ray structures of the TonB-dependent transporter FhuA shows this protein in complex with lipid A, suggestive of strong binding.[28] Thus, in future, the combination of experimental and simulation methodologies will undoubtedly yield further insights into the complex interactions that exist between native proteins and the local bacterial membrane environment.

Asymmetric Energetic Barriers for Permeation of Small Molecules

While the computational studies discussed above clearly demonstrate that molecular permeation across the LPS-containing outer leaflet is more difficult compared to permeation across the inner leaflet, given the former diffuses slower and is tightly cross-linked by divalent ions, comparison of the energetic differences in the permeation processes requires free-energy calculations. Carpenter et al. employed umbrella sampling with the weighted histogram analysis method (WHAM) to calculate the potential of mean force profiles for permeation of benzene, hexane, ethane, acetic acid, and ethanol across the Khalid outer membrane model.[5,29] This study revealed a marked difference in the energetic barriers presented by the headgroups of LPS compared to phospholipids. The energetic barrier at the LPS headgroup region was most pronounced for hexane, which is likely a consequence of the longer hydrophobic length of hexane compared to benzene and ethane. The maximum free energy in the LPS headgroup regions was calculated as ~6 kcal/mol for hexane, <5 kcal/mol for benzene, and ~3 kcal/mol for ethane; interestingly no appreciable barriers to the entry of any of the hydrophobic solutes into the head groups of the phospholipid leaflet were observed, whereas previously free-energy calculations of small solute permeation through symmetrical phospholipid bilayers reported small aliphatic chains such as propane to have similar barriers through the headgroup region of phosphatidylethanolamine lipid headgroups as benzene.[30,31] Visual inspection of the umbrella sampling windows corresponding to these regions revealed that the lack of barriers reported by Carpenter *et al.* are a consequence of the high mobility of the phospholipids, which are able to rearrange locally, such that the hydrophobic solutes become exposed to parts of the lipid tail regions, rather than the headgroups. It was noted that the previously reported simulations were too short to observe these rearrangements. Thus, this study provided scientific insights, but also a note of caution regarding the ever-present problem of molecular dynamics simulation timescales.

Larger Simulation Systems: Correlated Lipid Motion

While there have been a number of coarse-grain simulations of large bacterial membrane systems incorporating many copy numbers of outer membrane proteins embedded within symmetrical phospholipid bilayers, there are fewer such studies of larger systems reported in the literature in

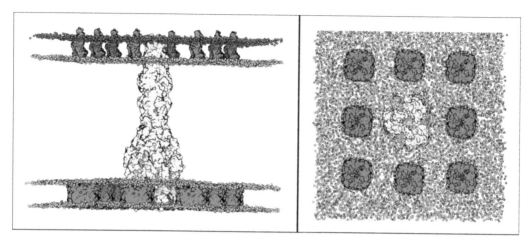

FIGURE 4 Two views of one of the simulation systems incorporating both the inner and outer membranes of E. coli reported by Hsu et al.[33] LPS headgroups are purple, phospholipid headgroups are cyan, AqpZ is red, OmpA is blue, and the TolC-AcrABZ efflux pump is yellow. (See color insert for the color version of this figure)

which LPS is included in the membrane model.[25,26,32] Hsu et al. reported simulations of the *E. coli* cell envelope in which the TolC/AcrABZ efflux pump spanned both membranes and also the periplasmic space, the latter was approximated as a water-filled region, separating the two membranes (Figure 4).[33] The results of these simulations revealed stark differences in the motion of LPS compared to phospholipids. While it had previously been shown that the motion of LPS is restricted compared to phospholipids, these larger simulations also revealed that LPS motion within the outer leaflet is highly correlated. Indeed, the LPS molecules are correlated with each other, but also with the proteins within the membrane: they all move in the same direction. In contrast, the phospholipids of the inner leaflet and those within both leaflets of the inner membrane display far less correlation in their motion. Furthermore, enrichment of specific lipid types around AcrB and Aqp1, two proteins of the inner membrane, was reported. Similar lipid 'fingerprints' of membrane proteins have been reported from coarse-grain, large-scale simulations of other membrane proteins (www.biorxiv.org/content/early/2017/09/20/191486), most notably from Tieleman and co-workers (Corradi et al., ACS Central Science, 2018, 4, 707–717). Communication between the two leaflets of the outer membrane were recently reported by Shearer *et al.*, once again from CG simulations which showed regions of high lipid disorder in the outer leaflet to be directly above regions of high order in the inner leaflet.[34] Thus, the studies of larger simulations enabled by CG models are opening up new channels of lipid–protein interactions; these studies are still very much in their infancy and not yet *de rigor* within the biomolecular simulation community, but show enormous promise for the future.

Summary and Outlook

In summary, molecular dynamics simulations of bacterial outer membranes have advanced enormously in the past decade in terms of the biochemical accuracy of the lipidic components and also the copy numbers of proteins that are embedded within the membranes. The development of coarse-grain models to complement their atomistic counterparts has further broadened the scope of molecular simulations of bacterial membranes, particularly when both levels of granularity are used to study different aspects of the same system; the details of specific interactions are provided by the atomistic simulations whereas properties such as diffusion and local membrane biophysics are provided by coarse-grain models. These simulation studies are now revealing many of the hitherto unexplored complexities of bacterial membrane structure–dynamics–function relationships. The coupled nature of protein and lipid motion, communication between the two leaflets of the outer membrane, the vast differences between the mobilities of the two leaflets of the outer membrane, and the ability of proteins to sort lipids in their vicinity are all aspects of bacterial membrane behavior in which large-scale molecular dynamics simulations have played a key role in advancing our understanding. The larger simulation studies described in this chapter are beginning to bridge the traditional gap between simulation and experimental system lengthscales and thus enable comparative studies. Given the breadth of bacterial membrane protein and lipid structures such large-scale, multidisciplinary studies will be essential for elucidating all of the areas of biological significance of protein–lipid interactions. While the two membranes have been the focus of much attention from the international biomolecular simulation community, the cell wall has been less well studied, although there have been one or two reports in the literature of atomistic models of peptidoglycan.[15,35] Given the importance of the cell wall in terms of mechanical strength imparted to the cell envelope and the number of proteins that reside within the periplasm, it is a component that requires inclusion in any models aiming to mimic the *in vivo* cell envelope, and given that simulations of peptidoglycan are now beginning to emerge, no doubt will be incorporated into future models of the 'virtual cell envelope.'

REFERENCES

1. Nikaido, H., Molecular basis of bacterial outer membrane permeability revisited. *Microbiol Molec Biol Rev* 2003, *67*, 593–656.
2. Erridge, C.; Bennett-Guerrero, E.; Poxton, I. R., Structure and function of lipopolysaccharides. *Microbes Infect* 2002, *4* (8), 837–851.
3. Lins, R. D.; Straatsma, T. P., Computer simulation of the rough lipopolysaccharide membrane of *Pseudomonas aeruginosa*. *Biophys J* 2001, *81*, 1037–1046.
4. Straatsma, T. P.; Soares, T. A., Characterization of the outer membrane protein OprF of Pseudomonas aeruginosa in a lipopolysaccharide membrane by computer simulation. *Proteins-Struct Funct Bioinf* 2009, *74* (2), 475–488.
5. Piggot, T. J.; Holdbrook, D. A.; Khalid, S., Electroporation of the E. coli and S. Aureus membranes: Molecular dynamics simulations of complex bacterial membranes. *J Phys Chem B* 2011, *115* (45), 13381–13388.
6. Oostenbrink, C.; Soares, T. A.; van der Vegt, N. F.; van Gunsteren, W. F., Validation of the 53A6 GROMOS force field. *Eur Biophys J* 2005, *34* (4), 273–284.
7. Wu, E. L.; Engstrom, O.; Jo, S.; Stuhlsatz, D.; Wildmalm, G.; Im, W., Molecular dynamics simulations of E. coli lipopolysaccharide bilayers. *Biophys J* 2013, *104* (2), 586a–586a.
8. Soares, T. A.; Straatsma, T. P., Assessment of the convergence of molecular dynamics simulations of lipopolysaccharide membranes. *Mol Simulat* 2008, *34* (3), 295–307.
9. Ma, H.; Irudayanathan, F. J.; Jiang, W.; Nangia, S., Simulating gram-negative bacterial outer membrane: A coarse grain model. *J Phys Chem B* 2015, *119* (46), 14668–14682.
10. Jefferies, D.; Hsu, P. C.; Khalid, S., Through the lipopolysaccharide glass: A potent antimicrobial peptide induces phase changes in membranes. *Biochem* 2017, *56* (11), 1672–1679.
11. Santos, D. E. S.; Pol-Fachin, L.; Lins, R. D.; Soares, T. A., Polymyxin binding to the bacterial outer membrane reveals cation displacement and increasing membrane curvature in susceptible but not in resistant lipopolysaccharide chemotypes. *J Chem Inf Model* 2017, *57* (9), 2181–2193.
12. Berglund, N. A.; Piggot, T. J.; Jefferies, D.; Sessions, R. B.; Bond, P. J.; Khalid, S., Interaction of the antimicrobial peptide polymyxin B1 with both membranes of E. coli: A molecular dynamics study. *PLoS Comput Biol* 2015, *11* (4), e1004180.
13. Piggot, T. J.; Holdbrook, D. A.; Khalid, S., Conformational dynamics and membrane interactions of the E. coli outer membrane protein FecA: A molecular dynamics simulation study. *Biochimica et biophysica acta* 2013, *1828* (2), 284–293.
14. Balusek, C.; Gumbart, J. C., Role of the native outer-membrane environment on the transporter BtuB. *Biophys J* 2016, *111* (7), 1409–1417.

15. Samsudin, F.; Boags, A.; Piggot, T. J.; Khalid, S., Braun's lipoprotein facilitates OmpA interaction with the Escherichia coli cell wall. *Biophys J* 2017, *113* (7), 1496–1504.

16. Samsudin, F.; Ortiz-Suarez, M. L.; Piggot, T. J.; Bond, P. J.; Khalid, S., OmpA: A flexible clamp for bacterial cell wall attachment. *Structure* 2016, *24* (12), 2227–2235.

17. Holdbrook, D. A.; Piggot, T. J.; Sansom, M. S.; Khalid, S., Stability and membrane interactions of an autotransport protein: MD simulations of the Hia translocator domain in a complex membrane environment. *Biochimica et biophysica acta* 2013, *1828* (2), 715–723.

18. Fleming, P. J.; Patel, D. S.; Wu, E. L.; Qi, Y.; Yeom, M. S.; Sousa, M. C.; Fleming, K. G.; Im, W., BamA POTRA domain interacts with a native lipid membrane surface. *Biophys J* 2016, *110* (12), 2698–2709.

19. Wu, E. L.; Fleming, P. J.; Klauda, J. B.; Fleming, K. G.; Im, W., A molecular dynamics simulation study of Outer Membrane Phospholipase A (OMPLA) structure and dynamics in an asymmetric lipopolysaccharide membrane. *Biophys J* 2014, *106* (2), 656a–656a.

20. Lee, J.; Patel, D. S.; Kucharska, I.; Tamm, L. K.; Im, W., Refinement of OprH-LPS interactions by molecular simulations. *Biophys J* 2017, *112* (2), 346–355.

21. Patel, D. S.; Re, S.; Wu, E. L.; Qi, Y.; Klebba, P. E.; Widmalm, G.; Yeom, M. S.; Sugita, Y.; Im, W., Dynamics and interactions of OmpF and LPS: Influence on pore accessibility and ion permeability. *Biophys J* 2016, *110* (4), 930–938.

22. Hoenger, A.; Pages, J. M.; Fourel, D.; Engel, A., The orientation of porin OmpF in the outer membrane of Escherichia coli. *J Mol Biol* 1993, *233* (3), 400–413.

23. Hoenger, A.; Ghosh, R.; Schoenenberger, C. A.; Aebi, U.; Engel, A., Direct in situ structural analysis of recombinant outer membrane porins expressed in an OmpA-deficient mutant Escherichia coli strain. *J Struct Biol* 1993, *111* (3), 212–221.

24. Jaroslawski, S.; Duquesne, K.; Sturgis, J. N.; Scheuring, S., High-resolution architecture of the outer membrane of the gram-negative bacteria Roseobacter denitrificans. *Mol Microbiol* 2009, *74* (5), 1211–1222.

25. Holdbrook, D. A.; Huber, R. G.; Piggot, T. J.; Bond, P. J.; Khalid, S., Dynamics of crowded vesicles: Local and global responses to membrane composition. *PLoS One* 2016, *11* (6), e0156963.

26. Goose, J. E.; Sansom, M. S., Reduced lateral mobility of lipids and proteins in crowded membranes. *PLoS Comput Biol* 2013, *9* (4), e1003033.

27. Arunmanee, W.; Pathania, M.; Solovyova, A. S.; Le Brun, A. P.; Ridley, H.; Basle, A.; van Den Berg, B.; Lakey, J. H., Gram-negative trimeric porins have specific LPS binding sites that are essential for porin biogenesis. *Proc Natl Acad Sci USA* 2016, *113* (34), E5034–43.

28. Ferguson, A. D.; Hofmann, E.; Coulton, J. W.; Diederichs, K.; Welte, W., Siderophore-mediated iron transport: Crystal structure of FhuA with bound lipopolysaccharide. *Science* 1998, *282*, 2215–2220.

29. Carpenter, T. S.; Parkin, J.; Khalid, S., The free energy of small solute permeation through the Escherichia coli outer membrane has a distinctly asymmetric profile. *J Phys Chem Lett* 2016, *7* (17), 3446–3451.

30. Bemporad, D.; Luttmann, C.; Essex, J. W., Computer simulation of small molecule permeation across a lipid bilayer: Dependence on bilayer properties and solute volume, size, and cross-sectional area. *Biophys J* 2004, *87* (1), 1–13.

31. Wennberg, C. L.; van der Spoel, D.; Hub, J. S., Large influence of cholesterol on solute partitioning into lipid membranes. *J Am Chem Soc* 2012, *134* (11), 5351–5361.

32. Fowler, P. W.; Helie, J.; Duncan, A.; Chavent, M.; Koldso, H.; Sansom, M. S., Membrane stiffness is modified by integral membrane proteins. *Soft Matter* 2016, *12* (37), 7792–7803.

33. Hsu, P. C.; Samsudin, F.; Shearer, J.; Khalid, S., It is complicated: Curvature, diffusion, and lipid sorting within the two membranes of Escherichia coli. *J Phys Chem Lett* 2017, *8* (22), 5513–5518.

34. Shearer, J.; Khalid, S., Communication between the leaflets of asymmetric membranes revealed from coarse-grain molecular dynamics simulations. *Sci Rep* 2018, *8* (1), 1805.

35. Gumbart, J. C.; Beeby, M.; Jensen, G. J.; Roux, B., Escherichia coli peptidoglycan structure and mechanics as predicted by atomic-scale simulations. *PLoS Comput Biol* 2014, *10* (2), e1003475.

Index